Springer Tracts in Modern Physics

Volume 289

Springer Tracts in Modern Physics provides comprehensive and critical reviews of topics of current interest in physics. The following fields are emphasized:

- Particle and Nuclear Physics
- Condensed Matter Physics
- Light Matter Interaction
- Atomic and Molecular Physics

Suitable reviews of other fields can also be accepted. The Editors encourage prospective authors to correspond with them in advance of submitting a manuscript. For reviews of topics belonging to the above mentioned fields, they should address the responsible Editor as listed in "Contact the Editors".

Shinji Watanabe · Kazumasa Miyake

Quantum Critical Phenomena of Valence Transition

Heavy Fermion Metals and Related Systems

Springer

Shinji Watanabe
Department of Basic Sciences
Kyushu Institute of Technology
Kitakyushu, Fukuoka, Japan

Kazumasa Miyake
Graduate School of Science
Center for Advanced High Magnetic Field Science
Osaka University
Toyonaka, Osaka, Japan

ISSN 0081-3869 ISSN 1615-0430 (electronic)
Springer Tracts in Modern Physics
ISBN 978-981-99-3517-8 ISBN 978-981-99-3518-5 (eBook)
https://doi.org/10.1007/978-981-99-3518-5

This Springer imprint is published by the registered company Springer Nature Singapore Pte Ltd.
The registered company address is: 152 Beach Road, #21-01/04 Gateway East, Singapore 189721, Singapore

Preface

Heavy fermion systems offer an interesting research field of strongly correlated electron systems. The 4f-electron at the rare-earth atom, such as Ce and Yb, has the localized character, which hybridizes with the conduction electrons in the crystal, giving rise to the heavy fermion state. The underlying mechanism is the Kondo effect, which is the many-body effect between the localized 4f magnetic moment and itinerant conduction electrons.

In heavy fermion systems, the phase transition between the paramagnetic metallic phase and the magnetically ordered phase often takes place by changing a control parameter of the system such as the pressure. The phase diagram has been understood on the basis of the so-called Doniach's phase diagram. However, anomalous phenomena such as the unconventional quantum critical phenomena and the superconductivity with the Cooper pair not mediated by magnetic fluctuations, which seem to be beyond the ordinary concept, have been observed experimentally in several Ce- and Yb-based compounds since the end of the last century.

In this monograph, it is discussed that the quantum critical phenomena originating from the quantum critical point (QCP) of the valence transition play a key role in understanding these anomalous phenomena. The valence transition in rare-earth-based compounds has long history of research since the discovery of the first-order transition in the Ce metal. This monograph intends to point out that the critical point of the first-order valence transition can be suppressed near absolute zero in the rare-earth-based heavy fermion systems such as Ce- and Yb-based compounds. The critical point located at zero temperature is defined as the QCP.

At the QCP, the critical valence fluctuations diverge, which not only cause the new type of the quantum criticality in the physical quantities at low temperatures but also trigger the electronic instability such as the superconductivity. This monograph emphasizes that the critical valence fluctuations are the key concept which brings about the paradigm shift from the ordinary concept based on the Doniach's phase diagram in heavy fermion systems.

The magnetic quantum critical phenomena have been well understood on the basis of the self-consistent renormalization (SCR) theory for spin fluctuations. The quantum criticality in the physical quantities at low temperatures shown by the SCR theory has been endorsed by the renormalization group theory. An advantageous point of the SCR theory is that it describes not only the quantum critical regime but also the higher temperature regime in a unified way. In this monograph, the SCR theory is explained in the modern form using the path-integral formalism, which is considered to give the most straightforward and coherent way of the derivation.

The microscopic origin of the valence transition is the inter-orbital Coulomb repulsion between 4f and 5d electrons at the rare-earth atom. Then, the universal model is the Anderson lattice model with the inter-orbital Coulomb repulsion, which is referred to as the extended Anderson lattice model. The theory of the quantum critical phenomena of valence fluctuations is constructed on the basis of the SCR theory starting from the extended Anderson lattice model. The emergence of the almost localized valence fluctuations mode due to the strong correlation effect is shown to be the key factor to give rise to the new type of quantum criticality.

In the realistic parameter regime in the extended Anderson lattice model, the interplay of the magnetic QCP and the valence QCP is shown to be realized. This provides the important viewpoint to clarify the mechanism of unresolved phenomena, which are hard to be understood merely by the viewpoint of the magnetic instability, as observed in several Ce- and Yb-based heavy fermion systems.

The universality class of critical valence fluctuations is applied not only to periodic crystals but also to quasicrystals. Qualitatively new phenomena such as the robust quantum criticality under pressure and non-divergent Grüneisen parameter at zero temperature observed in the Yb-based quantum critical quasicrystal are discussed to be explained by the theory of the critical Yb-valence fluctuations. Recently, the rare-earth-based quasicrystals have attracted much attention as a new research field of strongly correlated electron systems. The method of construction of the microscopic model in the quasicrystal is also concretely explained in Sect. 6.6.

In this monograph, starting from the brief review of the valence transition, the early history of critical valence fluctuations is overviewed. Fundamentals of the heavy fermion state are explained on the basis of Green's function formalism in the Anderson lattice model. The concept of the Fermi-liquid theory is explained from experimental and theoretical aspects. Then, the quantum critical phenomena of valence fluctuations are explained from both aspects of the theory and experiments in periodic crystals and quasicrystals.

This monograph is written as follows:

Watanabe: Sects. 4.1, 4.5, 4.6, 8.1.1, 8.1.2, Chaps. 5–7 and Appendices. Miyake: Chaps. 1–3, Sects. 4.2–4.4 and 8.1.3–8.1.6.

If this monograph contributes to understanding of researchers and students, it is authors' great pleasure.

Kitakyushu, Japan Shinji Watanabe
Osaka, Japan Kazumasa Miyake
March 2023

Acknowledgements

The authors acknowledge collaborations with following researchers: J. Flouquet, D. Jaccard, A. T. Holmes, G. Seyfarth, A. Tsuruta, Y. Onishi, O. Narikiyo, H. Maebashi, I. Sheikin, A. Huxley, D. Braithwaite, J.-P. Brison, A.-S. Rüetschi, K. Sengupta, M. Imada, A. Georges, K. Matsubayashi, T. Hirayama, T. Yamashita, S. Ohara, N. Kawamura, M. Mizumaki, N. Ishimatsu, K. Kitagawa, and Y. Uwatoko. The authors thank the following researchers for permission to use figures in the text: T. Mito, T. C. Kobayashi, Y. Ōnuki, Y. H. Matsuda, K. Ishida, A. T. Holmes, K. Umeo, R. Küchler, K. Kuga, K. Imura, H. Shishido, G. Knebel, D. Jaccard, and K. Matsubayashi. The authors thank valuable discussions with H. Harima, N. K. Sato, T. Ishimasa, K. Deguchi, K. Imura, S. Nakatsuji, K. Kuga, Y. Matsumoto, H. Kobayashi, Y. Kitaoka, T. Watanuki, E. Bauer, F. Steglich, J. D. Thompson, T. Park, and H. Q. Yuan. The authors are indebted for discussions with C. M. Varma and G. G. Lonzarich. One of the authors (Watanabe) acknowledges Y. Kuramoto who introduced him to the field of the heavy fermion systems. One of the authors (Miyake) acknowledges the warm encouragement of T. Kasuya at early stage of works discussed in this monograph. The authors acknowledge H. Shiba for recommending us to publish this monograph from Springer-Nature.

Contents

Chapter 1
Prologue

In this chapter, as a prologue, the concept and typical examples of conventional valence transition and the criticality associated with critical valence transition are introduced paying attention to the difference from the critical magnetic transition. Brief summary of the chapters discussed in this monograph is given at the end of the chapter.

1.1 Conventional Valence Transition

The valence transitions were observed in a series of compounds including rare-earth compounds, and attracted much attention since mid-1970s. Among those, the cases of Ce metal and SmS were discussed actively as prototypical ones both from experimentally [1] and theoretically [2] for example. Since then, there has been huge accumulation of researches on the valence-transition problem. In this section, as typical cases, valence transitions observed in γ-α transition of Ce metal and $YbInCu_4$ are discussed.

1.1.1 Ce Metal

The typical first-order valence transition is the γ-α transition of Ce metal as shown in Fig. 1.1 [3, 4] in which the valence-transition temperature T_v, between the γ and the α phases, is shown as a function of pressure P. The transition between γ and α phases is the *first* order and ceases at the critical point at finite temperature $T = T_c$. As this example, valence transitions are the *first order* in general.

S. Watanabe and K. Miyake, *Quantum Critical Phenomena of Valence Transition*, Springer Tracts in Modern Physics 289,
https://doi.org/10.1007/978-981-99-3518-5_1

In the γ-α transition, the valence of Ce changes discontinuously from $Ce^{+3.03}$ in the γ phase to $Ce^{+3.14}$ in the α phase at room temperature, maintaining the crystal structure of the face-centered cubic symmetry [4]. This means that number of 4f-electron per Ce ion changes from 0.97 to 0.86, accompanied by the reduction of system volume about 18% [3, 5]. This fact implies that the electronic state in the γ phase is in the so-called Kondo regime, while that in the α phase is in the so-called *valence-fluctuation* regime as discussed shortly.

Since both phases are the normal metallic phase with the same crystal symmetry, there exists a critical point of the first-order transition as in the case of liquid-gas transition. Around this critical point at $T = T_c$, critical valence fluctuations highly develop in general as in the case of density fluctuations around the liquid-gas critical point where the density fluctuations or the compressibility diverges. In the case where T_c is suppressed to $T_c = 0$, there arises singular T dependence of various physical quantities near $T = 0$ by the quantum critical valence fluctuations as discussed in Sect. 1.2.

According to the Clapeyron-Clausius relation along the valence transition line, the following relation holds:

$$\frac{dT}{dP} = \frac{V_\gamma - V_\alpha}{S_\gamma - S_\alpha} > 0. \tag{1.1}$$

Therefore, the entropy S_γ in the γ phase is larger than S_α in the α phase because $V_\gamma - V_\alpha < 0$ under the pressure in general. This suggests that, in the γ phase, the nearly localized $4f^1$-state, or the Kondo state, is realized so as to obtain the larger entropy, arising from the *spin* degrees of freedom of almost localized $4f^1$ electron, compared to the smaller entropy in the *valence fluctuation* regime in which the $4f^1$-state has more itinerant character with smaller entropy.

The reason why the slope of the valence transition is positive, $dT_v/dP>0$, in Fig. 1.1 is given as follows. First of all, at $T = 0$, it is noted that the valence transition occurs when the total energies of both phases, E_K in the Kondo regime and E_{VF} in the *valence-fluctuation* regime coincide at the *extrapolated* negative pressure $P_0(< 0)$ at $T = 0$ as imagined from Fig. 1.1. The physical reason why the *valence-fluctuation* state is stabilized by increasing P is that 4f-electrons are forced to move into the conduction band by the effect of P. As a result, the number of 4f-electrons decreases from nearly $4f^1$ (Kondo regime) state toward the mixed-valence state with less 4f-electrons number.

This means that the difference in energies (at $T = 0$), $\Delta E \equiv (E_K - E_{VF})$, increases from 0 (at $P = P_0$) to the positive value with increasing P from $P = P_0$ to the positive pressure region of $P > 0$. Therefore, the higher transition temperature $T_v \propto \Delta E$ becomes necessary to gain the entropy in the higher T region for realizing the Kondo regime in the γ-phase at higher temperature side by reducing the Helmholtz free energy $F = E - TS$. This explains why the slope dT_v/dP is positive.

The aspect discussed here should be modified in the Yb-based compound, $YbInCu_4$, discussed in the next Sect. 1.1.2. This is because the correlated electronic state, such as heavy fermion properties, can be understood on the basis of the hole picture of $4f^1_{hole}$ because original $4f^{13}$ electronic state of Yb can be described as the hole because fully occupied state of f-orbital is in the f^{14} configuration.

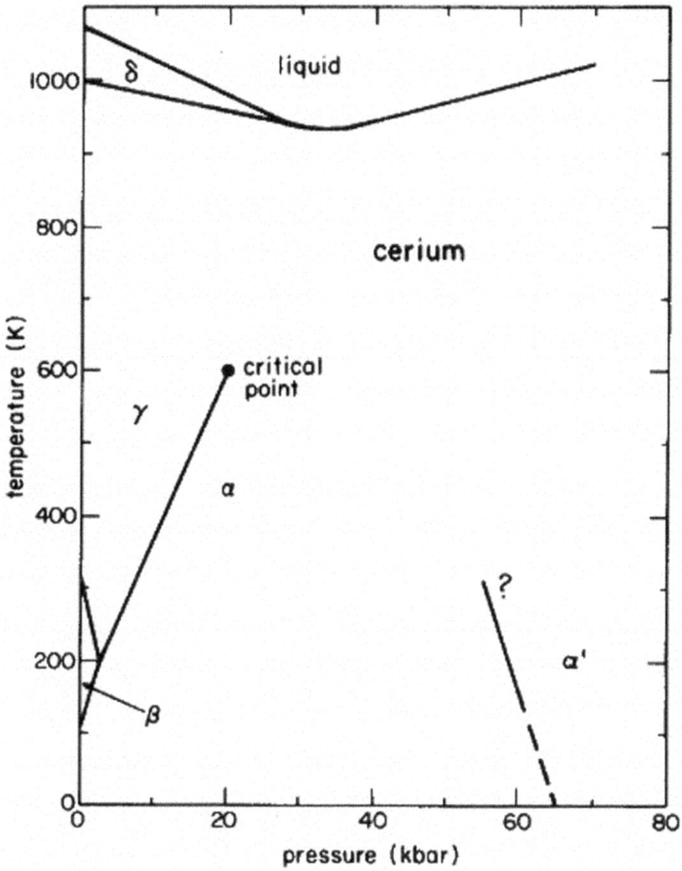

Fig. 1.1 Temperature-pressure phase diagram of Ce metal [3]. The first-order valence-transition line (solid line) between the γ phase and the α phase terminates at the critical point (filled circle). Reprinted with permission from Ref. [3]. Copyright 1970 American Physical Society

1.1.2 YbInCu$_4$

The valence transition and associated phenomena are observed also in Yb-based compounds. A typical example is $YbInCu_4$ as shown in Fig. 1.2. As mentioned at the end of the previous subsection, the electronic state of Yb is described essentially as $4f^1_{\rm hole}$ system, so that the aspect of valence transition in the T-P phase diagram is somewhat contrasting from that in Ce-based compounds, such as Ce metal discussed in the previous subsection.

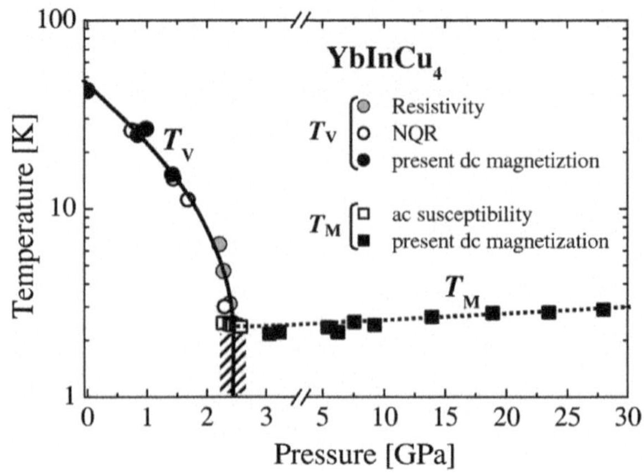

Fig. 1.2 Temperature-pressure phase diagram of $YbInCu_4$ [6]. T_v and T_M are transition temperatures of valence transition and ferromagnetic phase transition, respectively. The reason why the derivative dT_v/dP diverges as $T_v \to 0$ is understood by the Clapeyron-Clausius relation, Eq. (1.1). Reprinted with permission from Ref. [6]. Copyright 2007 American Physical Society

At ambient pressure, $YbInCu_4$ exhibits valence transition from $Yb^{+2.90}$ to $Yb^{+2.74}$ at $T = T_v = 42$ K as shown in Fig. 1.2 [7]. This implies that the $4f_{hole}$ number changes discontinuously from 0.90 to 0.74, accompanied by the reduction of the system volume of about 0.5% [8]. In contrast to the case of Ce metal, according to the recent report [9], the structure of the crystal changes slightly from the cubic to the tetragonal structure.

By the similar argument to the case of Ce metal, the Clapeyron-Clausius relation (1.1) implies that the higher T region ($T > T_v$) is in the Kondo regime in the hole picture, while the lower T region ($T < T_v$) is in the *valence-fluctuation* regime.

The reason why T_v is suppressed by the pressure P, i.e., $(dT_v/dP) < 0$, is understood in parallel to the above argument for Ce metal modifying it to the case of $4f^1_{hole}$ system. However, the crucial difference from the case of Ce metal is that $4f^1_{hole}$ state is stabilized by the pressure. This is understood physically as follows. Under the pressure, the $4f^1_{hole}$ state is stabilized against the $4f^0_{hole}$ state, because $4f^1_{hole}$ state ($4f^{13}$ state in electron picture) is stabilized under the pressure against $4f^0_{hole}$ state (fully occupied $4f^{14}$ state in electron picture).

For comparison with the phase diagram of Ce metal shown in Fig. 1.1, it is noted that the critical point of valence transition does not appear in the phase diagram of Fig. 1.2. The critical point of valence transition in $YbInCu_4$ is considered to be in the hypothetical negative pressure region ($P < 0$). However, there are some cases where the critical point exists around $T = 0$ giving rise to anomalous quantum critical phenomena as discussed later in Chap. 5.

1.2 Quantum Critical Valence Fluctuations

The quantum critical valence fluctuations (QCVF) develop when the critical temperature T_c is pulled down to zero as shown in Fig. 1.3 for global behavior and in Fig. 1.4 which focuses around the quantum critical point ($T_c = 0$). This QCVF gives rise to

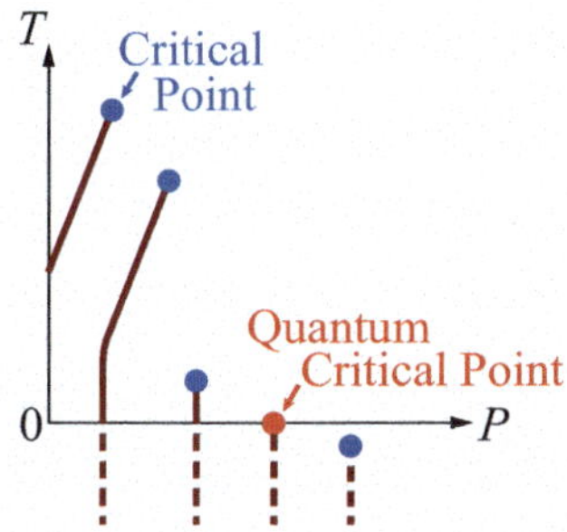

Fig. 1.3 A series of schematic behaviors of valence transitions in the P-T (pressure-temperature) plane in Ce-based compounds, including Ce metal, as shown in Fig. 1.1. The first-order valence-transition line (solid line) terminates at the critical point (filled circle). The dashed lines in the $T < 0$ region represent the hypothetical valence-transition line in the negative temperature region. Note that the derivative dT_v/dP, in the $T \to 0$ limit, diverges as explained in Fig. 1.2. The quantum critical valence fluctuations develop around the quantum critical point with $T_c = 0$

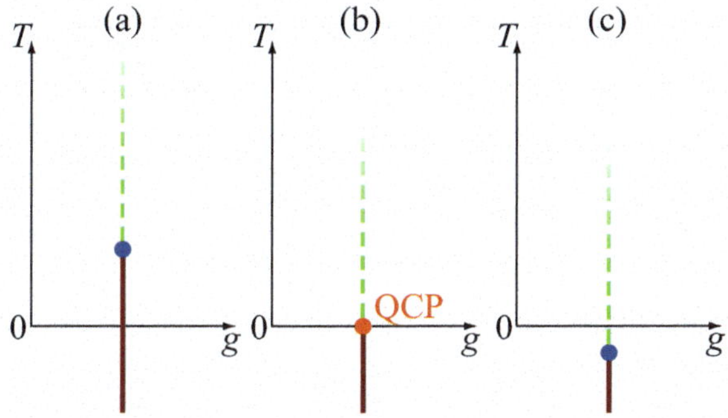

Fig. 1.4 Conceptual phase diagram for the Ce-based system in the T-g (control parameter) plane in the region $T \sim 0$ [10]. Solid line and broken line in each figure are first-order valence transition and valence crossover lies, and filled circles are critical points, respectively. **a**: $T_v > 0$, **b**: critical temperature $T_c = 0$, and **c** $T_v < 0$. QCP means the quantum critical point

the critical T dependence in a series of physical quantities as shown in Table 1.1 together with those based on the magnetic criticality, ferromagnetic (F) and antiferromagnetic (AF), and also the case of Fermi-liquid state for comparison. It is clear that T dependence based on the QCVF is totally different from those on magnetic criticality. Namely, the criticality based on the QCVF forms a *different* and new universality class of quantum criticality.

These theoretical predictions are consistent with a series of experimental observations as discussed in, e.g., Sects. 5.3.7, 6.1, and 6.5.2. In particular, it should be stressed that even non-periodic system such as quasicrystal compound $Yb_{15}Al_{34}Au_{51}$ belongs to the same universality class, as discussed in Sect. 6.6.6.

Table 1.1 Temperature (T) dependence in the limit $T \to 0$ of a series of physical quantities, at quantum critical points in three dimensions, i.e., the magnetic QCP for each class specified by FM ($z = 3$), AFM ($z = 2$), AFM ($z = 3$), and quantum critical valence fluctuations (QCVF), together with the Fermi liquid. Dimensionless parameter η measures the deviation from the criticality, $(\rho - \rho_0)$ the electrical resistivity, with ρ_0 being the residual part, C/T the specific-heat coefficient, χ the uniform susceptibility, and $(T_1T)^{-1}$ the NMR relaxation rate, respectively. The parameter ζ ($0.5 \le \zeta \le 0.7$ with weak T-dependence) is the critical exponent for critical valence fluctuations (CVF), and ($T^{1.5} \to T$) for $(\rho - \rho_0)$ implies that T-dependence crosses over from $T^{1.5}$ to T below the extremely small temperature scale T_0 compared to the highly renormalized effective Fermi energy T_F^* (See, Ref. [14] for the derivation of T_0 by the microscopic calculations)

Class	η	$(\rho - \rho_0)$	C/T	χ	$(T_1T)^{-1}$	Refs.
FM ($z = 3$)	$T^{4/3}$	$T^{5/3}$	$-\ln T$	$T^{-4/3}$	$T^{-4/3}$	[11]
AFM ($z = 2$)	$T^{3/2}$	$T^{3/2}$	const. $-T^{1/2}$	const. $-T^{1/4}$	$T^{-3/4}$	[12]
AFM ($z = 3$)	$T^{4/3}$	$T^{5/3}$	$-\ln T$	const. $-T^{1/3}$	$T^{-4/3}$	[13]
CVF	T^{ζ}	$T^{1.5} \to T$	$-\ln T$	$T^{-\zeta}$	$T^{-\zeta}$	[14]
Fermi liquid	–	T^2	const.	const.	const.	

1.3 Brief Summary of Chapters Discussed in This Monograph

The goal of this monograph is to present a comprehensive review on the critical valence fluctuations phenomena.

In Chap. 2, "Early History of Critical Valence Fluctuations", it is introduced how the existence of critical valence fluctuations (CVF) was realized experimentally in a canonical heavy fermion superconductor $CeCu_2Si_2$ under pressure and was interpreted as the phenomenon caused by CVF. Namely, unconventional properties, such as *sharp enhancement* of the superconducting transition temperature T_c and the residual resistivity ρ_0, together with *sharp decrease* of the A coefficient of T^2 terms in the resistivity (or effective mass m^* of quasiparticles), were interpreted theoretically as the phenomena caused by CVF.

In Chap. 3, "Fundamentals of Heavy Fermion State", fundamental properties of heavy fermion systems are discussed putting emphasis on the following points: (i) the temperature dependence of relevant physical quantities, (ii) the mechanism of huge mass enhancement, (iii) the so-called Kadowaki-Woods ratio, A/γ^2, with A and γ being the coefficient of T^2 term in the resistivity $\rho(T)$ and the Sommerfeld constant γ, respectively.

In Chap. 4, "Anomalous Phenomena due to Critical Valence Transition", anomalous phenomena due to CVF are discussed. In order to understand anomalous properties observed in a series of compounds in a unified way, the extended Anderson lattice model with the repulsive interaction among f- and conduction electrons is introduced as a minimal model for understanding. On this model, the valence-fluctuation mediated superconductivity and the unconventional non-Fermi-liquid behavior in the electrical resistivity are shown to be understood, together with the effects of magnetic field on the valence transition and valence fluctuations.

In Chap. 5, "Self-Consistent Renormalization Theory", the framework of the so-called self-consistent renormalization (SCR) theory for explaining the quantum critical phenomena of itinerant magnetism near the QCP is discussed, because it is also useful to understand the CVF phenomena. It is demonstrated that the enhanced magnetic fluctuations near the QCP cause a series of anomalous behaviors in the physical quantities distinct from the Fermi-liquid behavior expected in normal metallic state.

In Chap. 6, "Quantum Criticality of Valence Transition—Experiments and Theory", the quantum criticality of the valence transition is discussed from both aspects of the theory and experiments in detail especially for a series of Yb-based compounds including a quasicrystal $Yb_{15}Al_{34}Au_{51}$.

In Chap. 7, "Interplay Between Magnetic QCP and Valence QCP", the interplay between two QCPs, due to the magnetic transition (discussed in Chap. 5) and the valence transition (discussed in Chap. 6), is discussed. This interplay offers a key novel concept for understanding the heavy fermion systems. As a typical example, case of $CeRhIn_5$ is discussed in detail.

In Chap. 8, "Perspectives–Ubiquity of Critical Valence Fluctuations", the ubiquity of critical valence fluctuations is briefly discussed by introducing the cases already discussed in previous chapters and new cases where the relevance to these phenomena are realized recently. In particular, sharp and pronounced enhancement of the residual resistivity is emphasized as the characteristic of critical valence fluctuations.

In Appendices, derivations of mathematical relations and tables for physical quantities, used in the main text, are discussed for helping the understanding of readers.

References

1. J.M. Lawrence, P.S. Riseborough, R.D. Parks, Rep. Prog. Phys. **44**, 1 (1981)
2. C.M. Varma, Rev. Mod. Phys. **48**, 219 (1976)
3. K.A. Gschneidner, L. Eyring, *Handbook on the Physics and Chemistry of Rare Earths* (North-Holland, Amsterdam, 1978)
4. D. Wohlleben, J. Rhohler, J. Appl. Phys. **55**, 15 (1984)
5. J.W. Allen, R.M. Martin, Phys. Rev. Lett. **49**, 1106 (1982)
6. T. Mito, M. Nakamura, M. Otani, T. Koyama, S. Wada, M. Ishizuka, M.K. Forthaus, R. Lengsdorf, M.M.A. Elmeguid, J. L. Sarrao, Phys. Rev. B **75**, 134401 (2007)
7. H. Sato, K. Yoshikawa, K. Hiraoka, M. Arita, K. Fujimoto, K. Kojima, T. Muro, Y. Saitoh, A. Sekiyama, S. Suga, M. Taniguchi, Phys. Rev. (2004)
8. A.L. Cornelius, J.M. Lawrence, J.L. Sarrao, Z. Fisk, M.F. Hundley, G.H. Kwei, J.D. Thompson, C.H. Booth, F. Bridges, Phys. Rev. B **56**, 7993 (1997)
9. S. Tsutsui, K. Sugimoto, R. Tsunoda, Y. Hirose, T. Mito, R. Settai, M. Mizumaki, J. Phys. Soc. Jpn. **85**, 063602 (2016)
10. S. Watanabe, K. Miyake, J. Phys. Condens. Matter **24**, 294208 (2012)
11. T. Moriya, *Spin Fluctuations in Itinerant Electron Magnetism* (Springer, 1985)

12. T. Moriya, T. Takimoto, J. Phys. Soc. Jpn. **64**, 960 (1995)
13. M. Hatatani, O. Narikiyo, K. Miyake, J. Phys. Soc. Jpn. **64**, 3434 (1995)
14. S. Watanabe, K. Miyake, Phys. Rev. Lett. **105**, 186403 (2010)

Chapter 2
Early History of Critical Valence Fluctuations

In this chapter, it is introduced how the existence of the critical valence fluctuations (CVF) was realized experimentally and theoretically in a canonical heavy fermion superconductor $CeCu_2Si_2$ and its sister compound $CeCu_2Ge_2$, and its solid solution $CeCu_2(Si,Ge)_2$. Namely, unconventional properties, such as *sharp enhancements* of the superconducting transition temperature T_c and the residual resistivity ρ_0, together with the *sharp decrease* of the A coefficient of T^2 term in the resistivity (or effective mass m^* of quasiparticles) under pressure, were interpreted theoretically as the phenomena caused by CVF or enhanced valence fluctuations associated with the sharp valence transition.

2.1 Anomalous Response of Physical Quantities Under Pressure

The existence of the highly enhanced or critical valence fluctuations (CVF) is thought to be first realized through measurements of pressure (P) dependence of a series of physical quantities in $CeCu_2Ge_2$ [1] with the theoretical discussion [2]. Similar behaviors were also observed in its sister compound $CeCu_2Si_2$ by more systematic measurements as shown in Fig. 2.1 [3], where the anomalous (non-monotonic) P dependence of the superconducting transition temperature T_c, the Sommerfeld coefficient γ, the residual resistivity ρ_0, and the A coefficient of T^2 term in the resistivity $\rho(T)$ were reported. Note that T_1^{max} shown in the inset of Fig. 2.1c has considerable P dependence [1] so that T_1^{max} offers the measure of P.

S. Watanabe and K. Miyake, *Quantum Critical Phenomena of Valence Transition*, Springer Tracts in Modern Physics 289,
https://doi.org/10.1007/978-981-99-3518-5_2

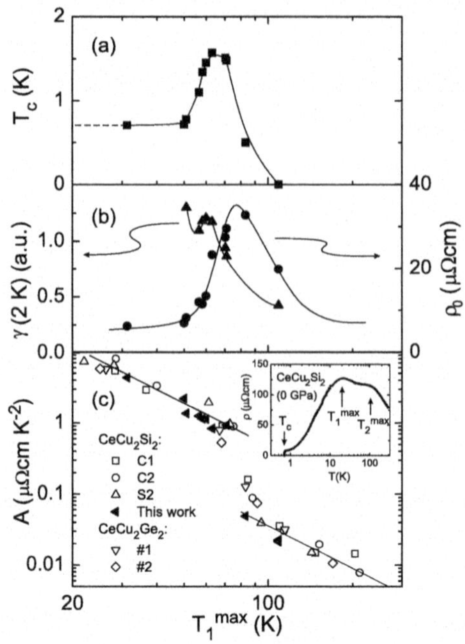

Fig. 2.1 Plotted against T_1^{max} (defined in the inset of (**c**)), a measure of the characteristic energy scale of the system, are **a** the bulk superconducting transition temperature T_c, **b** the residual resistivity ρ_0 and the γ coefficient of the electronic specific heat, and **c** the coefficient A of the $\rho - \rho_0 \simeq AT^2$ law of resistivity. Note that the straight lines represent the expected $A \propto (T_1^{max})^{-2}$ scaling. The maximum of T_c coincides with the start of the crossover between two straight lines, while the maximum in the residual resistivity is situated in the middle of the collapse in A. Pressure increases toward the right-hand side of the scale (high T_1^{max}) [3]. Reprinted from Ref. [3]. Copyright 2007 American Physical Society

2.1.1 *Drastic Decrement of A-Coefficient in Resistivity*

A drastic logarithmic scale decrement in the A coefficient of the T^2 term in the resistivity $[\rho(T) \simeq \rho_0 + AT^2]$ by about two orders of magnitude around the pressure $P = P_v \simeq 4.5$ GPa, as shown in Fig. 2.1c, suggests a sharp valence crossover occurs at around $P = P_v$. This is because the A coefficient is scaled by $(m^*)^2$, the effective mass of quasiparticles, which is a sharply decreasing function of pressure in the case of Ce-based compounds. Note here that T_1^{max} is an increasing function of pressure P and simulates the variation of P. The residual resistivity ρ_0 exhibits a sharp and pronounced enhancement there, as shown in Fig. 2.1b. This implies that the effective mass m^* of the quasiparticles also decreases sharply there, since A is scaled by $(m^*)^2$ [6]. This decrement of m^* implies in turn a *sharp change* in the valence of Ce ion, deviating from Ce^{3+}, considering the fact that the following approximate (but canonical) formula holds in the strong correlation limit [7, 8]:

$$\frac{m^*}{m_{\text{band}}} = \frac{1 - (n_{\text{f}}/2)}{1 - n_{\text{f}}}, \tag{2.1}$$

where m_{band} is the band mass without electron correlations, and n_{f} is the f-electron number per Ce ion.

2.1.2 *Sharp Crossover of Kadowaki-Woods Ratio* A/γ^2

A sharp crossover in the valence of Ce ion gives rise to that of the so-called Kadowaki-Woods (KW) ratio [6], A/γ^2, where γ is the Sommerfeld coefficient of the electronic specific heat. This is because the decrease of A coefficient in Fig. 2.1c is in the logarithmic scale while that of γ in Fig. 2.1b is in the linear scale. As shown in Fig. 2.2, there exist two classes of materials: one for *strongly* correlated systems with $A/\gamma^2 = 1.0 \times 10^{-5}$, and another for *weakly* correlated ones with $A/\gamma^2 = 0.4 \times 10^{-6}$. Note that γ^{-1} is related to the temperature $T_1^{\max}$, where the resistivity $\rho(T)$

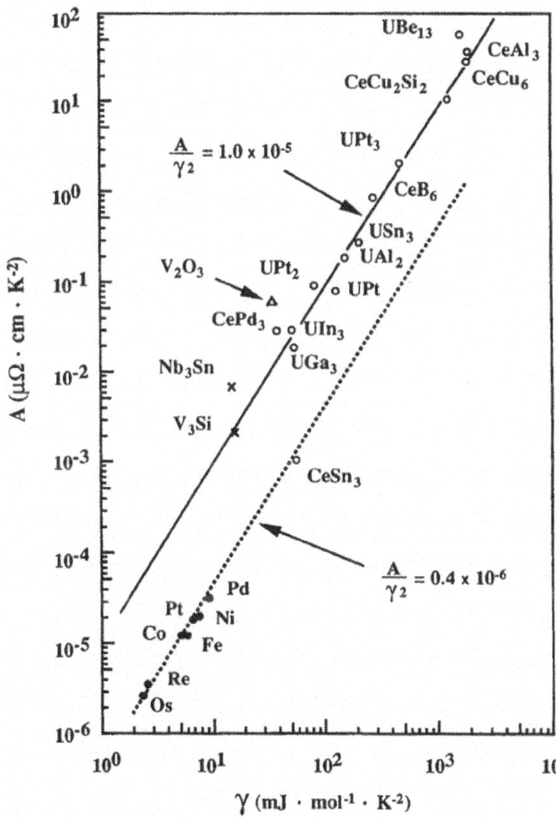

Fig. 2.2 Kadowaki-Woods plot in $\gamma - A$ plane [6, 9]. The line of $A/\gamma^2 = 1.0 \times 10^{-5}$ is for strongly correlated system as heavy fermion systems, and the line of $A/\gamma^2 = 0.4 \times 10^{-6}$ is for less correlated metals in logarithmic scale. Reprinted from Ref. [10]. Copyright 2013 The University of Nagoya Press

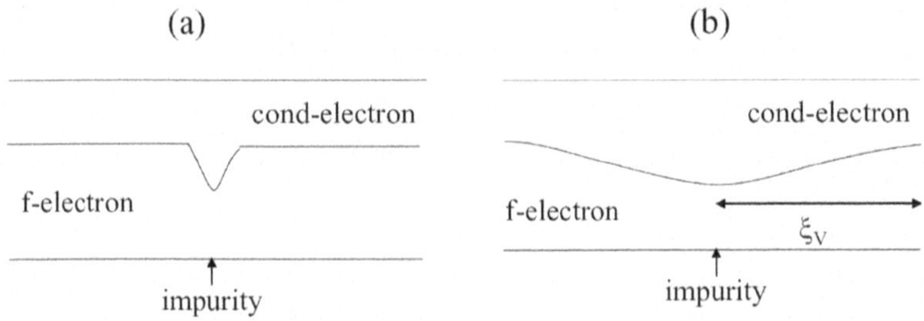

Fig. 2.3 Schematic view of charge distribution of f- and conduction electrons around impurity: **a** at far from $P = P_v$ where the effect of impurity remains as short-ranged, the residual resistivity ρ_0 is not enhanced; **b** at around $P = P_v$ where the effect of impurity extends to long-range, because the correlation length ξ_v of valence fluctuations diverges as $P \to P_v$, leading to the highly enhanced ρ_0. Reprinted from Ref. [11]. Copyright 2007 IOP Publishing

exhibits maximum, as shown in the inset of Fig. 2.1c. This indicates, as discussed in Ref. [9], that the mass enhancement due to the dynamical electron correlation is quickly lost at around $P = P_v$.

2.1.3 Huge Enhancement of Residual Resistivity

The huge enhancement of ρ_0 at $P \simeq P_v$, as shown in Fig. 2.1b, can be understood as the effect that the disturbance in the ratio of numbers of f- and conduction electrons around impurity site extends to the correlation length ξ_v which grows appreciably at around $P = P_v$, giving rise to the strong scattering of quasiparticles. The physical picture of this fact is illustrated in Fig. 2.3. The microscopic justification has been discussed in Ref. [10]. This is in contrast to the effect of AF critical fluctuations on ρ_0 which is rather moderate, as discussed in Ref. [11]. Thus, the critical pressure P_v can be clearly identified by the sharp maximum of ρ_0 as a function of pressure (P).

2.1.4 Direct Experimental Evidence of Sharp Valence Transition

A direct evidence of sharp crossover in the valence of Ce in $CeCu_2Si_2$ was obtained through the measurement of NQR frequency ν_Q as shown in Fig. 2.4 [4]. Leaving the details of analyses to Ref. [4], the amount of deviation $\Delta\nu_Q$ from $P = 3.9$ to $P = 4.5$, $\Delta\nu_Q \simeq -0.03$ MHz, which is the difference [ν_Q("$CeCu_2Si_2$") − ν_Q("$LaCu_2Si_2$")] $\simeq -0.77$ MHz at around $P = 3.9$–4.5 GPa. This negative value of $\Delta\nu_Q$ implies an increase in the itinerant character of 4f-electrons, and the ratio of

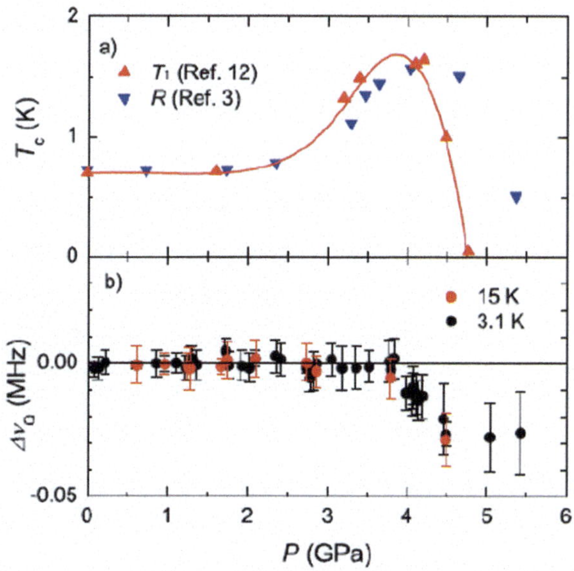

Fig. 2.4 Pressure dependence of the superconducting transition temperature T_c and deviation from the linear P dependence of background in the NQR frequency $^{63}\nu_Q$ in $CeCu_2Si_2$. Reprinted with permission from Ref. [4]. Copyright 2013 The Physical Society of Japan

4f-electrons of acquiring itinerant character is estimated as $0.03/0.77 \simeq 0.04$ of the localized component, suggesting $\Delta n_f \simeq 0.04$.

2.1.5 Birds Eye View of Early History

The birds eye view of those discussed in this chapter is illustrated in Fig. 2.5. Here, a salient property associated with the sharp crossover of the valence is that the temperatures T_i^{max} (i = 1, 2), corresponding to two maxima in the resistivity $\rho(T)$, merge at $P \simeq P_V$.

The existence of the two peaks in $\rho(T)$ and the emergence of them under pressure can be understood as an effect of smearing of the crystalline-electric-field (CEF) splitting by the increase of the c-f hybridization under pressure in general. This is because the splitting of the crystal-filed-level splitting, which is the origin of the separation of the two peaks in the T dependence of the resistivity $\rho(T)$, loses its meaning at $P > P_V$, as discussed in [12]. However, it is nontrivial that the emergence occurs at around $P = P$v. The reasoning for this remarkable fact was given by the discussion in Sect. 2 of Ref. [5]. Such a behavior was also observed in $CeCu_2Ge_2$ [1], $CeAu_2Si_2$ [13] and $CeAl_2$ [14], suggesting that the emergence is a generic property associated with the sharp valence crossover of Ce ion as argued in [3].

The non-monotonic P dependence of T_c shown in Fig. 2.5 suggests that there exist two different mechanisms of superconductivity. Namely, SC II is caused by CVF while SC I is promoted by the enhanced AF spin fluctuations associated with

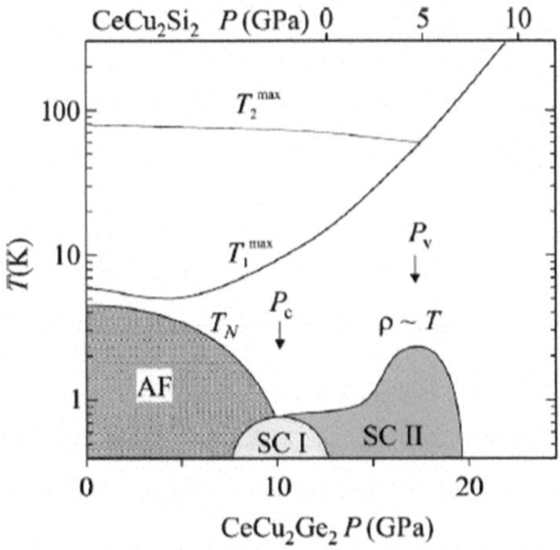

Fig. 2.5 Schematic P-T phase diagram for $CeCu_2(Si/Ge)_2$ showing the two critical pressures P_c and P_v [3]. At P_c, where the antiferromagnetic ordering temperature $T_N \to 0$, superconductivity in region SC I is mediated by antiferromagnetic spin fluctuations; around P_v, in the region SC II, valence fluctuations provide the pairing mechanism and the resistivity is linear in temperature. The characteristic temperatures T_1^{max} and T_2^{max}, defined in the inset of Fig. 2.1c, merge at a pressure $P \simeq P_v$. Reprinted from Ref. [3]. Copyright 2007 American Physical Society

the disappearance of AF order. This point will be discussed later in Sect. 5.3 in detail. This idea was demonstrated more comprehensively by the observation of two separated dome of T_c's in the sister compound $CeCu_2(Si_{0.9}Ge_{0.1})_2$ under P [15].

References

1. D. Jaccard, H. Wilhelm, K. Alami-Yadri, E. Vargoz, Phys. B **259–261**, 1 (1999)
2. K. Miyake, O. Narikiyo, Y. Onishi, Phys. B **259–261**, 676 (1999)
3. A.T. Holmes, D. Jaccard, K. Miyake, Phys. Rev. B **69**, 024508 (2004)
4. T.C. Kobayashi, K. Fujiwara, K. Takeda, H. Harima, Y. Ikeda, T. Adachi, Y. Ohishi, C. Geibel, F. Steglich, J. Phys. Soc. Jpn. **82**, 114701 (2013)
5. K. Miyake, S. Watanabe, Philos. Mag. B **97**, 3495 (2017)
6. K. Kadowaki, S.B. Woods, Solid State Commun. **58**, 507 (1986)
7. T.M. Rice, K. Ueda, Phys. Rev. B **34**, 6420 (1986)
8. H. Shiba, J. Phys. Soc. Jpn. **55**, 2765 (1986)
9. K. Miyake, T. Matsuura, C.M. Varma, Solid State Commun. **71**, 1149 (1989)
10. K. Miyake, H. Maebashi, J. Phys. Soc. Jpn. **71**, 1007 (2002)
11. K. Miyake, O. Narikiyo, J. Phys. Soc. Jpn. **71**, 867 (2002)
12. Y. Nishida, A. Tsuruta, K. Miyake, J. Phys. Soc. Jpn. **75**, 064706 (2006)
13. Z. Ren, G. Giriat, G.W. Scheerer, G. Lapertot, D. Jaccard, Phys. Rev. B **9**, 094515 (2015)
14. B. Barbara, J. Beille, B. Cheaito, J.M. Laurant, M.F. Rossignol, A. Waintal, S. Zemirli, J. Phys. (Paris) **48**, 635 (1987)
15. H.Q. Yuan, F.M. Grosche, M. Deppe, C. Geibel, G. Sparn, F. Steglich, Science **302**, 2104 (2003)

Chapter 3
Fundamentals of Heavy Fermion State

In this chapter, the fundamental concept of *Fermi liquid* is discussed, for understanding the electronic state of heavy fermion system which is closely related to the valence transition and critical valence fluctuations discussed in this monograph. What is the Fermi liquid? It is possible to describe the low-energy excited states in heavy fermion systems, such as electrons system based on f and d electrons, as collections of quasiparticles renormalized by strong interaction among constituent fermions. Effective mass of such quasiparticles is highly enhanced by the effect of the strong interaction among f-electrons.

3.1 Heavy Fermion Systems

3.1.1 Fermi Gas and Fermi Liquid

First, let us review the fundamental properties of free electron model before discussing heavy electron system. In the low enough temperature region than the Fermi temperature, $T \ll T_{\rm F} \equiv E_{\rm F}/k_{\rm B}$ where $E_{\rm F}$ is the Fermi energy and $k_{\rm B}$ is the Boltzmann constant, the specific heat at constant volume $C_{\rm V}$ and the magnetic susceptibility χ are given as follows:

$$C_V \simeq \frac{\pi^2}{3} D_{\rm F} k_{\rm B}^2 T, \tag{3.1}$$

$$\chi \simeq \mu_{\rm B}^2 D_{\rm F}. \tag{3.2}$$

Here, the density of states at the Fermi level $D_{\rm F}$ is given as

$$D_{\rm F} = V \frac{m k_{\rm F}}{\pi \hbar^2}, \tag{3.3}$$

S. Watanabe and K. Miyake, *Quantum Critical Phenomena of Valence Transition*, Springer Tracts in Modern Physics 289,
https://doi.org/10.1007/978-981-99-3518-5_3

where $k_{\rm F} \equiv (3\pi^2 N/V)^{1/3}$ is the Fermi wave number, and corresponding Fermi energy is given as $E_{\rm F} = \hbar^2 k_{\rm F}^2/2m$. In Eq. (3.2), $\mu_{\rm B}$ is the Bohr magneton.

The free electron model gives a correct qualitative description of real metals, even though there exists the Coulomb interaction among electrons, as far as the properties of phenomena at low-temperature region $T \ll T_{\rm F}^* \equiv (m/m^*)T_{\rm F}$ are concerned, where $T_{\rm F}^*$ is the effective Fermi temperature and m^* is the effective mass. Behind this fact, there exists the Fermi-liquid theory [4]. Namely, the specific-heat and magnetic susceptibility of interacting electron systems are given in the low-temperature region $T \ll T_{\rm F}^*$, in parallel to the free electron model, as follows:

$$C_V \simeq \frac{\pi^2}{3} D_{\rm F}^* k_{\rm B}^2 T, \tag{3.4}$$

$$\chi \simeq \frac{\mu_{\rm B}^2 D_{\rm F}^*}{1 + F_0^a}, \tag{3.5}$$

where the density of states at the Fermi level $D_{\rm F}^*$ is expressed with the use of the effective mass m^*, renormalized by the interaction among electrons as in the similar form of free electron model as

$$D_{\rm F}^* = V \frac{m^* k_{\rm F}}{\pi \hbar^2}. \tag{3.6}$$

The Fermi-liquid parameter F_0^a in Eq. (3.5) represents the effect which cannot be taken into account through the effective mass m^*, and is of the order of O(1). Near the ferromagnetic transition, $F_0^a \to -1$, leading to the divergence of χ.

3.1.2 Behaviors of Heavy Fermions at Low Temperatures

First, let us discuss the temperature dependence of $C_V(T)$ in a typical heavy fermion system $CeCu_6$ as shown in Fig. 3.1 [1, 2]. In the temperature region $T \lesssim 1$ K, the relation Eq. (3.4) holds. However, the magnitude is about 10^3-times larger than that of the typical ones in usual metals, i.e., $C_V/T \sim 1\,{\rm mJ/mole \cdot K^2}$. Since its electron number density is of the same order as normal metals, the effective mass m^* of $CeCu_6$ is about 10^3-times larger than normal metals. This means that the effective Fermi temperature is given as $T_{\rm F}^* \equiv (m/m^*)T_{\rm F} \sim 10\,{\rm K}$. Namely, heavy fermion state is realized in the temperature region $T < 1\,{\rm K} \ll T_{\rm F}^* \sim 10\,{\rm K}$, explaining the behavior shown in Fig. 3.1.

This situation can be physically understood as follows. The localized magnetic moment manifested in the high-temperature region $T > T_{\rm F}^*$, giving large entropy, suddenly decreases in the low-temperature region $T < T_{\rm F}^*$. Therefore, the Sommerfeld constant $\gamma \equiv C_V/T = ({\rm d}S/{\rm d}T)_V$ in the low-temperature region $T < T_{\rm F}^* (= T_0)$ is highly enhanced, as seen in Fig. 3.2.

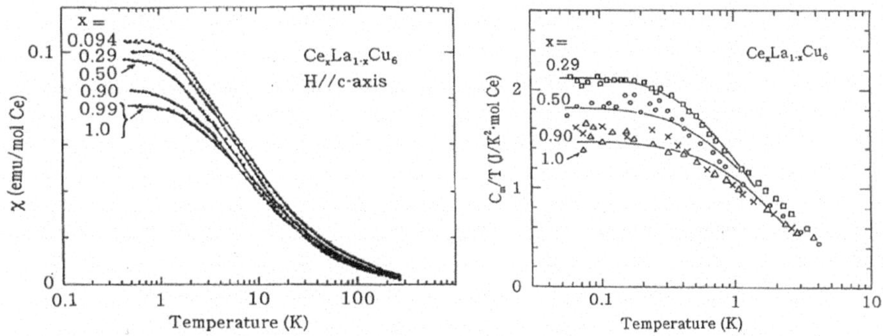

Fig. 3.1 Temperature dependence of χ and C_V/T of $Ce_xLa_{1-x}Cu_6$ [1, 2]. Reprinted with permission from Ref. [1]. Copyright 2013 The University of Nagoya Press

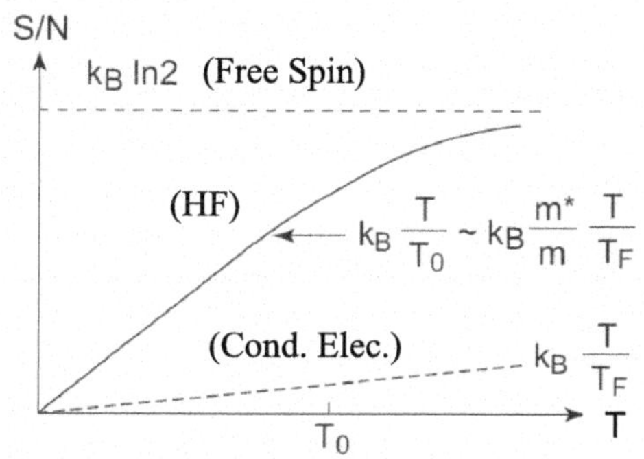

Fig. 3.2 Conceptual figure of T-dependence of the entropy per particle. Solid line represents the case of the heavy fermion system, and dashed line represents the case of the conduction electron band. The characteristic temperature T_0 is the same as $T_F^* = (m/m^*)T_F$ as seen in the Figure. It can be seen that the localized magnetic moment arises in the region $T > T_0 (= T_F^*)$ where $S/N \simeq k_B \ln 2$

The magnetic susceptibility shown in Fig. 3.1 follows the Curie law in wide temperature region $T \gtrsim 10$ K. Namely, in this T-region, the localized magnetic moment is manifested. If the degeneracy of the magnetic moment is assumed as 2, its entropy per particle is $k_B \ln 2$. In usual metals, rather high temperature of $O(T_F \sim 10^4\,\mathrm{K})$ is necessary to attain such high entropy state. On the other hand, in heavy fermion systems, in rather wide low T region of $T_F^* < T \ll T_F$, the localized magnetic moment is manifested, while in the low-temperature region, $T < T_F^* \ll T_F$, the localized moment vanishes and the heavy fermi liquid state is realized.

It can be seen from the T-dependence of the resistivity $\rho(T)$ in the low T-region ($T < 10$ K) that $CeCu_6$ is a good metal as shown in Fig. 3.3 in which $\rho(T)$'s of $Ce_xLa_{1-x}Cu_6$ [1] is shown. On the other hand, the resistivity exhibits T-dependence as $\rho(T) \propto -\ln T$ at high T-region ($T > T_0$ in Fig. 3.2) which is a typical behavior of the Kondo effect and implies the existence of the localized magnetic moment fluctuating independently at each site.

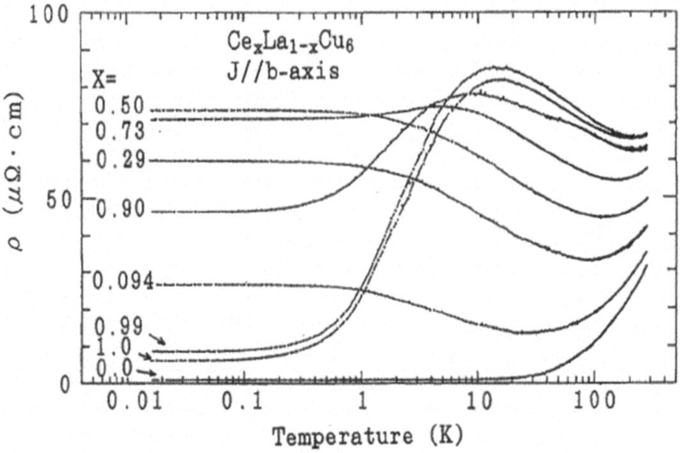

Fig. 3.3 Temperature dependence of the resistivity in $Ce_xLa_{1-x}Cu_6$ [2]. Reprinted with permission from Ref. [2]. Copyright 1986 The Physical Society of Japan

3.2 Origin of Heavy Electron State: Intuitive Understanding

What is the reason why the heavy electron state appears in Ce- or Yb-based compounds? Let us consider the case of Ce-based compounds (the case of Yb can be understood as a hole picture) in which the relatively simple physical picture is possible. The electronic configuration of Ce^{+3}, in its usual valence state, is considered to be $5s^2 5p^6$ closed shell of Xe+$4f^1$. The extent of wave function of $4f^1$ electron has a peak within the Bohr radius from the Ce nuclei, as shown in Fig. 3.4, so that the strong Coulomb repulsion of the order of $U_f \sim 1$ Ry ~ 13.6 eV works among 4f-electrons, forbidding the doubly occupied state of $4f^2$. This is the origin for the appearance of the localized magnetic moment within the $4f^1$ configuration in the higher temperature region, i.e., $T > T_0(= T_F^*) \simeq 10$ K, as seen from the T-dependence of the entropy shown in Fig. 3.2.

Another crucial point is that, as shown in Fig. 3.5, the energy levels of the states with angular momentum $\boldsymbol{J} \equiv \boldsymbol{\ell} + \boldsymbol{s}$ split into the states with $J = 7/2$ and $J = 5/2$ of the energy separation of O(3000K) by the strong spin-orbit interaction. Furthermore, by the relatively weak crystal field effect, the ground state of the sextet with $J = 5/2$ splits into three Kramers doublets with energies separation of O(100 K) as shown also in Fig. 3.5. This is an origin of why the entropy per Ce^{3+} ion remains as the large one, $S/N \simeq k_B \log 2$, down to the low temperature $T_0(T_F^*) \sim 10\,\mathrm{K} \ll 100\,\mathrm{K}$ as shown in Figs. 3.1 and 3.2.

The wave function of $4f^1$ electron of a specified Ce ion hybridizes with the conduction electrons of Ce at around the specified Ce or other ion(s) (Fig. 3.4). Therefore, its energy level broadens with the energy width Γ, which is given by the Kondo temperature T_K as $\Gamma = k_B T_K$ in the case of *impurity problem*. The Γ ($= k_B T_0$ or $k_B T_F^*$) in the lattice system is considered to remain in the same order of $k_B T_K$. If

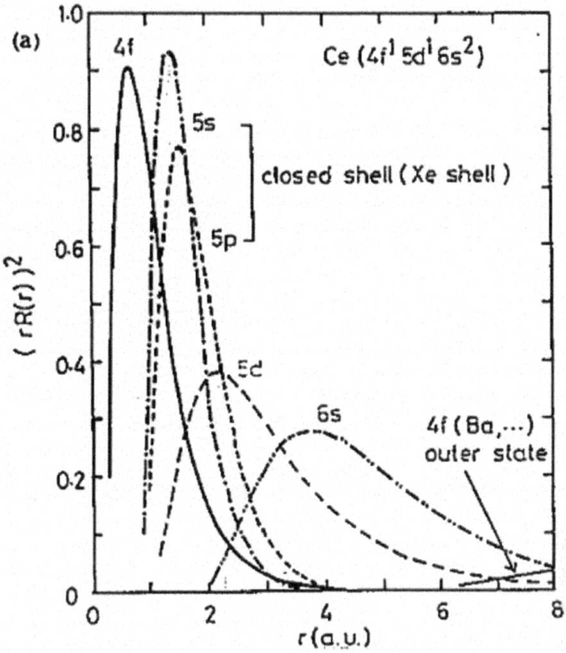

Fig. 3.4 Radial distribution of electron charge density for each atomic orbitals of Ce. Unit of the radius is the Bohr radius [3]. Reprinted with permission from Ref. [3]. Copyright 1991 AGNE Gijutsu Center Co., Ltd

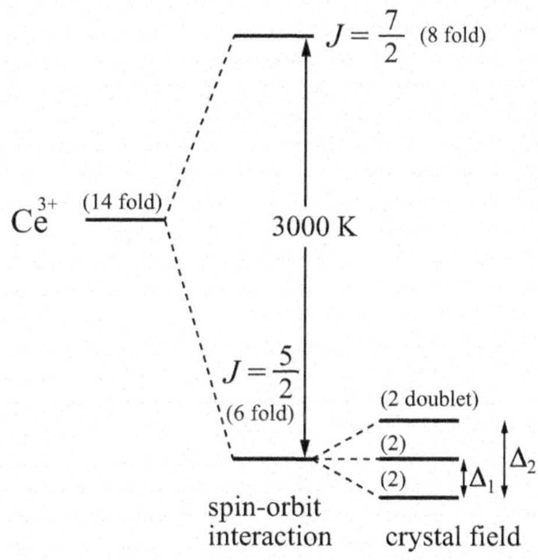

Fig. 3.5 Energy levels of Ce^{3+} in the crystal field

the inequality $\Gamma \ll \Delta_1$ holds, the double degeneracy of the lowest state in Fig. 3.5 is essentially preserved because $k_B T_K \ll \Delta_1$ in general.

For realizing such situation, the following two conditions are necessary:

$$E_f + \Gamma \ll E_F, \tag{3.7}$$

and

$$(E_f + U_f) - E_F \gg \Gamma. \tag{3.8}$$

where $E_F \equiv \mu(T = 0)$ is the *renormalized* chemical potential at $T = 0$. The first condition Eq. (3.7) implies that the $4f^1$ state is stabilized even considering the effect of c-f hybridization, while the second condition Eq. (3.8) implies that the energy of the doubly occupied $4f^2$ state is higher than E_F so that the $4f^2$ state is not realized.

3.3 Representation of Quasiparticles by Green Function

Properties of the heavy quasiparticles (heavy electrons) can be described by the one-particle retarded Green function $G^R_\sigma(\mathbf{k}, \omega)$. In order to obtain the Green function explicitly, detailed theoretical calculations are necessary as discussed in detail in [4].

The explicit form of $G^R_\sigma(\mathbf{k}, \omega)$ so obtained is given as

$$G^R_\sigma(\mathbf{k}, \omega) = \frac{1}{\omega - \xi_\mathbf{k} - \Sigma^R_\sigma(\mathbf{k}, \omega)}. \tag{3.9}$$

Here, Σ^R is called the self-energy and represents the effect of the interaction among constituent particles. The real part of $\Sigma^R_\sigma(\mathbf{k}, \omega)$ represents the effect of energy level shift of quasiparticles while the imaginary part gives the effect of damping in general, in particular, near the Fermi level or $k \sim k_F$, with k_F being the Fermi wave number, and at low-energy $|\omega| \ll E_F$. The pole of ω of $G^R_\sigma(\mathbf{k}, \omega)$ in ω plane represents the dispersion relation of quasiparticles. In order to see the explicit form of $\Sigma^R_\sigma(\mathbf{k}, \omega)$, it is expanded as

$$\Sigma^R(\mathbf{k}, \omega) \simeq \Sigma^R(\mathbf{k}_F, 0) + \frac{\partial \Sigma^R}{\partial \omega}\omega + \frac{\partial \Sigma^R}{\partial \mathbf{k}} \cdot (\mathbf{k} - \mathbf{k}_F) + i\mathrm{Im}\,\Sigma^R(\mathbf{k}_F, \omega). \tag{3.10}$$

Substituting this into Eq. (3.9), the Green function (Eq. (3.9)) is given at around $\omega \sim 0$ and $\mathbf{k} \sim \mathbf{k}_F$ as follows:

$$G^R_\sigma(\mathbf{k}, \omega) \simeq \frac{z_\mathbf{k}}{\omega - \tilde{\xi}_\mathbf{k} + i\tilde{\gamma}_\mathbf{k}}, \tag{3.11}$$

where

$$z_\mathbf{k} \equiv \left(1 - \frac{\partial \Sigma^R}{\partial \omega}\right)^{-1}, \tag{3.12}$$

$$\tilde{\xi}_\mathbf{k} \equiv z_\mathbf{k}\left[\xi_\mathbf{k} + \frac{\partial \Sigma^R}{\partial \mathbf{k}} \cdot (\mathbf{k} - \mathbf{k}_F)\right] \tag{3.13}$$

and

$$\tilde{\gamma}_{\mathbf{k}} \equiv -z_{\mathbf{k}} \mathrm{Im}\, \Sigma^{\mathrm{R}}(\mathbf{k}_{\mathrm{F}}, \tilde{\xi}_{\mathbf{k}}). \tag{3.14}$$

The chemical potential μ is determined as

$$\mu = \epsilon_{\mathbf{k}_{\mathrm{F}}} + \Sigma^{\mathrm{R}}(\mathbf{k}_{\mathrm{F}}, 0), \tag{3.15}$$

where the chemical potential μ in $\xi_{\mathbf{k}} \equiv (\epsilon_{\mathbf{k}} - \mu)$ includes the effect of the interaction.

The retarded Green function Eq. (3.11) is the generalized form of that for the free electron system. The factor in the numerator, $z_{\mathbf{k}}$, is called *renormalization factor*, and represents the ratio of the quasiparticles in the one-particle spectral weight. Namely, it represents the ratio which can be described by the *quasiparticles* in the bare particles with $\mathbf{k} \sim \mathbf{k}_{\mathrm{F}}$. Therefore, the distribution n_σ of bare particle $n_\sigma(\mathbf{k}) \equiv \langle a^\dagger_{\mathbf{k}\sigma} a_{\mathbf{k}\sigma} \rangle$ is expected to exhibit the discontinuity $z_{\mathbf{k}_{\mathrm{F}}}$ at $\mathbf{k}_{\mathrm{F}}$, even under the effect of interaction, as

$$n_\sigma(\mathbf{k}_{\mathrm{F}}^-) - n_\sigma(\mathbf{k}_{\mathrm{F}}^+) = z_{\mathbf{k}_{\mathrm{F}}}. \tag{3.16}$$

Here, $\mathbf{k}_{\mathrm{F}}^{+(-)}$ implies the wave vector with just above (below) the Fermi surface. Namely, the number of particles with wave vector $\mathbf{k}$ exhibits the discontinuity of $z_{\mathbf{k}_{\mathrm{F}}}$ which can be described as the quasiparticles [4]. This implies that the quasiparticles near the Fermi level follows the Fermi distribution function just as the non-interacting fermion system. This implies that the Fermi surface exits even in the interacting fermion systems as the curved surface in the $\mathbf{k}$ space as that exhibits the discontinuity of $n_\sigma(\mathbf{k})$. It was first assumed by the physical intuition of Landau in the Fermi-liquid theory that the Fermi wave number k_{F} is not renormalized by the interaction. Its validity was proved later by microscopic calculations using the Green function technique [4]. Indeed, it is shown that the particle number density (N/V) is given as

$$\frac{N}{V} = 2 \sum_k \theta(G^{\mathrm{R}}(k, 0)). \tag{3.17}$$

This relation implies that the volmue of $\mathbf{k}$ space satisfying the condition $\tilde{\xi}_{\mathbf{k}} < 0$ is not renormalized by the interaction. Namely, the Fermi wave number k_{F} is not renormalized by the interaction in the isotropic system. In the anisotropic systems as real metals, the Fermi wave vector $\mathbf{k}_{\mathrm{F}}$ is deformed in general while the volume in the $\mathbf{k}$-space is not renormalized. On this basis, the picture of quasiparticles satisfying the Fermi statistics is assured, so that the low-temperature properties of strongly correlated fermion systems can be understood by the picture of moderately interacting fermionic quasiparticles system.

3.4 Heavy Fermion System as Fermi Liquid

3.4.1 Periodic Anderson Model

In this subsection, the periodic Anderson model is introduced as a canonical model describing heavy fermion system because it is used in other chapters as an extended version.

The periodic Anderson model is given by the following Hamiltonian $H_{\rm PAM}$:

$$\begin{aligned} H_{\rm PAM} &= \sum_{\mathbf{k},m}\left[\xi_{\mathbf{k}} c^{\dagger}_{\mathbf{k}m} c_{\mathbf{k}m} + E_{\rm f} f^{\dagger}_{\mathbf{k}m} f_{\mathbf{k}m} + (V_{\mathbf{k}} c^{\dagger}_{\mathbf{k}m} f_{\mathbf{k}m} + {\rm h.c.})\right] \\ &\quad + \frac{U}{N}\sum_{\mathbf{k},\mathbf{k}',\mathbf{q}} f^{\dagger}_{\mathbf{k}-\mathbf{q}+} f^{\dagger}_{\mathbf{k}'+\mathbf{q}-} f_{\mathbf{k}'-} f_{\mathbf{k}+}, \end{aligned} \tag{3.18}$$

where $c_{\mathbf{k}m}$ and $f_{\mathbf{k}m}$ are the annihilation operators of the conduction- and 4f-electron with the wave vector $\mathbf{k}$ and the crystal field level specified by the label $m(=\pm)$, respectively. This is possible if the lowest crystal field level is well separated from other Kramers doublet states and doubly degenerate. The quantities, $\xi_{\mathbf{k}}$, $E_{\rm f}$, $V_{\mathbf{k}}$, and U, represent the dispersion of the conduction electrons, the 4f-level energy, the c-f hybridization, and the onsite Coulomb repulsion among 4f-electrons, respectively.

In the case without the onsite Coulomb interaction ($U = 0$), The Hamiltonian Eq. (3.18) is easily diagonalized and the hybridization band is formed as shown in Fig. 3.6. The Green function of f-electron $G_{\rm f}^{(0)}$ is given explicitly as

$$G_{\rm f}^{(0)}(\mathbf{k},\omega) = \frac{1}{\omega - E_{\rm f} - \dfrac{|V_{\mathbf{k}}|^2}{\omega - \xi_{\mathbf{k}}}}. \tag{3.19}$$

The pole of this Green function $G_{\rm f}^{(0)}(\mathbf{k},\omega)$ gives the energy dispersion of quasiparticles as

$$E_{\mathbf{k}}^{\pm} = \frac{1}{2}\left[E_{\rm f} + \xi_{\mathbf{k}} \pm \sqrt{(E_{\rm f} - \xi_{\mathbf{k}})^2 + 4|V_{\mathbf{k}}|^2}\right]. \tag{3.20}$$

3.4.2 Heavy Electrons as Quasiparticles

The effect of the Coulomb repulsion U is expressed as the correction of the 4f-level due to the self-energy $\Sigma_{\rm f}$. Explicitly, the Green function of f-electron is modified as

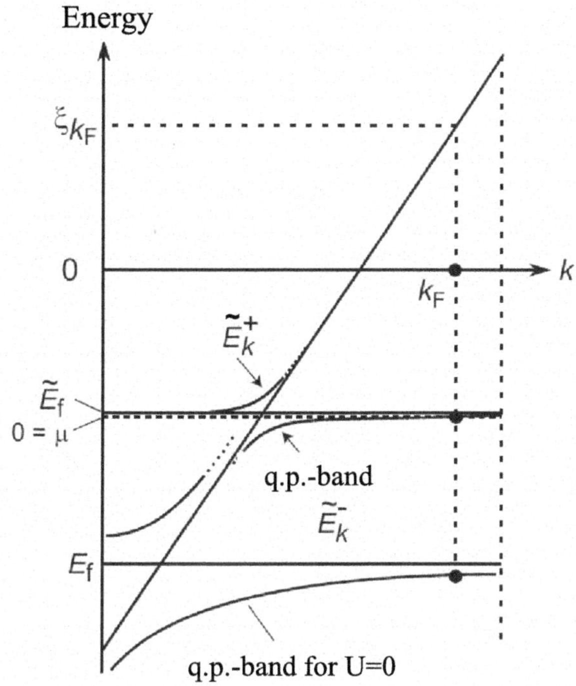

Fig. 3.6 Hybridization band and heavy quasiparticle band

$$G_{\mathrm{f}}^{\mathrm{R}}(\mathbf{k},\omega) = \frac{1}{\omega - E_{\mathrm{f}} - \Sigma_{\mathrm{f}}(\mathbf{k},\omega) - \dfrac{|V_{\mathbf{k}}|^2}{\omega - \xi_{\mathbf{k}}}}. \tag{3.21}$$

Expanding Σ_{f} as in Eq. (3.10), the Green function $G_{\mathrm{f}}^{\mathrm{R}}(\mathbf{k},\omega)$ is obtained as

$$G_{\mathrm{f}}^{\mathrm{R}}(\mathbf{k},\omega) \simeq \frac{a_{\mathrm{f}}}{\omega - \tilde{E}_{\mathrm{f}} + i\tilde{\gamma}_{\mathbf{k}}(\omega) - \dfrac{a_{\mathrm{f}}|V_{\mathbf{k}}|^2}{\omega - \xi_{\mathbf{k}}}}, \tag{3.22}$$

where

$$a_{\mathrm{f}} \equiv \left(1 - \frac{\partial \Sigma_{\mathrm{f}}}{\partial \omega}\right)^{-1}, \tag{3.23}$$

$$\tilde{E}_{\mathrm{f}} \equiv a_{\mathrm{f}}[E_{\mathrm{f}} + \Sigma_{\mathrm{f}}(\mathbf{k},0)], \tag{3.24}$$

and

$$\tilde{\gamma}_{\mathbf{k}}(\omega) \equiv -a_{\mathrm{f}} \mathrm{Im}\,\Sigma_{\mathrm{f}}(\mathbf{k},\omega). \tag{3.25}$$

Note that $\tilde{\gamma}_{\mathbf{k}_{\mathrm{F}}}$ is equal to Γ appearing in Eqs. (3.7) and (3.8).

Comparing Eq. (3.22) with Eq. (3.19) for $U = 0$, the effect of the Coulomb repulsion U appears as the renormalization of the 4f-level near the Fermi level. The pole of Eq. (3.22) gives the dispersion of quasiparticles as follows if the effect of damping $\tilde{\gamma}$ is neglected.

$$\tilde{E}_{\mathbf{k}}^{\pm} = \frac{1}{2}\left[\tilde{E}_{\mathrm{f}} + \xi_{\mathbf{k}} \pm \sqrt{(\tilde{E}_{\mathrm{f}} - \xi_{\mathbf{k}})^2 + 4a_{\mathrm{f}}|V_{\mathbf{k}}|^2}\right]. \tag{3.26}$$

According to Eq. (3.26), the necessary condition for appearance of heavy quasiparticles with extremely small energy dispersion is given as

$$a_{\mathrm{f}} \ll 1. \tag{3.27}$$

This can be understood qualitatively from the dispersion shown in Fig. 3.6. Namely, the renormalized c-f hybridization ($a_{\mathrm{f}}^{1/2}V_{\mathbf{k}}$) is necessary to be suppressed by strong repulsion U among 4f-electrons. On the other hand, the renormalized 4f-level ($\tilde{E}_{\mathrm{f}}$) must be determined such that the number of 4f-electron number at each site satisfies the condition $n_{4\mathrm{f}} \simeq 1$.

3.4.3 One-Particle Spectral Weight of 4f Electron

Behavior of one-particle spectral weight of f-electron $\rho^{\mathrm{f}}(\omega)$ defined below is useful to grasp the physical picture of what is going on in strongly correlated metals. In particular, it is useful for discussing the relation between conduction and magnetism in heavy fermion system.

$$\rho^{\mathrm{f}}(\omega) \equiv -\frac{1}{\pi N}\sum_{\mathbf{k}} \mathrm{Im}G_{\mathrm{f}}^{\mathrm{R}}(\mathbf{k}, \omega + i\delta). \tag{3.28}$$

In the region $\omega \sim 0$, with the use of Eq. (3.23) and the condition $a_{\mathrm{f}} \ll 1$, $\rho^{\mathrm{f}}(\omega)$ is approximated as

$$\rho^{\mathrm{f}}(\omega) \simeq \sum_{\mathbf{k}} a_{\mathrm{f}}\left[\delta(\omega - \tilde{E}_{\mathbf{k}}^{+}) + \delta(\omega - \tilde{E}_{\mathbf{k}}^{-})\right]. \tag{3.29}$$

This value is almost unchanged from that in the case without the repulsive interaction, i.e., $U = 0$. Namely, it is unchanged from the non-interacting case, while the density of states N_{F}^* is of quasiparticles is highly enhanced. This is because the increase in N_{F}^* is almost cancelled by the small factor $a_{\mathrm{f}} \ll 1$. Since the integrated value of $\rho^{\mathrm{f}}(\omega)$, Eq. (3.29), is tiny one of the order of a_{f} ($\ll 1$), its spectral weight of the quasiparticles is also tiny in the case of Ce (or Yb-)-based heavy fermion systems. Indeed, according to the sum-rule for the spectral weight, the following relation holds [4]:

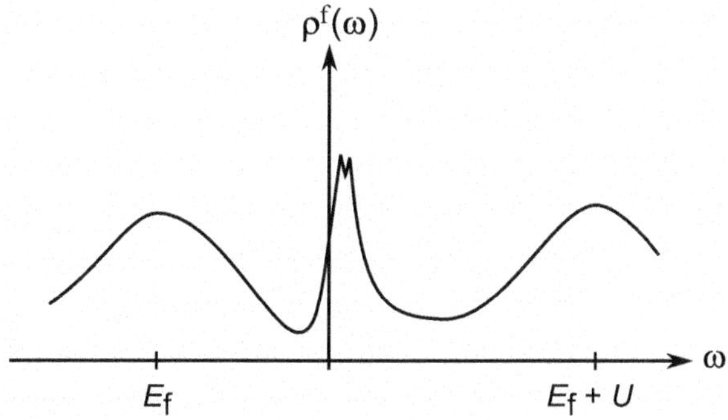

Fig. 3.7 Qualitative behavior of spectral weight $\rho^{\mathrm{f}}(\omega)$ in heavy fermion system with f^1 configuration

$$\int_{-\infty}^{+\infty} d\omega\big[\rho^{\mathrm{f}}_{\uparrow}(\omega) + \rho^{\mathrm{f}}_{\downarrow}(\omega)\big] = 2, \tag{3.30}$$

where the double degeneracy of the ground state of crystal field level is taken into account.

As discussed in Sect. 3.2, the condition for the heavy fermion state to appear is that the energies of $4\mathrm{f}^1$- and $4\mathrm{f}^2$-state are well separated about the energy of the Fermi level. Namely, a conceptual pattern of $\rho^{\mathrm{f}}(\omega)$ should be the form shown in Fig. 3.7. The sharp peak around $\omega = 0$ is related to the quasiparticles around the Fermi level or the Fermi-liquid state. On the other hand, the broad peak around E_{f} represents the state *singly* occupied by 4f-electron, while the broad peak around E_{f} represents the *doubly* occupied f^2 state. This type of structure of $\rho^{\mathrm{f}}(\omega)$ expresses the itinerant property around $\omega = 0$ and the magnetic property of singly occupied state associated with the local magnetic moment. In this sense, it symbolizes the property of "itinerant-localized duality" of f^1-based heavy fermion systems. The former degrees of freedom is related with Fermi-liquid properties such as the transport, while the latter one is related with the appearance of magnetic orders.

3.5 Kadowaki-Woods Relation

Temperature dependence of the resistivity $\rho(T)$ of heavy fermion systems is given, in the low-temperature limit, by the following relation

$$\rho(T) \simeq \rho_0 + AT^2 + \cdots . \tag{3.31}$$

It was pointed out in [5] that the coefficient A varies from one compound to another, while it takes larger value as the effective mass m^* is larger. Kadowaki and Woods organized this fact systematically for huge number of compounds, and pointed out

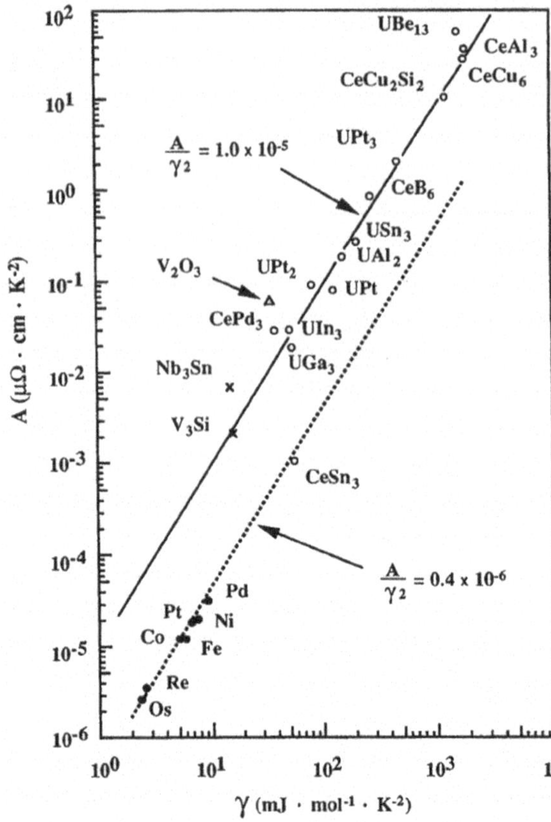

Fig. 3.8 Relation between A and γ^2 (Kadowaki-Woods relation) [8]. Reprinted with permission from Ref. [1]. Copyright 2013 The University of Nagoya Press

[6] that there exists a scaling relation between A and the coefficient γ of the T linear term of the molar specific heat as

$$\frac{A}{\gamma^2} \simeq 1.0 \times 10^{-5}[\mu\Omega\text{cm}(\text{mole} \cdot \text{K}/\text{mJ})^2]. \tag{3.32}$$

Namely, the values of A/γ^2 remains as almost constant universal value, even though the A and γ are very different from compounds to compounds (see Fig. 3.8).

In heavy fermion systems, the resistivity is proportional to the scattering probability given by Eq. (3.25). Then, except for the numerical constant of $O(1)$, it is given by the imaginary part of the self-energy $\text{Im}\,\Sigma_\text{f}$. On the other hand, the specific-heat coefficient $\gamma \equiv \lim_{T\to 0}(C/T)$ is enhanced in proportion to the inverse of the renormalization factor a_f, Eq. (3.23), or the ω dependence of the real part of the self-energy $\text{Re}\,\Sigma_\text{f}$. As a result the specific-heat coefficient γ in heavy fermion states is related with γ_band given by the band structure calculations (without strong correlation effect) as follows:

$$\gamma = \gamma_{\rm band}\left(1 - \frac{\partial {\rm Re}\Sigma(\omega)}{\partial \omega}\right) \equiv \gamma_{\rm band} + \gamma_{\rm cor}. \tag{3.33}$$

Here, the second term $\gamma_{\rm cor}$ represents the mass enhancement due to the strong correlation effect, and is defined as $\gamma_{\rm cor} \equiv -\gamma_{\rm band}[\partial {\rm Re}\Sigma(\omega)/\partial\omega]$.

Since the self-energy $\Sigma(\omega)$ is the analytic function in the upper half of complex ω plane as the Retarded Green function $G^{\rm R}(\omega)$, the real part and imaginary part of the self-energy $\Sigma^{\rm R}(\omega)$ satisfies the Kramers-Krönig relation [7]. Therefore, the following relation holds:

$${\rm Re}\Sigma(\mathbf{k}, \omega + i\delta) = \frac{\rm P}{\pi}\int_{-\infty}^{+\infty} dx \frac{{\rm Im}\Sigma(\mathbf{k}, x + i\delta)}{x - \omega}. \tag{3.34}$$

In the case the electron correlation effect is strong $|\partial {\rm Re}\Sigma(\omega)/\partial\omega| \gg 1$ as in heavy fermion systems, i.e., $\gamma_{\rm cor} \gg \gamma_{\rm band}$, the following approximate relation holds:

$$\frac{A}{\gamma^2} \simeq \frac{A}{\gamma_{\rm cor}^2}. \tag{3.35}$$

Since A is given by ${\rm Im}\Sigma(\omega)$ and $\gamma_{\rm cor}$ is given by ${\rm Re}\Sigma(\omega)$, it is easily imagined that the Kadowaki-Woods relation holds through the relation, Eq. (3.34), irrespective of the largeness of the effective mass m^*, as explicitly shown in the discussions below. Indeed, ${\rm Im}\Sigma(\omega; T)$ exhibits the following dependence on ω and T:

$${\rm Im}\Sigma(\omega; T) = -\left(\frac{1}{2\tau_0} + s\frac{\tilde{\omega}^2}{T_{\rm F}^{*2}}\right)\theta(T_{\rm F}^* - \tilde{\omega}) - \left(\frac{1}{2\tau_0} + s\right)F\left(\frac{\tilde{\omega}}{T_{\rm F}^*}\right)\theta(\tilde{\omega} - T_{\rm F}^*). \tag{3.36}$$

Here, $\tilde{\omega} \equiv [\omega^2 + (\pi T)^2]^{1/2}$, and $F(y)$ is the slowly varying and decreasing function satisfying the relations $F(1) = 1$ and $F(\infty) = 0$. The timescale τ_0 represents the lifetime due to the impurity scattering, and the parameter s is determined so that the approximate relation, $-a_{\rm f}{\rm Im}\Sigma(\omega) \simeq T_{\rm F}^*$ (condition for the unitarity scatterg), holds at $\omega \simeq T_{\rm F}^*$.

The resistivity is given by the Drude-Lorentz formula as

$$\rho = \frac{m^*}{ne^2}\frac{1}{\tau^*}. \tag{3.37}$$

where $1/\tau^* = -a_{\rm f}{\rm Im}\Sigma(T; T)$ is the damping rates of thermally excited quasiparticles, and m^* is the effective mass of the quasiparticles. Therefore, with the use of the relation (Eq. (3.36)), the A-coefficient of T^2 term in the resistivity is proportional to the square of the effective Fermi temperature $T_{\rm F}^*$:

$$A \propto T_{\rm F}^{*-2}. \tag{3.38}$$

On the other hand, substituting Eqs. (3.36) into (3.34), the $\mathrm{O}(\omega)$ term of $\mathrm{Re}\Sigma(\omega)$ is estimated easily giving the following relation:

$$\gamma \propto T_{\mathrm{F}}^{*-1}. \tag{3.39}$$

Here, it is crucial to note that the proportionality coefficients both in Eqs. (3.38) and (3.42) do not include the small renormalization amplitude $a_{\mathrm{f}} \propto (m/m^*) \ll 1$, but are determined essentially by the band structure calculations in which the essential character of heavy fermion state is not taken into account. Since the properties of the hybridization band without the strong short range repulsion U are insensitive to the composition of heavy fermion compounds, the Kadowaki-Woods relation is considered to hold for a variety of heavy fermion compounds, as discussed explicitly in [8].

On the other hand, in the *weakly correlated case* $|\partial\mathrm{Re}\Sigma(\omega)/\partial\omega| \sim 1$, where the band structure calculations are valid, the situation is different as discussed below. First of all, the Drude-Lorentz formula, Eq. (3.40), is valid also for this case:

$$\rho(T) = \frac{m_{\mathrm{band}}}{ne^2}\frac{1}{\tau^*(T)}. \tag{3.40}$$

Here, the T-dependence of the damping rate $1/\tau^*(T)$ is given approximately by

$$\frac{1}{\tau^*(T)} \simeq \frac{1}{\tau_{\mathrm{imp}}} + \frac{T^2}{T_{\mathrm{F}}^{\mathrm{band}}}, \tag{3.41}$$

where $\tau_{\mathrm{imp}} \equiv \tau^*(0)$ is the lifetime due to *only* the impurity scattering. This is because the resistivity $\rho(T)$ reaches the unitarity limit at $T \simeq T_{\mathrm{F}}^{\mathrm{band}}$ (the Fermi energy given by band structure calculation), as discussed in Eq. (3.36). Therefore, the T-dependence of the resistivity $\rho(T)$ in the low-temperature region is approximately given as

$$\rho(T) \simeq \rho_0 + \frac{m_{\mathrm{band}}}{ne^2}\frac{T^2}{T_{\mathrm{F}}^{\mathrm{band}}} + \cdots. \tag{3.42}$$

Namely, the coefficient A is given as

$$A = \frac{m_{\mathrm{band}}}{ne^2}\frac{1}{T_{\mathrm{F}}^{\mathrm{band}}}. \tag{3.43}$$

On the other hand, the specific-heat coefficient γ is given as

$$\gamma = \frac{\pi^2}{2}N\frac{k_{\mathrm{B}}}{T_{\mathrm{F}}^{\mathrm{band}}}. \tag{3.44}$$

On the basis of detailed calculations in [8], the observed value, $A/\gamma^2 \simeq 0.4 \times 10^{-6}$, for weakly correlated metals, is approximately reproduced, by choosing an appropriate and reasonable set of parameters.

As mentioned in Chap. 2, in the systems, $CeCu_2(Si,Ge)_2$, the value of A/γ^2 really crosses over under applied pressures, exhibiting the crossover from the strongly correlated class to that described by the conventional band structure calculation [8].

References

1. A. Sumiyama, Y. Oda, H. Nagano, Y. Ōnuki, K. Shibutani, T. Komatsubara, J. Phys. Soc. Jpn. **55**, 1294 (1986)
2. K. Satoh, T. Fujita, Y. Maeno, Y. Ōnuki, T. Komatsubara, J. Phys. Soc. Jpn. **58**, 1012 (1989)
3. T. Kasuya, Butsuri (Membership Journal of The Physical Society of Japan) **42**, 722 (1987); A. Hasegawa, private communication
4. A.A. Abrikosov, L.P. Gor'kov, I.E. Dzyaloshinskii, *Method of Quantum Field Theroy in Statistical Physics*, 2nd ed., (Pergamon, 1965)
5. N. Sato, A. Sumiyama, S. Kunii, H. Nagano, T. Kasuya, J. Phys. Soc. Jpn. **54**, 1923 (1985)
6. K. Kadowaki, S.B. Woods, Solid State Commun. **58**, 507 (1986)
7. L.D. Landau, I.M. Lifshitz, *Statisitical Physics*, 3rd. ed., (Elsevier 1984). Sect. 123
8. K. Miyake, T. Matsuura, C.M. Varma, Solid State Commun. **71**, 1149 (1989)

Chapter 4
Anomalous Phenomena Due to Critical Valence Transition

In this chapter, anomalous phenomena due to critical valence fluctuations are discussed. In Sect. 4.1, the extended Anderson lattice model is introduced as a minimal model for understanding the anomalous phenomena associated with the valence transition in a unified way. In Sect. 4.2, anomalies in the resistivity and specific heat are discussed from a theoretical point of view. In Sect. 4.3, trends of unconventional superconductivity are introduced and discussed both from the mechanism on the critical antiferromagnetic and valence fluctuations. In Sect. 4.4, microscopic model calculations for the superconducting transition are given on the basis of the extended Anderson lattice model, and the valence-fluctuation mediated superconductivity is discussed. In Sect. 4.5, results on the valence transition obtained by the DMRG (Density-Matrix Renormalization Group) are discussed. In Sect. 4.6, the effects of the magnetic field on the valence transition and valence fluctuations are discussed.

4.1 Extended Anderson Lattice Model for Valence Transition

According to the first principles calculation for the electronic state in Ce, the energy bands are mainly composed of the 4f and 5d orbitals [1]. As the effective model to describe the electronic states in the Ce- and Yb-based heavy fermion systems, we consider the extended Anderson lattice model

$$
\begin{aligned}
H = & \sum_{k\sigma} \varepsilon_k c^{\dagger}_{k\sigma} c_{k\sigma} + \sum_{k\sigma} \left(V_k f^{\dagger}_{k\sigma} c_{k\sigma} + h.c. \right) \\
& + \varepsilon_{\rm f} \sum_{i\sigma} n^{\rm f}_{i\sigma} + U_{\rm ff} \sum_{i} n^{\rm f}_{i\uparrow} n^{\rm f}_{i\downarrow} + U_{\rm fc} \sum_{i\sigma\sigma'} n^{\rm f}_{i\sigma} n^{\rm c}_{i\sigma'},
\end{aligned}
\tag{4.1}
$$

S. Watanabe and K. Miyake, *Quantum Critical Phenomena of Valence Transition*, Springer Tracts in Modern Physics 289,
https://doi.org/10.1007/978-981-99-3518-5_4

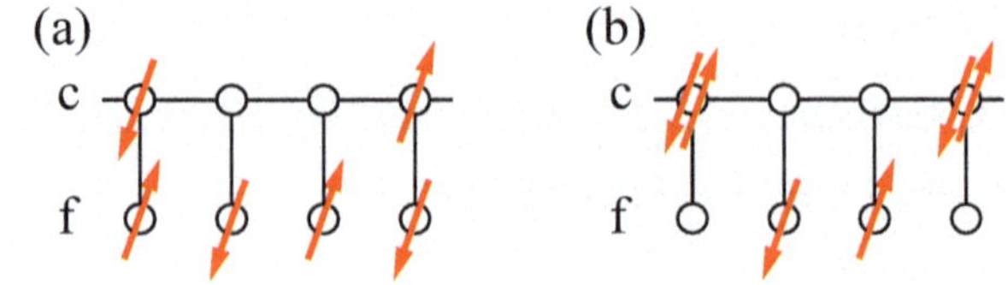

Fig. 4.1 In the extended Anderson lattice model, **a** Kondo state and **b** valence-fluctuation state. In **a**, $\bar{n}_{\rm f} \approx 1$ is realized while in **b** $\bar{n}_{\rm f} < 1$ is realized

where $\varepsilon_{\boldsymbol{k}}$ is the conduction band, $V_{\boldsymbol{k}}$ is the hybridization between f and conduction electrons,[1] $\varepsilon_{\rm f}$ is the ground energy level of the CEF, $U_{\rm fc}$ is the Coulomb repulsion between f and conduction electrons [2–11]. $\sigma = \uparrow, \downarrow$ denotes the Kramers states. This model describes the electronic state where the CEF ground doublet of the 4f-electron shown in Fig. 3.5 hybridizes with conduction electrons. In the Ce-based systems, we take the electron picture, while in the Yb-based systems, we take the hole picture. The model except for the $U_{\rm fc}$ term is known as the Anderson lattice model, which has been studied as the canonical model for the heavy fermion systems. The model (4.1) is "extended" to include the Coulomb repulsion between 4f and conduction electrons $U_{\rm fc}$ as the microscopic origin of the valence transition and the quantum critical point as well as the valence crossover. If we set $V_{\boldsymbol{k}} = 0$ in Eq. (4.1), the model reduces to the Falicov-Kimball model, which was the canonical model to study the valence transitions in 1970s [12]. Thus, it turns out that the model (4.1) is regarded as the extended model in order to describe the heavy fermion state in real rare-earth-based compounds. The general model including the model (4.1) was proposed by C. Varma to discuss the physics of valence fluctuations [13]. The similar model was also proposed to discuss the exciton mechanism of the high-temperature superconductivity in the cuprates [14].

In the model (4.1) for $U_{\rm fc} = 0$, in case that the f-level $\varepsilon_{\rm f}$ is located at a deep position, f-electrons can earn the energy of the double occupancy by $\uparrow$ and $\downarrow$ electrons. However, the strong onsite Coulomb repulsion prohibits such a doubly occupied state so that up to the single occupancy on average is realized. This state with

$$\bar{n}_{\rm f} \equiv \frac{1}{N_{\rm s}} \sum_{i} \sum_{\sigma=\uparrow,\downarrow} \langle n^{\rm f}_{i\sigma} \rangle \approx 1, \tag{4.2}$$

where $N_{\rm s}$ is the number of lattice, is referred to as the Kondo state (see Fig. 4.1a). On the other hand, in case that the f-level $\varepsilon_{\rm f}$ increases to approach the Fermi level, the hybridization strength becomes larger, which gives rise to the state with $\bar{n}_{\rm f} < 1$. This state is referred to as the valence-fluctuation state (see Fig. 4.1b). As will be remarked below Eq. (4.5), at the quantum critical point (QCP) of the valence transition, critical valence fluctuations diverge. To avoid the confusion between the ordinary "valence fluctuation" regime and "critical valence fluctuation", we refer to the former as the mixed-valence regime, following the papers published by the present authors. Since

[1] In case that there exists spatial inversion symmetry at the rare-earth site, there is no onsite hybridization but is offsite hybridization. This gives the momentum dependence of the hybridization $V_{\boldsymbol{k}}$.

$\bar{n}_f$ changes continuously as a function of ε_f, the Kondo state and the mixed-valence state change each other via the crossover.

Then, what happens for finite $U_{fc}(\geq 0)$? For large U_{fc}, in case that the f-level is located at deep enough position, the Kondo state with $\bar{n}_f \approx 1$ is realized as shown in Fig. 4.1a. On the other hand, in case that the f-level increases, the energy loss due to the Coulomb repulsion U_{fc} becomes larger than the energy benefit gained by the depth of the f-level. To avoid this, the f-electron is transferred to the energy band at the Fermi level via the f-c hybridization, giving rise to the state with $\bar{n}_f < 1$ shown in Fig. 4.1b. Namely, this abrupt change implies that the first-order transition from the Kondo state to the mixed-valence state occurs with a jump in the valence of the rare-earth ion, which is the first-order valence transition.

Within the mean-field picture, the condition is given by

$$\varepsilon_f + U_{fc}\bar{n}_c = E_F \tag{4.3}$$

where $\bar{n}_c$ is the number of conduction electrons per site

$$\bar{n}_c \equiv \frac{1}{N_s}\sum_i \sum_{\sigma=\uparrow,\downarrow} \langle n^c_{i\sigma} \rangle \tag{4.4}$$

(see Fig. 3.6) and E_F is the Fermi level. As U_{fc} decreases, the jump at the first-order valence transition $\Delta\bar{n}_f$ becomes smaller and vanishes at the critical point. At the critical point, the valence susceptibility diverges as

$$\chi_v \equiv -\frac{\partial \bar{n}_f}{\partial \varepsilon_f} = \infty. \tag{4.5}$$

Namely, valence fluctuation diverges critically. The critical point at $T = 0$ is the QCP, as shown in Fig. 1.4b. This critical valence fluctuation gives rise to the quantum critical phenomena of valence transition.[2] As U_{fc} further decreases, the Kondo state and the mixed-valence state changes via the crossover.

The model (4.1) has been studied in the spatial dimension $d = 1, 2, 3$, and ∞ by the slave-boson mean-field theory, the density-matrix renormalization group method, and the dynamical mean-field theory [2, 3, 5, 6, 8, 11]. The phase diagram is shown schematically in Fig. 4.2. As understandable in Eq. (4.3), the valence transition is caused by the onsite Coulomb repulsion locally, the detail of the phase diagram does not depend on the spatial dimensionality. In the phase diagram of the f-level ε_f, inter-orbital Coulomb repulsion U_{fc}, and temperature T, the first-order valence-transition surface terminates at the critical line connecting the critical points, from which the valence-crossover surface extends. The point where the critical line touches the $T = 0$

[2] The critical valence fluctuation diverges at the quantum critical point of the valence transition. It is necessary not to confuse the valence fluctuation which exists even for $U_{fc} = 0$ in the model (4.1). To distinguish them, in the papers of Refs. [3, 6–9], the former is denoted as critical valence fluctuation and the latter is denoted as mixed valence.

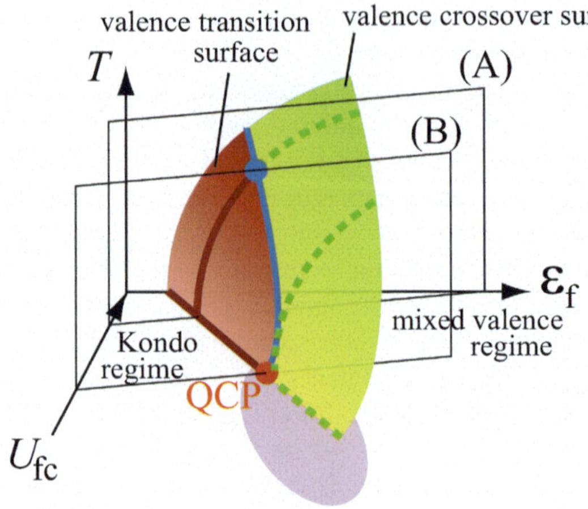

Fig. 4.2 The phase diagram of temperature T, f-level ε_f, and inter-orbital Coulomb repulsion U_{fc} of the Anderson lattice model. The point where the critical line dividing the first-order valence transition surface and valence-crossover surface touches the $T = 0$ plane is the quantum critical point (QCP) of the valence transition. Superconducting correlation is enhanced by critical valence fluctuations mainly in the Kondo regime near the QCP (pink region)

plane is the QCP of the valence transition. In the Kondo regime for smaller U_{fc} near the QCP, the superconducting correlation due to the critical valence fluctuations is enhanced [2, 3], which will be discussed in detail in Sects. 4.4 and 4.5.

In Fig. 4.2, the reason why the valence-transition surface and the valence-crossover surface are placed over the valence-fluctuation regime is that in general the free-energy gain is larger in the Kondo regime at high temperatures due to the entropy of the Kramers degeneracy of the localized limit of the 4f-electron state (see Fig. 3.2).

In the ε_f-U_{fc} plane at $T = 0$, the Kondo regime and the mixed-valence regime are smoothly connected detouring the first-order valence-transition line and the QCP. Hence, in both the regimes, the large Fermi surface is realized, whose volume contains both numbers of f-electrons and conduction electrons according to Luttinger's theorem.[3].

In general, by applying pressure to the Ce (Yb)-based compounds, the f-level ε_f increases (decreases) and the hybridization strength $|V_{\boldsymbol{k}}|$ increases. It is noted that Fig. 4.2 is the phase diagram for a certain hybridization strength. Although by applying pressure, the hybridization strength increases so that the position of the first-order valence-transition surface as well as the valence-crossover surface shifts, the cross section (A) corresponds to Figs. 1.1 and 1.2 and the cross section (B)

[3] At $T = 0$, there exists the momentum where the momentum distribution function becomes discontinuous. The Fermi surface is defined by the surface of the volume enclosed by the momentum. The volume enclosed by the Fermi surface only depends on the number of electrons, which is not affected by the many-body effect. This is referred to as Luttinger's theorem or Luttinger's sum rule [15]. In the uniform Fermi liquid, the Fermi wave number $\boldsymbol{k}_F$ is not affected by the many-body effect, which provides the key concept of Landau's Fermi-liquid theory.

corresponds to Fig. 1.4b where the horizontal axis g is set as pressure P. Namely, it turns out that the microscopic parameter which controls Fig. 1.4a–c is the inter-orbital Coulomb repulsion U_{fc}. Indeed, in the Ce metal where the γ-α transition takes place, the main component of the conduction electron is the 5d orbital so that the inter-orbital Coulomb repulsion U_{fc} is onsite interaction, which cannot be neglected. On the other hand, in the Ce-based compounds and Yb-based compounds, in general, conduction electrons except for the Ce or Yb atom also contribute to the conduction band (i.e., ξ_k in Fig. 3.6). In these systems, as the contribution from the 5d orbital at the Ce or Yb atom to the energy band located at the Fermi level decreases, the cross section illustrated in Fig. 4.2 tends to be located at the smaller-U_{fc} position. Then, the cross sections (A) and (B) are considered to correspond to Fig. 1.4a, b, respectively.

When the cross section in Fig. 4.2 is located at the position smaller U_{fc} than that of the cross section (B), it corresponds to Fig. 1.4c. In this case, the valence-crossover temperature T_v^* appears in the T-g phase diagram (g is the control parameter such as pressure P, magnetic field, and chemical doping) as illustrated as the dashed line in Fig. 1.4c. Indeed, the valence-crossover temperature was observed in several Ce- and Yb-based compounds, as discussed in Sects. 4.6.3.2, 6.5.2, 7.2, and 8.1.1.

4.2 Anomalies in Resistivity and Specific Heat

4.2.1 *T-linear Resistivity and Enhanced Sommerfeld Coefficient at $P = P_v$*

In this subsection, the reason why the nearly T-linear temperature dependence of the resistivity is observed in a narrow pressure region around $P \sim P_v$ (as shown in the inset of Fig. 4 of Ref. [43]), on the basis of a semi-phenomenological theory. More solid treatment on the basis of the mode coupling theory (or self-consistent renormalization theory) will be discussed in Chap. 6 in which the T-dependence of the resistivity is given in Sect. 6.4 as shown in Fig. 6.3.

However, a rough but intuitive understanding would be possible by this simplified approach. The origin of the d-wave superconductivity induced by CVF is the locality of pairing interaction as discussed in Sects. 4.4.2.1 and 4.4.2.2. Namely, the static limit of the effective interaction $\Gamma^{(0)}(q, \mathrm{i}\omega_m)$ between quasiparticles is enhanced greatly around $P \sim P_v$, and is *almost independent* of q, the momentum transfer, up to $\sim$3/2 of p_F. This must be due to the local nature of CVF whose response function, $\chi_v(q, \omega)$, is also almost q-independent in the low frequency (ω) region. On this observation, we present here a semi-phenomenological theory explaining the nearly T-linear resistivity and the enhancement of the Sommerfeld coefficient γ around $P \sim P_v$.

We adopt an exponentially decaying phenomenological form for the valence-fluctuation propagator (dynamical valence susceptibility) χ_v:

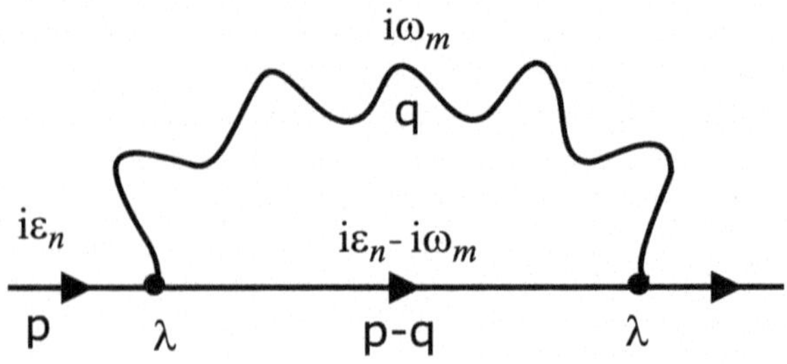

Fig. 4.3 Feynman diagram for the self-energy due to the one-fluctuation mode exchange process. λ denotes the coupling between the valence-fluctuation mode and the quasiparticles

$$\chi_{\rm v}(q,\omega) \equiv {\rm i}\int_0^\infty {\rm d}t e^{{\rm i}\omega t}\langle[n_{\rm f}(q,t), n_{\rm f}(-q,0)]\rangle \tag{4.6}$$

$$= \frac{K}{\omega_{\rm v} - {\rm i}\omega}, \text{ for } q < q_{\rm c} \sim p_{\rm F} \tag{4.7}$$

where $n_{\rm f}(q)$ is the Fourier component of the number of f-electron per Ce site, K is a constant of $O(1)$, and the parameter $\omega_{\rm v}$ parameterizes the closeness to criticality. $\omega_{\rm v}$ is inversely proportional to the valence susceptibility $\chi_{\rm v}(0,0) = -(\partial n_{\rm f}/\partial\epsilon_{\rm f})_\mu$, $\epsilon_{\rm f}$ being the atomic level of f-electron of Ce ion, and μ the chemical potential.

The retarded self-energy $\Sigma^{\rm R}_{\rm vf}(p, \epsilon + {\rm i}\delta)$ gives a measure of the quasiparticle effective mass and lifetime, respectively, in its real and imaginary parts. The explicit form of $\Sigma^{\rm R}_{\rm vf}(p, \epsilon + {\rm i}\delta)$ corresponding to the Feynman diagram Fig. 4.3 can be calculated in a standard way.

In typical limiting cases, ${\rm Im}\,\Sigma^{\rm R}_{\rm vf}(p_{\rm F}, \epsilon)$ can be straightforwardly calculated in the approximation on the dispersion of the quasiparticles, $\xi_{\mathbf{p}-\mathbf{q}} \simeq -vq\cos\theta$, where θ is the angle between **p** and **q**, v is the quasiparticle velocity, and p is assumed to be on the Fermi surface, i.e., $p = p_{\rm F}$.

In the case, $T = 0$ and $\epsilon \neq 0$:

$${\rm Im}\,\Sigma^{\rm R}_{\rm vf}(p_{\rm F}, \epsilon) \simeq -\frac{|\lambda|^2 K q_{\rm c}^2}{32\pi^2 v}\ln\left(1 + \frac{\epsilon^2}{\omega_{\rm v}^2}\right), \tag{4.8}$$

where $q_{\rm c}$ is the cutoff wavenumber of the order of $k_{\rm F}$.

In the case, $\epsilon = 0$ and $0 < T \ll \epsilon_{\rm F}$:

$${\rm Im}\,\Sigma^{\rm R}_{\rm vf}(p_{\rm F}, 0) \simeq -\frac{|\lambda|^2 K q_{\rm c}^2}{4\pi^2 v}\frac{T}{\omega_{\rm v}}\tan^{-1}\frac{T}{\omega_{\rm v}}. \tag{4.9}$$

This leads to the relation, ${\rm Im}\,\Sigma_{\rm vf}(p_{\rm F}, \epsilon = 0) \propto T/\omega_{\rm v}$ for $T \gg \omega_{\rm v}$, which implies that almost all the critical valence-fluctuation modes can be regarded as classical at $T > \omega_{\rm v}$. As a result, T-linear dependence stems from the asymptotic form of the Bose distribution function $g(\omega) = 1/(e^{\omega/T} - 1) \simeq T/\omega$.

In the limit $\omega_v \ll v p_F$, the real part of the self energy can be also calculated easily at $T = 0$ and $\epsilon \sim 0$, leading to

$$\mathrm{Re}\Sigma^{R}_{vf}(p_F, \epsilon) \simeq -\frac{|\lambda|^2 K q_c^2}{2\pi^2 v} \frac{\epsilon}{\omega_v} \int_0^1 du \frac{1-u^2}{u^2+1} \ln\left|\frac{1}{u}\right| \tag{4.10}$$

$$\propto -\frac{\epsilon}{\omega_v}, \tag{4.11}$$

where $u = vqt/\omega_v$.

The T-linear dependence of $\mathrm{Im}\Sigma^{R}_{vf}(p, 0)$, for $T > \omega_v$, Eq. (4.9), implies T-linear resistivity, as the quasiparticles are subject to the large angle scattering by the almost q-independent critical valence-fluctuation modes. These are effective in a wide region in the Brillouin zone due to their local nature and easily couple to the Umklapp process of quasiparticle scattering. This result is consistent with the experimental fact that nearly T-linear resistivity is observed in a narrow pressure region around P_v which is considered to correspond to a nearly critical transition of the valence of Ce ion.

Such a T-linear dependence has been discussed in the context of high-T_c cuprates with a marginal Fermi-liquid (MFL) assumption, [17] and charge-transfer fluctuations were once considered as an origin for MFL, [17, 18] while further theoretical models have been put forth up to now by Varma et al. [19]. Excepting the T-linear resistivity, the present result is different from MFL behavior. The self-energy exhibits different energy dependence, while the idea for the origin of our singular behavior shares aspects similar to the first idea of a charge-transfer mechanism for high-T_c cuprates. Indeed, $\Sigma(\epsilon)$ in the MFL model is given as $\Sigma(\epsilon) \propto (\epsilon \ln \epsilon - \mathrm{i}|\epsilon|)$ [17] which is indeed different from the present case (Eqs.(4.8) and (4.11)). In any case, it is noted that the T-linear resistivity is accompanied by the peak of T_c in both systems, high-T_c cuprates and $CeCu_2Si_2$.

The result, Eq. (4.11), implies that the mass enhancement $(1 - \partial\mathrm{Re}\Sigma^{R}_{vf}(\epsilon)/\partial\epsilon)$ is highly increased around $P \sim P_v$. Namely, the effective mass is given by

$$m^* \propto \bar{m}\frac{1}{\omega_v}, \tag{4.12}$$

where $\bar{m}$ is the effective mass renormalized by the conventional correlation effect, leading to heavy electrons, i.e., not including the effect of CVF. This latter effective mass $\bar{m}$ exhibits a drastic decrease around $P \sim P_v$, while the second factor of Eq. (4.12) is enhanced. Both effects should be reflected in the Sommerfeld coefficient γ, so that the peak of $\gamma \propto m^*$ is shifted to the lower pressure (larger $\bar{m}$) side, and the anomaly of γ due to the valence fluctuations may be smeared to some extent. Nevertheless, some traces should be observed. (The shift of peak of γ can be understood as the superposition of the two trends using a model P-dependence of $\bar{m}$ and ω_v.) Indeed, the experimental result presented in Fig. 2.1 may be explained by this effect. A similar trend has been reported in a series of solid solutions of Ce-based intermetallic compound [20].

4.3 Trends of Unconventional Superconductivity

Since the discovery of paramagnon mechanism of the Cooper pair formation in the superfluid ^{3}He in early 70s [21, 22], the electronic mechanism of superconductivity has been widely recognized in 80s through the discovery of unconventional superconductivity in heavy fermion compound $CeCu_2Si_2$ which later turned out to be located near the antiferromagnetic (AF) phase boundary [23]. In particular, it was proposed first in heavy fermion superconductivity that AF spin fluctuation mechanism works in Ce-based heavy fermions [24, 25], and later on that it works also in high-T_c cuprates [26, 27], and some organic metals which are all in proximity to the AF ordered state [28].

In 90s, the pressure-induced superconducting phase has been identified near the AF quantum critical point (QCP) in a series of Ce-based heavy fermion compounds, such as $CeIn_3$ [29], $CePd_2Si_2$ [30], $CeRh_2Si_2$ [31] as shown in Fig. 4.4, where AF spin fluctuation mechanism is considered to also work [32]. Let us call these compounds class (I).

In contrast with these developments, apparently different behavior of T_c against the pressure P was found in $CeCu_2Si_2$ and $CeCu_2Ge_2$. In particular, for $CeCu_2Si_2$, this was found even at quite early stage of research [34]. Both compounds exhibit a pronounced dome of T_c at P far above P_c which corresponds to the AF-QCP. Namely, T_c exhibits broad mountain at around $P \simeq 5$ GPa for $CeCu_2Si_2$ [34, 35], and at around $P \simeq 16$ GPa for $CeCu_2Ge_2$ [36–38]. These two isostructural compounds have a similar phase diagram in P-T plane and similar physical properties, if the origin of the pressure is shifted by about 12 GPa [39]. The schematic behavior of T_c against the pressure P is illustrated in Fig. 4.5. Around the pressure, corresponding to the maximum of T_c, they exhibit a drastic decrease of the coefficient A of the T^2 term of the resistivity [$\rho(T) = \rho_0 + AT^2$] and sharp enhancement of the residual resistivity ρ_0. Let us call these compounds class (II).

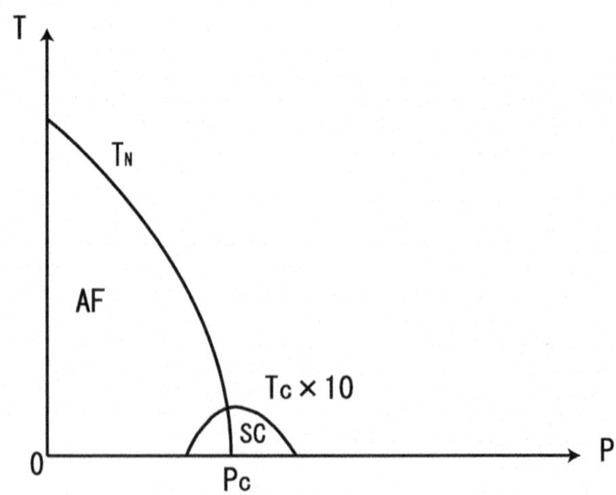

Fig. 4.4 Schematic P-T phase diagram for CeT_2Si_2 (T=Pd, Rh) and $CeIn_3$ showing rather low T_c in very narrow range of $P \sim P_c$. The T-dependence of the resistivity follows the non-Fermi-liquid law $\rho(T) \sim T^n$, with $n \simeq 1.5$. It is noted that the scale of T_c is expanded by 10 times. Reprinted from Ref. [65]. Copyright 2007 IOP Publishing

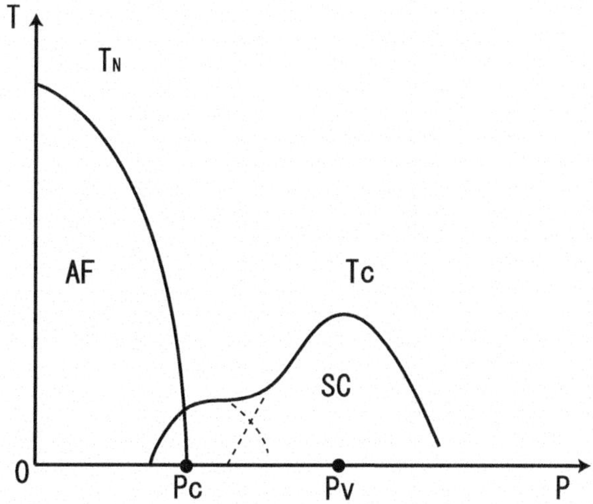

Fig. 4.5 Schematic *P*-*T* phase diagram for $CeCu_2(Si/Ge)_2$ showing the two critical pressures P_c and P_v. At P_c, where the Néel temperature $T_N \to 0$, superconductivity in region *SC I* is mediated by AF spin fluctuations; around P_v, in the region *SC II*, critical valence fluctuations provide the pairing mechanism. At $T > T_{SC\ II}$, the resistivity is linear in temperature. Reprinted from Ref. [65]. Copyright 2007 IOP Publishing

It turned out gradually that these phenomena, observed in materials of class (II), have a single origin of critical valence transition (CVT) of Ce ion at $P = P_v$ [2, 40, 41], while it was suggested that the enhancement of T_c is related to the valence transition of Ce ion from the first stage of research [34]. The nearly T-linear resistivity is observed in a wide temperature region up to a few 10 K except for very low-temperature region, which clearly shows that the universality class of QCP at $P = P_v$ is different from that of AF-QCP [37, 38]. It is also noted that T_c of the canonical class (I), shown in Fig. 4.4, is far lower than the present case shown in Fig. 4.5.

This novel point of view was reinforced by the following two experiments. First, the two separate domes of T_c was observed in $CeCu_2(Si_{0.9}Ge_{0.1})_2$, one at around $P \sim P_c$ and one $P \sim P_v$ [42], suggesting the existence of two different mechanism related to fluctuations associated with two different QCPs, one for AF-QCP and another for QCP associated with CVT. It was also revealed that T_c around P_c is far lower than that around P_v by about half. Second, the detailed and fine experiments of $CeCu_2Si_2$ under pressures was performed including measurements of the specific heat in addition to the resistivity [43], and their results were successfully explained by the theory which allows the CVT and fluctuations [2, 41]. Thus, the critical-valence-fluctuation mechanism, new mechanism of superconductivity of electronic origin, was identified both experimentally and theoretically.

4.4 Results on Microscopic Treatments of Extended PAM

In this section, a microscopic model calculation for the superconducting transition is given on the basis of the extended Anderson lattice model discussed in the Sect. 4.1 [66].

4.4.1 Superconductivity Induced by Critical Valence Fluctuations

In Ref. [2] the model Hamiltonian Eq. (4.1) with a spherical conduction band was treated by the mean-field approximation in the slave bosons formalism to discuss the possibility of valence transition and by the Gaussian fluctuation theory around the mean-field solution to discuss a possible superconducting state. As shown in Fig. 4.6, central results of Ref. [2] are summarized as follows:

(1) Sharp valence change is caused by the effect of $U_{\rm fc}$ with moderate strength of the order of D, the bandwidth of conduction electrons, when the f-level $\epsilon_{\rm f}$ is tuned, representing the effect of the pressure.
(2) The superconducting state is induced by the process of exchanging the slave-boson fluctuations for the values of $\epsilon_{\rm f}$ around which the sharp valence change occurs.
(3) The symmetry of so induced superconducting state is d-wave if the spherical model is adopted for conduction electrons. However, as seen in the argument below, this anisotropic pairing is induced by the CVF modes due to their *almost local* nature.

It is noted that aspects of the pressure dependence of $T_{\rm c}$ and A-coefficient of the resistivity observed experimentally [37], shown in Fig. 5 in Ref. [43], are well

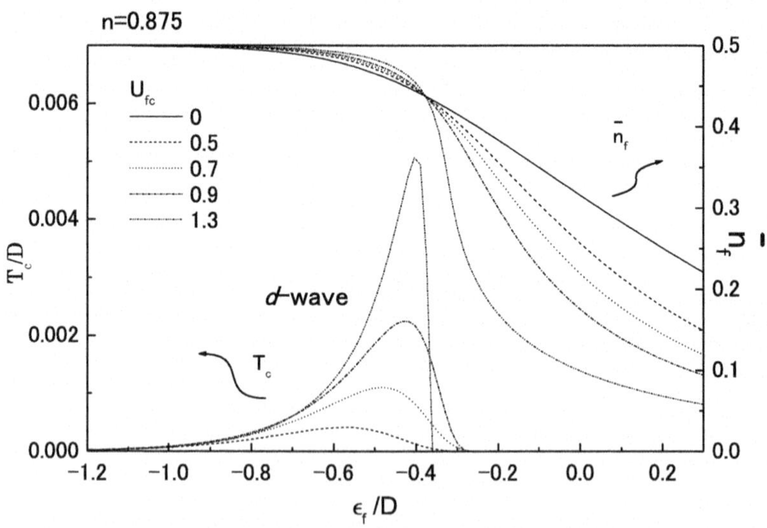

Fig. 4.6 $T_{\rm c}$ for d-wave channel and $\bar{n}_{\rm f}$, f-electron number per site and "spin", as a function of $\epsilon_{\rm f}$. Total number of electrons per site and "spin" is set as $n = 0.875$, and the c–f hybridization is set as $V = 0.5D$. The unit of energy is D, the Fermi energy of parabolic conduction band when it was decoupled from f-electrons: the dispersion $\epsilon_{\rm k} = k^2/2m - D$ is adopted for the conduction electrons. Reprinted from Ref. [2]. Copyright 2000 The Physical Society of Japan

reproduced since ϵ_f simulates the pressure variation. The derivation of these results will be briefly shown below.

4.4.2 Mean-Field Solution for Valence Change

The extended Anderson model Hamiltonian, Eq. (4.1), was treated by the *slave-boson* and large-N expansion approach [45] which has been shown to be effective for studying thermodynamic properties, the superconducting transition temperature, transport properties, etc. of the conventional periodic Anderson model (PAM) [45, 47, 48] as far as the low-temperature region $T \ll T_K$ is concerned. Following these methods, the effect of the f-c Coulomb interaction U_{fc} in Eqs. (4.1) and (4.2) was taken into account on the level of a Gaussian fluctuation approximation.

The Hamiltonian Eq. (4.1) with the constraint given by Eq. (4.2) is generalized to the case where the N-fold degeneracy exists, including both spin and orbital degrees of freedom. For this purpose, the slave-boson operators b's are introduced to exclude the doubly occupied f-electron state, assuming $U_{ff} = \infty$. Thus, the Hamiltonian is expressed as follows:

$$H = \sum_m \left[\sum_k \left(\varepsilon_k c^\dagger_{km} c_{km} + \varepsilon_f f^\dagger_{km} f_{km} \right) + \frac{V}{\sqrt{N_L}} \sum_{k,q} \left(c^\dagger_{kqm} f_{k+qm} b^\dagger_q + \text{h.c.} \right) \right.$$
$$\left. + U_{fc} \sum_{il} f^\dagger_{il} f_{il} c^\dagger_{im} c_{im} \right], \quad (4.13)$$

where b's are the slave-boson operators b's describing f^0 state with the constraint at each i-th site, i.e.,

$$Q_i = \sum_m f^\dagger_{im} f_{im} + b^\dagger_i b_i = 1, \quad (4.14)$$

in order to maintain the equivalence of the truncated Hilbert space and the original one, and m represents the component of the N-fold degeneracy of the spin-orbit state.

To generate $1/N$-expansion variables, the following parameterization is introduced:

$$\begin{cases} Q_i \to q_0 N \\ b \to b\sqrt{N} \\ V \to V/\sqrt{N} \\ U_{fc} \to U_{fc}/\sqrt{N} \end{cases} \quad (4.15)$$

In the calculations below, the radial gauge was adopted following Ref. [45]. (Although the radial and the Cartesian gauge formulations are ultimately equivalent, spurious infrared divergences do not appear in the radial gauge approach. Millis and Lee [48])

After the local gauge transformation, $b_i(\tau) = \rho_i(\tau)e^{\mathrm{i}\theta_i(\tau)}$, $f_{im}(\tau) = f'_{im}(\tau)e^{\mathrm{i}\theta_i(\tau)}$ and $\lambda'_i(\tau) \equiv \lambda_i + \dot{\theta}_i(\tau)$, was performed, the partition function is given as

$$Z = \int D(cc^\dagger ff^\dagger \rho\lambda)\exp(-S), \tag{4.16}$$

where the action S is given as

$$S = \int_0^\beta \mathrm{d}\tau \Bigg[\sum_{kk'm} f^\dagger_{ksm}(\tau)\{(\partial_\tau + \epsilon_\mathrm{f})\delta_{kk} + \frac{1}{\sqrt{N_\mathrm{L}}}\mathrm{i}\lambda(k-k';\tau)\} f_{k'm}(\tau) \tag{4.17}$$

$$+ \sum_{km} c^\dagger_{km}(\tau)(\partial_\tau + \epsilon_k)c_{km}(\tau) \tag{4.18}$$

$$+ \frac{V}{\sqrt{N_\mathrm{L}}}\sum_{kk'm}\{c_{km}(\tau)f_{km}(\tau)\rho(k-k';\tau) + \mathrm{h.c.}\} + \frac{U_\mathrm{fc}}{N}\sum_{ilm} n^\mathrm{f}_{il}n^\mathrm{c}_{im} \tag{4.19}$$

$$+ \mathrm{i}\frac{N}{\sqrt{N_\mathrm{L}}}\sum_{kk'}\rho(-k';\tau)\lambda(k'-k;\tau)\rho(k;\tau) - \mathrm{i}q_0 N\sqrt{N_\mathrm{L}}\lambda(\mathbf{0};\tau)\Bigg]. \tag{4.20}$$

In the above expressions, variables f' and λ' have been rewritten as f and λ, respectively, and the Jacobian factor $\prod_{i\tau}\rho_i(\tau)$ have been neglected following Ref. [45].

By introducing two kinds of Stratonovich-Hubbard fields ϕ^f and ϕ^c for U_fc, the functional integral over the Fermion fields can be performed in Eqs. (4.16) with (4.20). Then, the partition function Eq. (4.16) can be transformed as follows:

$$Z = \int D(\rho\lambda\phi^\mathrm{f}\phi^\mathrm{c})\exp(-S), \tag{4.21}$$

$$\begin{aligned} S = &- N\mathrm{Tr}(\ln\hat{A}) + \mathrm{i}\frac{N}{\sqrt{N_\mathrm{L}}}T^2\sum_{\mathbf{k}\mathbf{k}'}\rho(-\mathbf{k})\lambda(\mathbf{k}'-\mathbf{k})\rho(\mathbf{k}) \\ &- \mathrm{i}q_0 N\sqrt{N_\mathrm{L}}\int \mathrm{d}\tau\lambda(\mathbf{0};\tau) - \frac{NU_\mathrm{fc}}{4}T\sum_{\mathbf{k}}\phi^\mathrm{f}(\mathbf{k})\phi^\mathrm{c}(-\mathbf{k}), \end{aligned} \tag{4.22}$$

where the abbreviation $\mathbf{k} = (\mathbf{k}, \mathrm{i}\omega_n)$, etc. has been introduced and matrix $\hat{A}$ is defined as

$$A_{\mathrm{kk}'} = \begin{bmatrix} (-\mathrm{i}\omega_n + \epsilon_\mathbf{k})\delta_{\mathrm{kk}'} + \frac{U_\mathrm{fc}}{2\sqrt{z}}T\phi^\mathrm{f}(\mathrm{k}-\mathrm{k}') & T\frac{V}{\sqrt{N_\mathrm{L}}}\rho(\mathrm{k}-\mathrm{k}') \\ T\frac{V}{\sqrt{N_\mathrm{L}}}\rho^*(\mathrm{k}-\mathrm{k}') & (-\mathrm{i}\omega_n + \epsilon_\mathrm{f})\delta_{\mathrm{kk}'} + \frac{T}{\sqrt{N_\mathrm{L}}}\mathrm{i}\lambda(\mathrm{k}-\mathrm{k}') + \frac{TU_\mathrm{fc}}{2\sqrt{N_\mathrm{L}}}\phi^\mathrm{c}(\mathrm{k}-\mathrm{k}') \end{bmatrix}. \tag{4.23}$$

With the use of the mean-field solution for ρ, λ, ϕ^f, and ϕ^c, which optimizes the action (4.22), the f-electron number (per "spin"), $\bar{n}_\mathrm{f} \equiv n_\mathrm{f}/2$, as a function of ϵ_f, is given for a series of values of U_fc/D, D being the Fermi energy of the spherical conduction band, as shown on the right vertical axis of Fig. 4.6. Here we set the hybridization as $V = 0.5D$ and total electron number per "spin" as $n \equiv \bar{n}_{\rm f} + \bar{n}_{\rm c} = 0.875$. We adopted the dispersion of conduction electrons $\epsilon_{\mathbf{k}} = k^2/2m - D$, where the bottom of the conduction band was set as $-D$ and assumed that the density of states $N_0(\epsilon)$ per conduction electron with particular "spin" is given such that $\int_{-D}^{D} {\rm d}\epsilon N_0(\epsilon) = 1$.

These results are consistent with those of the previous works [49–51], and also with a physical picture that the rapid valence change occurs at the condition Eq. (4.3) which is fulfilled in the case where the energies of $\rm f^0$- and $\rm f^1$-state are degenerate. For much larger values of $U_{\rm fc}$ or smaller values of V than those presented in Fig. 4.6, there occurs a first-order like discontinuous transitions although they are not shown there. (See Ref. [2, 3].) In the mean-field level of approximation for slave-boson Hamiltonian, our treatment of the effect of $U_{\rm fc}$ is just like that in the Hartree-Fock approximation [49–51]. It is noted that the valence change occurs more sharply if we estimate it in a much more proper approximation on the extended Gutzwiller variational wave function [2]. In this sense, the sharpness of the valence change may be underestimated by the present treatment.

Using the mean-field solution, $\bar{\rho}$, $\bar{\lambda}$, $\bar{\varphi}^{\rm f}$, and $\bar{\varphi}^{\rm c}$, the dispersion of quasiparticles is given as

$$E_{\boldsymbol{k}}^{\pm} = \frac{1}{2}[\bar{\epsilon}_{\rm f} + \bar{\epsilon}_{\mathbf{k}} \pm \sqrt{(\bar{\epsilon}_{\rm f} - \bar{\epsilon}_{\mathbf{k}})^2 + 4\bar{V}^2}], \tag{4.24}$$

where

$$\bar{\epsilon}_{\boldsymbol{k}} \equiv \epsilon_{\boldsymbol{k}} + U_{\rm fc}\bar{n}_{\rm f},$$

$$\bar{V} \equiv V\bar{\rho}/\sqrt{N_{\rm L}}, \tag{4.25}$$

where $\bar{n}_{\rm f} \equiv \langle n_{\rm f}\rangle_{\rm MF}$, the f-electron numbers per site and "spin". Then, the density of states $\rho(\mu)$ of quasiparticles at the Fermi energy is calculated, and the Sommerfeld constant γ of the specific heat is given by the relation

$$\gamma = \frac{\pi^2}{3} D\rho(\mu)N, \tag{4.26}$$

where N (now $N = 2$) is the degeneracy of the quasiparticles. The so-called Kondo temperature, or the characteristic temperature of the present model, is defined simply as $(\bar{\epsilon}_{\rm f} - \mu)$.

In Fig. 4.7, we can see that the relation $T_{\rm K} \propto \gamma^{-1}$ holds rather nicely. Note also that the relation $T_{\rm K}$ versus γ with different value of $U_{\rm fc}$ are lying on a line exhibiting a kind of universality. It implies that the effect of $U_{\rm fc}$ can be absorbed into the other parameters of the conventional PAM at least in the mean-field level approximation, like in the case of the single impurity Anderson model. However, in the lattice case,

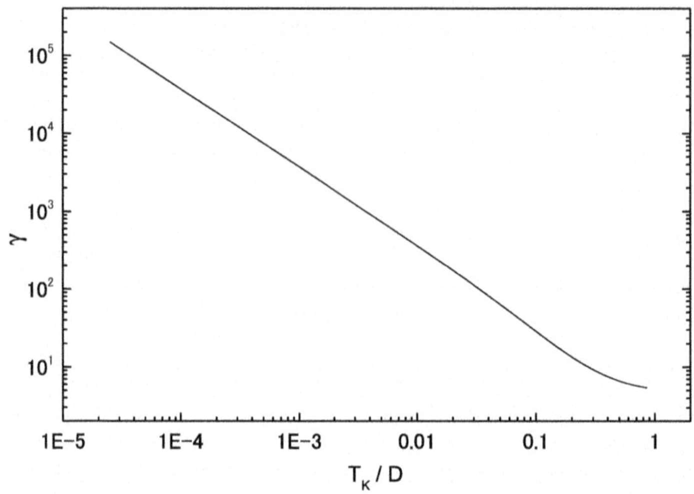

Fig. 4.7 Sommerfeld constant γ versus Kondo temperature T_K, defined by $T_K = \bar{\epsilon}_f - \mu$. The parameters used are the same as those in Fig. 4.6. Results for different values of U_{fc}/D lie on the same line, exhibiting a *universal* behavior. Reprinted from Ref. [2]. Copyright 2000 The Physical Society of Japan

the first-order like transition occurs through the effect of U_{fc} even in the mean-field level approximation. This is considered to be a distinctive effect of U_{fc} in the PAM.

4.4.2.1 Superconducting Transition Temperature [2]

In this subsection, a possible type of the superconducting state is discussed near the region where the rapid valence change occurs owing to the f-c Coulomb interaction U_{fc}. In the conventional PAM, problems of determining the superconducting transition temperature T_c have been studied by several authors within the slave-boson and $1/N$-expansion method [46, 47], in which the Gaussian fluctuations of relevant parameters around the mean-field solution are taken into account. Following their method, the Gaussian fluctuations of $\tilde{\rho} \equiv \rho - \bar{\rho}$, $\tilde{\lambda} \equiv \lambda - \bar{\lambda}$, $\tilde{\varphi}^f \equiv \varphi^f - \bar{\varphi}^f$, and $\tilde{\varphi}^c \equiv \varphi^c - \bar{\varphi}^c$, are taken into account. The transition temperature T_c is obtained

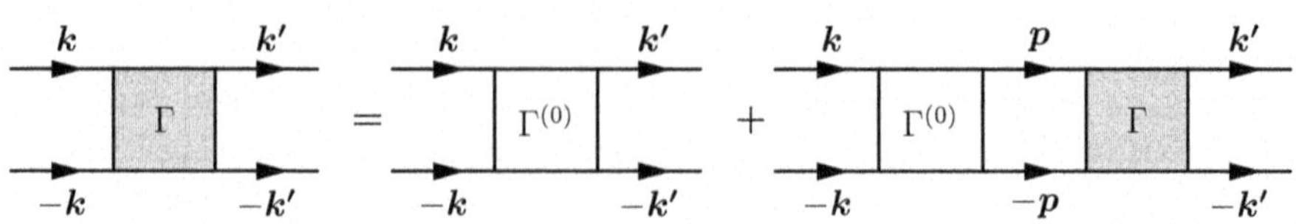

Fig. 4.8 Feynman diagram representing the integral equations for the scattering amplitude of two quasiparticles Γ from (**k**, −**k**) to (**k**′, −**k**′). Reprinted from Ref. [2]. Copyright 2000 The Physical Society of Japan

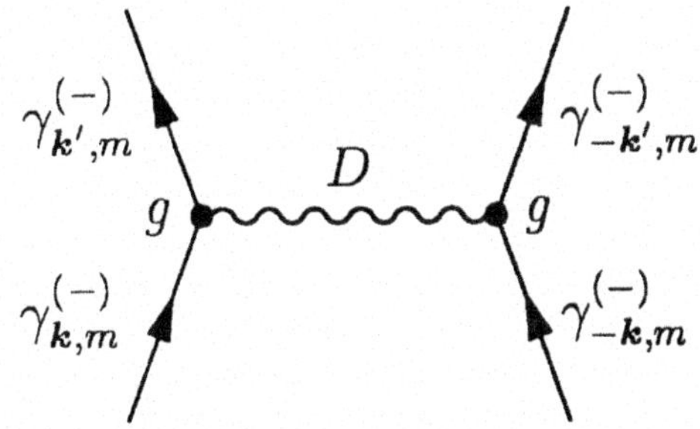

Fig. 4.9 Feynman diagram for the irreducible vertex part $\Gamma^{(0)}$ to leading order in $1/N$. The explicit form of the fluctuation propagators D is given by Eq. (4.29) and its interaction vertex g with quasiparticles arises from the matrix $\hat{A}$ given in Eq. (4.23). Reprinted from Ref. [2]. Copyright 2000 The Physical Society of Japan

from the integral equation of the particle-particle scattering amplitude, Γ, for two quasiparticles with opposite momentum near the Fermi surface. (see Fig. 4.8.)

The quasiparticle operators can be represented by a unitary transformation in terms of f- and conduction electron operators as

$$\begin{bmatrix} \gamma_{\mathbf{k}}^{(+)} \\ \gamma_{\mathbf{k}}^{(-)} \end{bmatrix} = \begin{bmatrix} u_{\mathbf{k}} & v_{\mathbf{k}} \\ -v_{\mathbf{k}} & u_{\mathbf{k}} \end{bmatrix} \begin{bmatrix} \mathrm{f}_{\mathbf{k}} \\ \mathrm{c}_{\mathbf{k}} \end{bmatrix}, \tag{4.27}$$

where $\gamma^{(\pm)}$ corresponds to the eigen values $E^{\pm}$, Eq. (4.24), respectively. The scattering amplitude Γ is obtained from two-quasiparticle correlation function,

$$\langle \mathrm{T}_\tau \gamma^{(-)}_{\mathbf{k}',m}(\tau_1)\gamma^{(-)}_{-\mathbf{k}',m'}(\tau_2)\gamma^{(-)\dagger}_{-\mathbf{k},m'}(\tau_3)\gamma^{(-)\dagger}_{\mathbf{k},m}(\tau_4)\rangle,$$

by removing the external legs.

To the leading order in $1/N$, the irreducible vertex part $\Gamma^{(0)}$ includes only single-fluctuations exchange processes as shown as the Feynman diagram in Fig. 4.9. With the use of relation, Eq. (4.27), the analytic expression of for the pairing interaction $\Gamma^{(0)}$ is given as follows [2]:

$$\begin{aligned} \Gamma^{(0)} = {} & v^4 \left\{ D_{\lambda\lambda} + \frac{U_\mathrm{c}}{2} D_{\lambda\phi^\mathrm{c}} + \frac{U_\mathrm{c}}{2} D_{\phi^\mathrm{c}\lambda} + (\frac{U_\mathrm{c}}{2})^2 D_{\phi^\mathrm{c}\phi^\mathrm{c}} \right\} - 4uv^3 \left\{ V D_{\lambda\rho} + \frac{U_\mathrm{c}}{2} D_{\phi^\mathrm{c}\rho} \right\} \\ & + u^2 v^2 \left[2 \left\{ \frac{U_\mathrm{c}}{2} D_{\lambda\phi^\mathrm{f}} + (\frac{U_\mathrm{c}}{2})^2 D_{\phi^\mathrm{c}\phi^\mathrm{f}} \right\} + 4V^2 D_{\rho\rho} \right] \\ & - u^3 v 4V \frac{U_\mathrm{c}}{2} D_{\phi^\mathrm{f}\rho} + u^4 (\frac{U_\mathrm{c}}{2})^2 D_{\phi^\mathrm{f}\phi^\mathrm{f}}, \end{aligned} \tag{4.28}$$

where fluctuation propagators D's are defined as

$$D_{\alpha\beta}(\boldsymbol{k};\tau) \equiv -\langle \mathrm{T}_\tau \alpha(\boldsymbol{k},\tau)\beta^\dagger(\boldsymbol{k},0)\rangle, \tag{4.29}$$

where the subscripts α and β stand for $\tilde{\rho}$, $\tilde{\lambda}$, $\tilde{\varphi}^\mathrm{f}$ and $\tilde{\varphi}^\mathrm{c}$.

With the use of (4.28), T_c is calculated in the weak coupling theory in which the external momenta are set on the Fermi surface, i.e., $|\boldsymbol{k}|$, $|\boldsymbol{k}|' \to k_\mathrm{F}$, and the static limit, $\omega \to 0$, is taken in the boson propagators. Then, the linearized gap equation, represented by Fig. 4.8, is solved by decoupling the scattering amplitude into the Legendre polynomial as follows: The scattering amplitude Γ_ℓ corresponding to the channel with relative angular momentum ℓ is given by

$$\Gamma_\ell = \frac{\Gamma_\ell^{(0)}}{1 + \rho(\mu)\Gamma_\ell^{(0)} \ln(\frac{T_K}{T})}, \tag{4.30}$$

where $\Gamma_\ell^{(0)}$'s are related to $\Gamma^{(0)}(\hat{\boldsymbol{k}} \cdot \hat{\boldsymbol{k}}')$'s by the formula

$$\Gamma_\ell^{(0)} = \frac{1}{2}\int_{-1}^{1} d(\cos\theta)\Gamma^{(0)}(\theta)P_\ell(\cos\theta). \tag{4.31}$$

Then, the transition temperature T_c for the ℓ-wave channel is given by

$$T_c = T_K \exp\left[\frac{1}{\rho(\mu)\Gamma_\ell^{(0)}}\right], \tag{4.32}$$

where the energy cutoff, corresponding to the Debye frequency, is given by $T_K \equiv \bar{\epsilon}_f - \mu$, the bandwidth of quasiparticles.

A non-negligible T_c is obtained for the d-wave ($\ell = 2$) channel as far as the channels, $\ell = 0$, 1 and 2, are concerned. The results of T_c versus ϵ_f are shown in Fig. 4.6 for a series of values of U_{fc}, the repulsion between f- and conduction electrons. There exists a sharp peak of T_c at around ϵ_f where n_f starts to show a rapid decrease. Its tendency becomes more drastic as U_{fc} increases making the valence change sharper. In the region where the f-electron number is decreased enough, T_c is strongly suppressed.

It is analyzed which term in Eq. (4.28) plays a crucial role in pairing interaction. As seen in Fig. 4.10, it turns out that the major part of $\Gamma^{(0)}(q)$ ($= V_{\mathbf{k},\mathbf{k}'}$ with $\mathbf{q} = \mathbf{k} - \mathbf{k}'$) is induced by the scattering process (f,f)→(f,c) or (f,c)→(f,f), in which the valence of f-electrons is changed directly. Here, it is remarked that the pairing interaction $\Gamma^{(0)}(q)$ is almost q-independent up to $q < 3k_F/2$ reflecting the local nature of CVF while the dimensionless coupling $\Gamma^{(0)}(q)\rho(0)$ at the Fermi level is greatly enhanced near the CVT point. It is this locality that is the origin of the d-wave pairing as in the case of two-dimensional isotropic Fermi liquid near the magnetic phase boundary [2]. Namely, the q-independent part of $\Gamma^{(0)}$ gives a short range repulsion of s-wave component, while the decrement (at $q > (3/2)k_F$) from the q-independent part gives us an attractive component with large wave number transfer of $O(2k_F)$, promoting the pairing with higher angular momentum, the d-wave in the present case. This locality should be maintained for the case where the conduction band is subject to the effect of lattice periodicity. In the latter case, the attractive interaction works among quasiparticles located at adjacent sites and p- or d- or f-wave like pairing is induced depending on the shape of the Fermi surface because the attraction due to CVF does not have "spin" dependence unlike spin fluctuation mechanism [24].

The reason for the d-wave pairing can be seen more clearly by inspecting a real space picture of the paring interaction

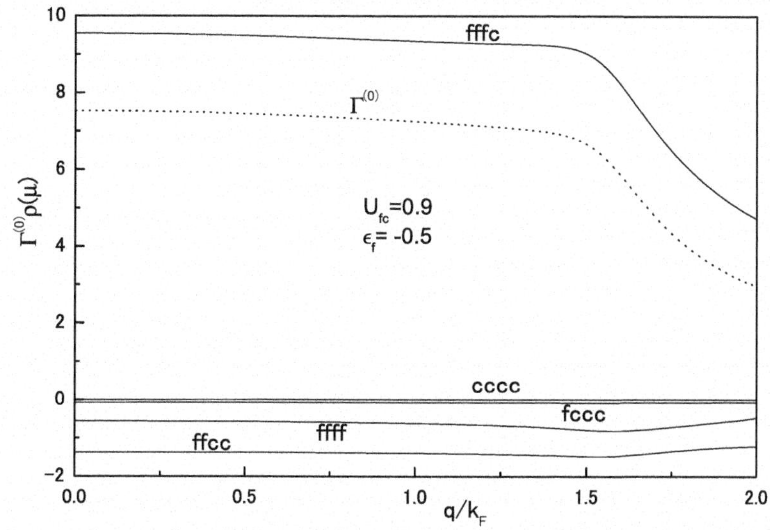

Fig. 4.10 $\Gamma^{(0)}(q)$ as a function of momentum transfer q for each scattering channel included in Eq. (4.28). For example, "ffff" means the v^4 term of Eq. (4.28). The others are represented similarly. Reprinted from Ref. [2]. Copyright 2000 The Physical Society of Japan

$$\Gamma^0(r) = \sum_{\mathbf{q}} \Gamma^0(q) e^{i\mathbf{q}\cdot\mathbf{r}}. \tag{4.33}$$

The result for $\Gamma^0(r)$ is shown in Fig. 4.11 for the same parameter set as Fig. 4.10. This clearly shows the existence of the extended attraction together with the onsite strong repulsion. If we assume $k_F \sim \pi/a$, a being the lattice constant in typical metals, the attraction works between the nearest-neighbor sites. Then, according to the discussion above, the d-wave pairing is promoted. However, the p-wave "spin-triplet" pairing is also possible *in principle* in this case because the attraction mediated by the valence fluctuation has no spin dependence.

4.4.2.2 Physical Picture of Critical-Valence-Fluctuation Mechanism

The physical interpretation of the superconductivity induced by the critical valence fluctuations (CVF) is obtained from Fig. 4.12 in which the reason why the nearest-neighbor pairing is induced by the CVF is demonstrated.

First, let us consider the onsite problem as shown in the left figures. If the condition, Eq.(4.3), is fulfilled, the *local* f^0-hole state is created, as shown in the left lower figure. This *positively* charged f^0-hole would attract the excess conduction electrons to its vicinity.

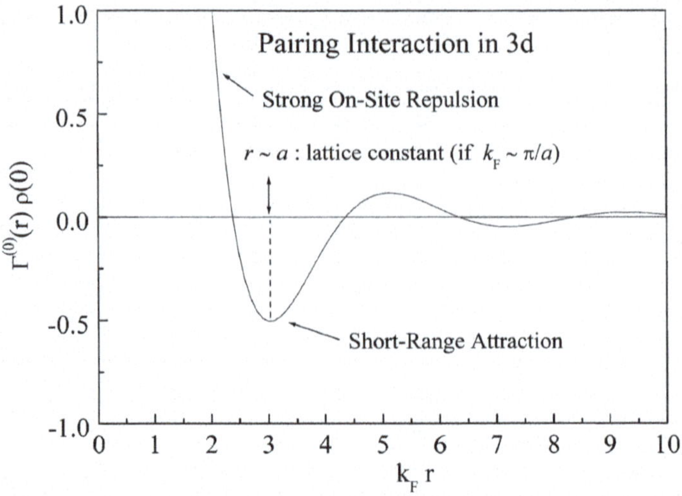

Fig. 4.11 Real-space pairing interaction $\Gamma^0(r)\rho(0)$. The parameter set adopted is the same as that in Fig. 4.10. Reprinted from Ref. [65]. Copyright 2007 IOP Publishing

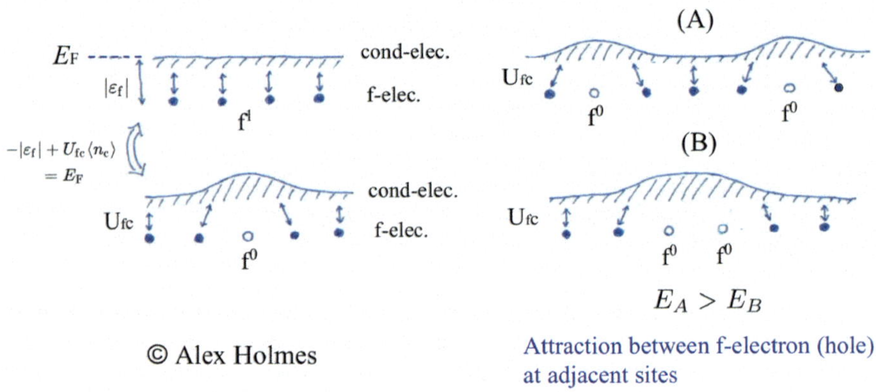

Fig. 4.12 The vertical scales are double, one is for the energy and the other for the number density of conduction electrons. E_F is the (bare) Fermi energy of conduction electrons, and ε_f is the energy level of the single f-electron measured from E_F, and U_{fc} is the local f-c repulsion. Filled and open circles represent the 4f-electrons and 4f-holes, respectively. This figure (original handwritten version was supplied by Alex Holmes) offers an intuitive view to understand the origin of the superconductivity induced by the CVF

Then, the right two figures exhibit a physical picture of the two holes problem in which two cases, (A) two holes are far separated, and (B) two holes are located at adjacent sites. It is apparent that $E_B < E_A$. This is because the effect of the f-c repulsion, U_{fc}, is about 2-times larger (in the present one-dimensional model) in the case (A) than in the case (B). This reflects the difference of the effect of the U_{fc}, between excess conduction electrons around f^0-holes. Namely, the excess potential energy in the case (A) is about 2-times larger than the case (B).

The present mechanism of the superconductivity is in contrast to the conventional mechanism based on the electron-phonon interaction discussed in [44]. In that case, the retardation effect was crucial and the paring interaction extends in time but is essentially local in real space, leading to *essentially* the s-wave pairing.

4.5 DMRG Calculation

In this section, the nature of the first-order valence transition as well as the electronic states near the QCP is studied by the numerical calculation [3]. It is shown that the superconducting correlation develops in the Kondo regime near the sharp valence crossover on the basis of the density-matrix renormalization group (DMRG) [52] calculation for the one-dimensional extended Anderson lattice model

$$\begin{aligned} H = &-t\sum_{i\sigma}(c^{\dagger}_{i\sigma}c_{i+1\sigma} + c^{\dagger}_{i+1\sigma}c_{i\sigma}) + \varepsilon_{\rm f}\sum_{i\sigma} n^{\rm f}_{i\sigma} \\ &+ V\sum_{i\sigma}\left(f^{\dagger}_{i\sigma}c_{i\sigma} + c^{\dagger}_{i\sigma}f_{i\sigma}\right) + U\sum_{i=1}^{N} n^{\rm f}_{i\uparrow}n^{\rm f}_{i\downarrow} \\ &+ U_{\rm fc}\sum_{i=1}^{N} n^{\rm f}_{i}n^{\rm c}_{i}. \end{aligned} \tag{4.34}$$

Here, the notation is standard and the number operator is defined by $n^{\rm a}_{i\sigma} = a^{\dagger}_{i\sigma}a_{i\sigma}$ and $n^{\rm a}_{i} = n^{\rm a}_{i\uparrow} + n^{\rm a}_{i\downarrow}$ for a = f and c. The last term is the Coulomb repulsion between f and conduction electrons, which is important in causing the valence transition [2, 53, 54].

The filling n is defined by $n = (n_{\rm f} + n_{\rm c})/2$ with $n_{\rm a} = \sum_{i=1}^{N}\langle n^{\rm a}_{i}\rangle/N$ for a = f and c with N being the number of lattice sites. We note that half filling is realized when $n = 1$ in this definition. Here, we consider the case $n = 7/8$, $t = 1$, $V = 0.1$ and $U = 100$. In the DMRG calculation, we have kept the number of states up to 1500 and the system sizes up to $N = 80$ with the open-boundary condition.

Figure 4.13a shows the f-electron-number $n_{\rm f}$ vs the f-level $\varepsilon_{\rm f}$ for $U_{\rm fc} = 7.0$, which has been obtained by a linear extrapolation to $N \to \infty$ using the data for $N =$ 24, 32, 40, 48, 56, 64, 72 and 80. We see that a jump in $n_{\rm f}$ appears at $\varepsilon_{\rm f} = -5.6395$, which indicates the first-order quantum phase transition, since a jump in $n_{\rm f}$ results in the level crossing of the ground states by the relation $n_{\rm f} = \partial\langle H\rangle/\partial\varepsilon_{\rm f}$. The first-order

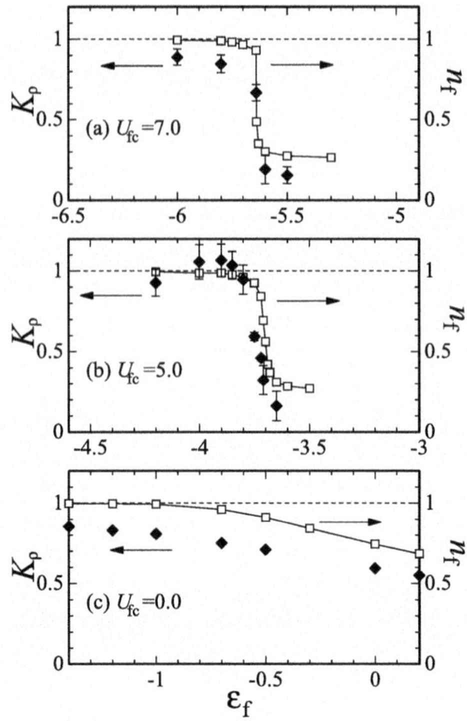

Fig. 4.13 n_f versus ε_f (open square) and K_ρ versus ε_f (solid diamond) for $t = 1$, $V = 0.1$ and $U = 100$ at $n = 7/8$ for **a** $U_{fc} = 7.0$, **b** 5.0 and **c** 0.0 [3]. Reprinted from Ref. [3]. Copyright 2006 The Physical Society of Japan

valence transition (FOVT) between the Kondo state with $n_f \simeq 1$ and the mixed-valence (MV) state with $n_f \simeq 1/4$ $(= 2(1 - n)$ for $1/2 \leq n \leq 1)$ is caused by U_{fc}, since a large U_{fc} forces the electrons to pour into either the f-level or the conduction band.

When U_{fc} decreases, the magnitude of the jump in n_f decreases. We see that the jump is invisible at $U_{fc} = 5.0$ and the sharp crossover from the Kondo to MV states appears in Fig. 4.13b. These results imply that the QCP at which both the jump disappears and the valence susceptibility $\chi_v \equiv -\partial^2\langle H\rangle/\partial\varepsilon_f^2 = -\partial n_f/\partial\varepsilon_f$ diverges, exists between $U_{fc} = 5.0$ and 7.0. As U_{fc} is set to be smaller than 5.0, the change in n_f becomes smoother and finally a gradual crossover appears at $U_{fc} = 0$ as seen in Fig. 4.13c.

To determine the location of the QCP, we have calculated the jump of the f-electron number, $\Delta n_f \equiv n_f^{K} - n_f^{MV}$, at the FOVT point as a function of ε_f for several U_{fc} values. Here, the superscripts K and MV specify the Kondo and mixed-valence states, respectively. Then, the QCP is estimated at $(\varepsilon_f^{QCP}, U_{fc}^{QCP}) \simeq (-4.5, 5.9)$ at which the jump disappears. In the slave-boson mean-field (MF) theory [33] the QCP is identified at $(\varepsilon_{f\,MF}^{QCP}, U_{fc\,MF}^{QCP}) = (0.19, 0.98)$.

We summarize the ground-state phase diagram in the U_{fc}-ε_f plane in Fig. 4.14. The solid line with solid diamonds represents the FOVT line determined by the DMRG, and the dashed line with open squares represents the point at which χ_v has

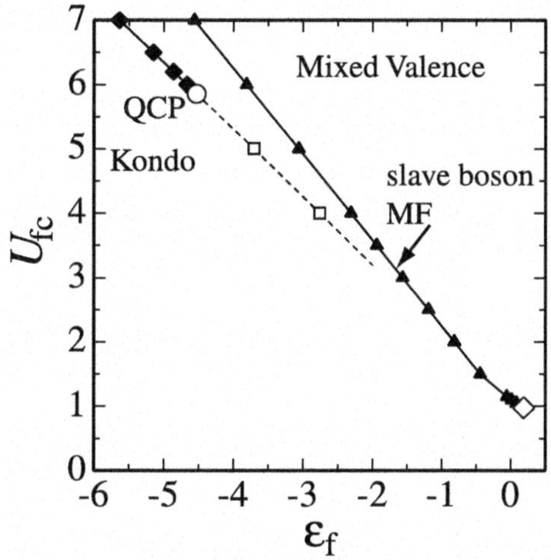

Fig. 4.14 Ground-state phase diagram in U_{fc}-ε_f plane for $t = 1$, $V = 0.1$, $U = 100$ at $n = 7/8$. The first-order valence-transition line (solid line with solid diamonds) with the quantum critical point (open circle) is determined by the DMRG. The dashed line with open squares represents the crossover determined by the point where χ_v has its maximum value. The first-order valence-transition line (solid line with solid triangles) with the QCP (open diamond) is determined by the slave-boson mean-field theory [3]. Reprinted from Ref. [3]. Copyright 2006 The Physical Society of Japan

a maximum, which indicates the crossover with enhanced valence fluctuation. Here, we note that the slope of the FOVT line is expressed as the ratio of the jump in n_f and $\langle n_i^f n_i^c \rangle$ at the transition by the "Claudius-Clapeyron relation" [55],

$$\frac{\delta U_{fc}}{\delta \varepsilon_f} = -\frac{n_f^K - n_f^{MV}}{C_{fc}^K - C_{fc}^{MV}}, \tag{4.35}$$

where C_{fc} is defined as $C_{fc} \equiv \sum_{i=1}^{N} \langle n_i^f n_i^c \rangle / N$. For $U_{fc} \gg U_{fc}^{QCP}$, $\delta U_{fc}/\delta \varepsilon_f$ is shown to be $-1/n_c^K$ in the MF theory [33], which coincides with the equation of the energy valance of the one-body f-electron energy at the FOVT; $\varepsilon_f + U_{fc} n_c^K \sim \mu$, with μ being the chemical potential [2] (see Eq. (4.3)). However, it should be stressed that the correct slope for $U_{fc} \gg U_{fc}^{QCP}$ is given as $\delta U_{fc}/\delta \varepsilon_f = -1$ by inserting the atomic-limit values of $n_f^K = 1, n_f^{MV} = 1/4, C_{fc}^K = 3/4$ and $C_{fc}^{MV} = 0$ into the thermodynamic relation. This discrepancy is due to the insufficiency of the description of the MV state by the MF theory.

Here, we give the following rigorous remarks on the valence transition: For general filling of $1/2 \leq n \leq 1$, the maximum value of the jump of n_f, which is realized for large values of $|\varepsilon_f|$ and U_{fc}, is given by $\Delta n_f|_{max} = 2n - 1$. This is in sharp contrast to the MF result, which always predicts the $4f^1(Ce^{3+})$ to $4f^0(Ce^{4+})$ transition; i.e., $\Delta n_f|_{max} = 1$. The slope of the FOVT line for large values of $|\varepsilon_f|$ and U_{fc} is expressed by $\delta U_{fc}/\delta \varepsilon_f = -1$, irrespective of total filling n. It is noted that these results hold even in the square- and cubic-lattice systems.

We also point out that the reason why the FOVT is difficult to be realized in the Ce compounds can be understood by the smallness of U_{fc}: In Ce metal, U_{fc} is large because of the onsite 4f-5d interaction, whereas in Ce compounds U_{fc} is considered to be smaller even if the coordination number between Ce and surrounding atoms is taken into account, since U_{fc} corresponds to the intersite 4f-(d, or p) interaction.

We have confirmed that on both sides of the FOVT, the peak structure appears at $q = \pi/4$ in the Fourier component of $\langle S_i^{az} \rangle$ for a=f and c when we apply a small magnetic field as a perturbation at the edge sites. The above period of the Friedel oscillations corresponds to $2k_F^{large} = \pi/4$, since $k_F^{large} = (1 + 3/4)\pi/2 = 7\pi/8$. The large Fermi surfaces detected in the Kondo and MV states are consistent with the existence of the QCP, since by detouring around the QCP, the Kondo and MV states can be adiabatically connected, with Luttinger's sum rule satisfied.

In Fig. 4.14, the solid line with solid triangles represents the FOVT line determined by the slave-boson MF theory. It is evident that the QCP (solid circle) identified by the DMRG is shifted from the QCP (open diamond) by the MF to the large-$|\varepsilon_f|$ and large-U_{fc} direction. We also find no evidence of the phase separation at the FOVT, such as negative compressibility, and an inhomogeneous charge distribution in the DMRG results for the ground state. This result is due to the fact that the order parameter n_f is not a conserving quantity, which is in sharp contrast to the MF theory where the FOVT is accompanied by the phase separation. These results indicate that the phase separation is destabilized by quantum fluctuation and electron correlation, and as a result, the crossover regime from the Kondo to MV states is strongly stabilized. In connection to the experiments of the Ce compounds such as $CeCu_2Ge_2$, this sharp crossover regime with strong fluctuation is most interesting, since the superconductivity is observed in this regime.

To clarify this point, we have calculated the superconducting correlation functions $\langle P_{m\pm}^{ab}(i) P_{m\pm}^{cd\,\dagger}(i+x) \rangle$ with

$$P_{m\pm}^{ab}(i) = \frac{1}{\sqrt{2}}(a_{i\uparrow} b_{i+m\downarrow} \pm a_{i\downarrow} b_{i+m\uparrow}) \tag{4.36}$$

for the singlet (−) and triplet (+) pairing. Here, a, b, c and d denote the f and/or conduction electron operators and we have calculated the nearest-neighbor (NN) pairing ($m = 1$) and also the onsite pairing ($m = 0$). To avoid the effect of the open boundary, we here fix the position of the operator at the central site, i.e., $i = N/2$ for the onsite pairing and $i = N/2 - 1$ for the NN pairing. Figure 4.15 shows the superconducting correlations for $\varepsilon_f = -3.9$ at $U_{fc} = 5.0$, which is close to the sharp valence change (see Fig. 4.13b). A remarkable point here is that the power of the decay of the correlations is smaller than 2 as clearly seen, for example, in the data of $|\langle P_{0-}^{cf} P_{1-}^{cc\,\dagger} \rangle|$ and $|\langle P_{0-}^{cf} P_{0-}^{cf\,\dagger} \rangle|$ with dashed lines indicating x^{-2}. We have confirmed that all correlations of $|\langle P_{m-}^{ab}(i) P_{m-}^{cd\,\dagger}(i+x) \rangle|$ show nearly the same power which is smaller than 2. On the other hand, for $\varepsilon_f = -4.2$, which is in the Kondo regime, we find that the power of the decay of the pairing correlation is larger than 2. It is also found that the pairing correlation decays rapidly as $\sim x^{-5}$ for $\varepsilon_f = -3.71$, where

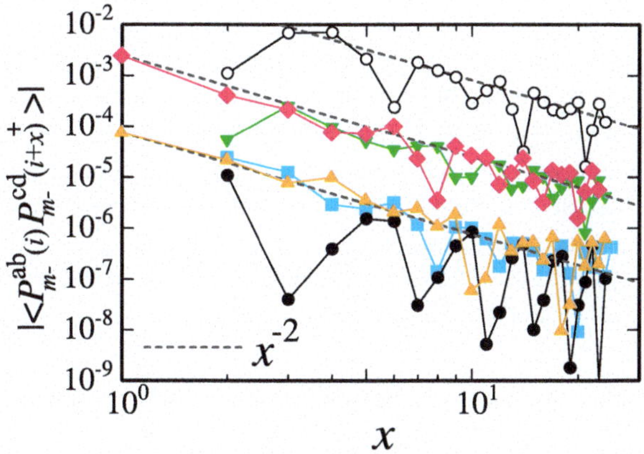

Fig. 4.15 Singlet pairing correlation functions $|\langle P^{\rm ab}_{m-}(i) P^{\rm cd}_{m-}{}^{\dagger}(i+x)\rangle|$ with $P^{\rm ab}_{m-}(i) = (a_{i\uparrow}b_{i+m\downarrow} - a_{i\downarrow}b_{i+m\uparrow})/\sqrt{2}$ (see text) for $\varepsilon_{\rm f} = -3.9$ and $U_{\rm fc} = 5.0$ at $t = 1, V = 0.1, U = 100$ and $n = 7/8$: $|\langle P^{\rm cc}_{1-} P^{\rm cc}_{1-}{}^{\dagger}\rangle|$ (black open circle), $|\langle P^{\rm ff}_{1-} P^{\rm ff}_{1-}{}^{\dagger}\rangle|$ (black solid circle), $|\langle P^{\rm cf}_{0-} P^{\rm cc}_{1-}{}^{\dagger}\rangle|$ (pink diamond), $|\langle P^{\rm ff}_{1-} P^{\rm cc}_{1-}{}^{\dagger}\rangle|$ (green triangle), $|\langle P^{\rm cf}_{0-} P^{\rm cf}_{0-}{}^{\dagger}\rangle|$ (orange triangle) and $|\langle P^{\rm ff}_{1-} P^{\rm cf}_{0-}{}^{\dagger}\rangle|$ (blue square). Here, $i = N/2$ is set for $m = 0$ and $i = N/2 - 1$ for $m = 1$. The dashed line $\sim x^{-2}$ is drawn for comparison [3]. Reprinted from Ref. [3]. Copyright 2006 The Physical Society of Japan

the sharp valence crossover occurs. We confirmed that these results are not changed by the calculations at several system sizes.

For $U = U_{\rm fc} = 0$, the f-level and conduction band are hybridized and μ is located at the lower hybridized band with $k_{\rm F}^{\rm large}$. We confirmed that the charge and spin gaps are closed with $k_{\rm F}^{\rm large}$ in the bulk limit in both the Kondo, MV and crossover regimes. The system is described by the single-band Tomonaga-Luttinger (TL) liquid. For the single-component TL liquid without the spin and charge gaps, the singlet pairing correlation decays as x^{-1-1/K_ρ} [56]. Since the charge and spin correlations decay as x^{-1-K_ρ} [56], the above result implies that $K_\rho > 1$, namely, that the superconducting correlation becomes dominant.

In order to estimate the TL parameter K_ρ systematically, we made a least-squares fit for the charge, spin, and superconducting correlations defined by $\langle \bar{O}_i \bar{O}_j^{\dagger}\rangle$ for $\bar{O}_i \equiv O_i - \langle O_i\rangle$ and $O_i = n_i^{\rm c}$, $S_i^{\rm cz}$ and $P^{\rm cc}_{m-}(i)$, respectively, since the conduction electrons have a large amplitude in the pairing correlation as seen in Fig. 4.15, which seems to give a reliable estimate. By assuming the asymptotic forms of the correlation functions of the single-component TL liquid [56], we obtained three K_ρ's from the singlet-superconducting (SS), the charge-density-wave (CDW) and the spin-density-wave (SDW) correlation functions. Then, we take the middle point between their maximum and minimum values as a mean value, with the error defined by the difference between the middle point and the maximum (minimum) value.

The K_ρ values evaluated in this manner are shown in Fig. 4.13 as the solid diamonds. Here, for $\varepsilon_{\rm f} \geq -3.75$ at $U_{\rm fc} = 5.0$ and for $\varepsilon_{\rm f} \geq -5.6$ at $U_{\rm fc} = 7.0$, K_ρ is estimated from $4k_{\rm F}$CDW and SS correlations. For $U_{\rm fc} = 5.0$, K_ρ appears to exceed 1 at around $\varepsilon_f = -3.9$ in Fig. 4.13b, suggesting that the superconducting correlation is dominant, which is consistent with the result presented in Fig. 4.15. For $U_{fc} = 7.0$, K_ρ jumps at the FOVT as in Fig. 4.13a. For $U_{fc} = 0$, K_ρ estimated from the SS correlation remains smaller than 1, indicating that the pairing correlation does not develop as in Fig. 4.13c. These results show that the superconducting correlation is enhanced in the Kondo regime near the sharp valence crossover, which is close to the QCP of the valence transition.

To get further insight into the non-monotonic behavior of K_ρ around $\varepsilon_f = -3.9$ at $U_{fc} = 5.0$, we calculated the charge compressibility

$$\kappa \equiv \frac{\partial(2n)}{\partial\mu}, \tag{4.37}$$

since the TL parameter is expressed by the charge velocity v_c and κ as

$$K_\rho = \frac{\pi v_c \kappa}{2}. \tag{4.38}$$

In Fig. 4.16, we plot the compressibility extrapolated to the $N \to \infty$ limit in the form of

$$\kappa = \frac{2}{N\Delta_c} \tag{4.39}$$

with the charge gap

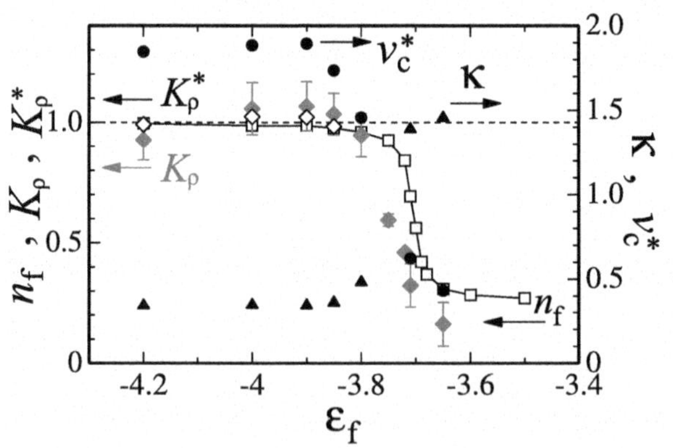

Fig. 4.16 Compressibility (solid triangle) obtained by $\kappa = \lim_{N\to\infty} 2/(N\Delta_c)$ and charge velocity (solid circle) obtained by $v_c^* = \lim_{N\to\infty} \Delta_c/\Delta q$ for $t = 1$, $V = 0.1$, $U = 100$ and $U_{fc} = 5.0$ at $n = 7/8$. The K_ρ (shaded diamond) obtained by the least-squares fit of correlation functions, $K_\rho^* = \pi v_c^* \kappa/2$ (open diamond), and n_f (open square) are shown [3]. Reprinted from Ref. [3]. Copyright 2006 The Physical Society of Japan

$$\Delta_c \equiv \frac{E(2nN+2,0)+E(2nN-2,0)-2E(2nN,0)}{2}, \tag{4.40}$$

where $E(N_e, S)$ is the ground-state energy for the electron number N_e and total spin S. As ε_f increases, κ increases, since total charge fluctuation becomes large. We see that κ does not show any diverging enhancement at the sharp valence crossover, $\varepsilon_f \sim -3.71$, in contrast to the enhanced χ_v. This is due to the cancellation of the leading terms between $\langle \bar{n}_i^a \bar{n}_j^a \rangle$ for a = f, c and $\langle \bar{n}_i^f \bar{n}_j^c \rangle$ in $\langle (\bar{n}_i^f + \bar{n}_i^c)(\bar{n}_j^f + \bar{n}_j^c) \rangle$, since the diagonal and off-diagonal charge correlations have the opposite signs for $U_{fc} > 0$.

A remarkable result here is that the compressibility stays constant around $\varepsilon_f = -3.9$, where K_ρ has a maximum. This indicates that the enhancement of K_ρ is caused by the enhancement of the charge velocity, but not by the compressibility. This is quite different from the ordinary models, which exhibits $K_\rho > 1$ [57]. In order to get further insight into this finding, we calculated the charge velocity defined by

$$v_c^* \equiv \lim_{N\to\infty} \frac{\Delta_c}{\Delta q} \tag{4.41}$$

with $\Delta q = \pi/(N+1)$ instead of v_c [58], since the numerical accuracy is not enough for the second-lowest eigenvalue in the Lanczos diagonalization. We see that the maximum of v_c^* appears around $\varepsilon_f = -3.9$ and this suggests that the enhancement of v_c is the origin of the enhancement of K_ρ. In Fig. 4.16, we show $K_\rho^* = \pi\kappa v_c^*/2$ for $\varepsilon_f \le -3.85$ by open diamonds and note that K_ρ^* coincides with the K_ρ obtained by the least-squares fit of the correlation functions within the error bars. Since K_ρ^* is obtained by v_c^* and κ after the extrapolation to $N \to \infty$, this suggests that K_ρ is evaluated properly within the finite sizes calculated in this study. A more accurate determination of K_ρ in the bulk limit is left for future study.

The increase in v_c^* toward $\varepsilon_f \simeq -3.9$ in the Kondo regime is considered to be due to the decrease in the density of states at the Fermi level, i.e., the broadening of the Kondo peak. As ε_f further increases, v_c^* is suppressed, since U_{fc} interrupts the coherent motion of the electrons in the MV state. This is the reason why v_c^* has a maximum near the sharp valence increase. The enhanced v_c^* implies the enlargement of the effective band width and thus the coherent motion of electrons with enhanced valence fluctuation is considered to be the origin of the development of the pairing correlation. The enhancement of K_ρ around $\varepsilon_f = -3.9$ also agrees with the following naive expectation: In the deep Kondo regime, the SDW correlation is dominant by the RKKY interaction and at the sharp valence-crossover point the CDW correlation becomes dominant with enhanced χ_v. Between the two regimes, the superconducting correlation can be dominant, since the regime midway between the Kondo state and the sharp valence crossover is unfavorable for both the SDW and CDW correlations.

To summarize, we have clarified the novel nature of the electronic states near the QCP of the valence transition. Since valence fluctuations are basically ascribed to atomic origin, we believe that the basic nature of the valence transition and effect of its fluctuation revealed in this study is universal in two- and three-dimensional systems as well.

4.6 Effect of Magnetic Field on Valence Transition and Valence Crossover

Valence transition and valence fluctuations are the phenomena, which are ascribed to charge degree of freedom. Hence, it is highly nontrivial how the magnetic field affects the valence transition and valence fluctuations. These interesting issues are addressed in this section. In Sect. 4.6.1, magnetic field dependence of the first-order valence transition is discussed. In Sect. 4.6.2, the effect of the magnetic field on the quantum critical point (QCP) of the valence transition is discussed. Theoretical analysis based on the extended periodic Anderson model shows that the field-induced QCP of the valence transition gives rise to metamagnetism, which offers an essentially new mechanism.

4.6.1 *Magnetic Field Dependence of First-Order Valence Transition*

First, we analyze how the critical point of the valence transition is suppressed to absolute zero by applying the magnetic field [4]. Let us consider the Claudius-Clapeyron relation for the first-order valence-transition temperature $T_{\rm v}$:

$$\frac{\delta T_{\rm v}}{\delta h} = -\frac{m_{\rm K} - m_{\rm MV}}{S_{\rm K} - S_{\rm MV}}, \tag{4.42}$$

where m and S denote the magnetization and the entropy, respectively. Here, K represents the Kondo regime where the f-electron (hole) density per site $\langle n_{\rm f}\rangle$ is close to 1 in the Ce (Yb) system, i.e., $\text{Ce}^{3+}(4\text{f}^1)$ and $\text{Yb}^{3+}(4\text{f}^{13})$, and MV represents the mixed-valence regime with $\langle n_{\rm f}\rangle < 1$.[4] Since the magnetization as well as the entropy in the Kondo regime are larger than those in the MV regime, as observed in the specific heat and the uniform susceptibility, it turns out that $T_{\rm v}$ is suppressed by applying h (see Fig. 4.17a). Then, the critical point is suppressed to $T = 0$ K by h.

Furthermore, the field dependence of $T_{\rm v}$ in the zero-temperature limit is also derived by using the above relation: For $T \to 0$, the entropy shows the T-linear behavior in both the Kondo and MV regimes and $S_{\rm K} - S_{\rm MV}$ is approximated to be proportional to $T_{\rm v}$ in case that $T_{\rm v}$ is smaller than the characteristic energy scales in the Kondo and MV regimes. Noting the temperature independent $m_{\rm K} - m_{\rm MV}$, we have $\delta T_{\rm v}/\delta h = -C_1/T_{\rm v}$, leading to $T_{\rm v} = C_2\sqrt{h_{\rm v} - h}$ with constants C_1 and C_2, which explains well the observed behavior in the Ce metal [59] and YbInCu_4 [60, 61] (see Fig. 4.17b). We stress here that our analysis not only provides firm

[4] We refer to spatially uniform and quantum-mechanically valence-fluctuating state with $\langle n_{\rm f}\rangle < 1$ as mixed valence.

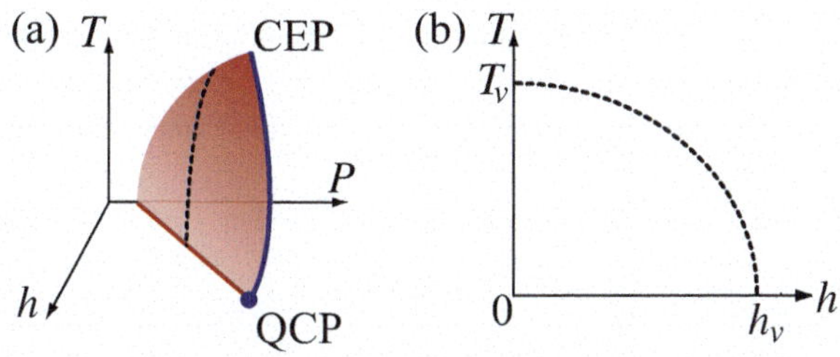

Fig. 4.17 **a** Schematic phase diagram, showing the FOVT surface in the T-P-h space, where P represents a control parameter (e.g., pressure, chemical concentration, etc.). The critical end point (CEP)s form a continuous transition line touched at $T = 0$ as the QCP. **b** FOVT line $(h/h_{\rm v})^2 + (T/T_{\rm v})^2 = 1$ [59–62] in the T-h plane for a fixed P, corresponding to the dashed line in **a** [4]. Reprinted from Ref. [4]. Copyright 2008 American Physical Society

ground for small-$T_{\rm v}$ behavior by considering the coherence of electrons which is essential for low temperature, but also interpolates high $T_{\rm v}$ satisfying the relation [59–62]

$$\left(\frac{h}{h_{\rm v}}\right)^2 + \left(\frac{T}{T_{\rm v}}\right)^2 = 1 \tag{4.43}$$

to zero temperature, since this relation was derived by assuming the isolated atomic entropy [59], which is justified only at the high-temperature regime.

4.6.2 *Field Induced Quantum Critical Point of the Valence Transition*

Although we have shown that the h dependence of the critical point can be understood from the viewpoint of the free-energy gain by the larger entropy in the Kondo regime, it is highly nontrivial how the QCP of the first-order valence transition is controlled by h at $T = 0$. To proceed with our analysis, we consider the extended periodic Anderson model which describes the essential part of the electronic state in the Ce and Yb systems [4]:

$$H = H_{\rm c} + H_{\rm f} + H_{\rm hyb} + H_{U_{\rm fc}} - h\sum_i (S_i^{\rm fz} + S_i^{\rm cz}), \tag{4.44}$$

where $H_{\rm c} = \sum_{k\sigma} \varepsilon_k c^\dagger_{k\sigma} c_{k\sigma}$, $H_{\rm f} = \varepsilon_{\rm f} \sum_{i\sigma} n^{\rm f}_{i\sigma} + U \sum_{i=1}^N n^{\rm f}_{i\uparrow} n^{\rm f}_{i\downarrow}$, $H_{\rm hyb} = V \sum_{i\sigma} \left(f^\dagger_{i\sigma} c_{i\sigma} + c^\dagger_{i\sigma} f_{i\sigma}\right)$ and $H_{U_{\rm fc}} = U_{\rm fc} \sum_{i=1}^N n^{\rm f}_i n^{\rm c}_i$. The $U_{\rm fc}$ term is the Coulomb repulsion between the f and conduction electrons, which is considered to play an important role in the valence transition [2, 3, 49–51, 63]. For example, in the case of Ce metal which exhibits the γ-α transition, the 4f- and 5d-electron bands are located at the Fermi level [64]. Since both the orbitals are located on the same Ce site, this term cannot be neglected. In the case of YbInCu$_4$, $H_{U_{\rm fc}}$ also plays a crucial role for the FOVT in the hole picture of Eq. (4.44) [63].

We apply the slave-boson-mean-field theory [2] to Eq. (4.44). To describe the state for $U=\infty$, we consider $Vf^{\dagger}_{i\sigma}b_i c_{i\sigma}$ instead of $Vf^{\dagger}_{i\sigma}c_{i\sigma}$ in Eq. (4.44) by introducing the slave-boson operator b_i at the i-th site to describe the f^0 state and require the constraint $\sum_i \lambda_i \left(\sum_\sigma n^{\rm f}_{i\sigma} + b^{\dagger}_i b_i - 1\right)$ with λ_i being the Lagrange multiplier. For $H_{U_{\rm fc}}$ in Eq. (4.44), we employ the mean-field decoupling as $n^{\rm f}_i n^{\rm c}_i \simeq n_{\rm f} n^{\rm c}_i + n_{\rm c} n^{\rm f}_i - \frac{1}{2} n_{\rm f} n_{\rm c}$. By approximating mean fields as uniform ones, i.e., $b = \langle b_i \rangle$ and $\bar{\lambda} = \lambda_i$, the set of mean-field equations are obtained by $\partial\langle H\rangle/\partial\lambda = 0$ and $\partial\langle H\rangle/\partial b = 0$:

$$\bar{\lambda} = \frac{V^2}{N}\sum_{\mathbf{k}\sigma} \frac{f(E^-_{k\sigma}) - f(E^+_{k\sigma})}{\sqrt{(\bar{\varepsilon}_{\rm f\sigma} - \bar{\varepsilon}_{\mathbf{k}\sigma})^2 + 4\bar{V}^2}}, \tag{4.45}$$

$$1 - |\bar{b}|^2 = \frac{1}{2N}\sum_{k\sigma,\pm}\left[1 \pm \frac{\bar{\varepsilon}_{\rm f\sigma} - \bar{\varepsilon}_{k\sigma}}{\sqrt{(\bar{\varepsilon}_{\rm f\sigma} - \bar{\varepsilon}_{k\sigma})^2 + 4\bar{V}^2}}\right] f(E^{\pm}_{k\sigma}), \tag{4.46}$$

and the following equation holds for the total electron number:

$$\bar{n}_{\rm f} + \bar{n}_{\rm c} = \frac{1}{N}\sum_{k\sigma}\left[f(E^-_{k\sigma}) + f(E^+_{k\sigma})\right]. \tag{4.47}$$

Here, $f(E)$ is the Fermi distribution function and $E^{\pm}_{k\sigma}$ are the lower $(-)$ and upper $(+)$ hybridized bands for the quasi particle with spin σ, respectively:

$$E^{\pm}_{k\sigma} = \frac{1}{2}\left[\bar{\varepsilon}_{\rm f\sigma} + \bar{\varepsilon}_{k\sigma} \pm \sqrt{(\bar{\varepsilon}_{\rm f\sigma} - \bar{\varepsilon}_{k\sigma})^2 + 4\bar{V}^2}\right], \tag{4.48}$$

where $\bar{\varepsilon}_{k\sigma}, \bar{\varepsilon}_{\rm f\sigma}$ and $\bar{V}$ are defined by $\bar{\varepsilon}_{k\sigma} \equiv \varepsilon_{\mathbf{k}} + U_{\rm fc}\bar{n}_{\rm f} - \frac{h\sigma}{2}, \bar{\varepsilon}_{\rm f\sigma} \equiv \varepsilon_{\rm f} + \bar{\lambda} + U_{\rm fc}\bar{n}_{\rm c} - \frac{h\sigma}{2}$ and $\bar{V} \equiv V|\bar{b}|$. The dispersion of the conduction electrons is taken as

$$\varepsilon_k = \frac{\boldsymbol{k}^2}{2m} - D \tag{4.49}$$

with D being the bottom of the conduction band and the density of states $N_0(\varepsilon)$ is set to satisfy the normalization condition, $\int_{-D}^{D} d\varepsilon N_0(\varepsilon) = 1$ per spin in three dimensions. We take D as the energy unit and show the results for $V = 0.5$ and the total filling $n = (\bar{n}_{\rm f} + \bar{n}_{\rm c})/2 = 7/8$.

At $h = 0$ and $T = 0$, a jump in $\bar{n}_{\rm f}$ appears as a function of $\varepsilon_{\rm f}$ for large $U_{\rm fc}$, which indicates the first-order quantum phase transition, since a jump in $\bar{n}_{\rm f}$ results in the level crossing of the ground states by the relation

$$\bar{n}_{\rm f} = \frac{\partial \langle H \rangle}{\partial \varepsilon_{\rm f}}. \tag{4.50}$$

The FOVT between the Kondo state and the MV state is caused by $U_{\rm fc}$, since a large $U_{\rm fc}$ forces the electrons to pour into either the f-level or the conduction band. The QCP in the $\varepsilon_{\rm f}$-$U_{\rm fc}$ plane is identified to be $(\varepsilon_{\rm f}^{\rm QCP}, U_{\rm fc}^{\rm QCP}) = (0.356, 1.464)$, at which the jump in $\bar{n}_{\rm f}$ disappears and the valence susceptibility

$$\chi_{\rm v} \equiv -\frac{\partial^2 \langle H \rangle}{\partial \varepsilon_{\rm f}^2} = -\frac{\partial \bar{n}_{\rm f}}{\partial \varepsilon_{\rm f}} \tag{4.51}$$

diverges. The characteristic energy scale of the system, the so-called Kondo temperature, which is defined as $T_{\rm K} \equiv \bar{\varepsilon}_{{\rm f}\sigma} - \mu$ is estimated to be $T_{\rm K} = 0.074$ at the QCP.

Note that the surface of the first-order valence transition exists in the parameter space of $\varepsilon_{\rm f}$, $U_{\rm fc}$ and V and a trajectory line is drawn in the space for corresponding experimental parameter such as pressure (i.e., P in Fig. 4.17a).

A remarkable result is found under the magnetic field in the valence-crossover regime for $U_{\rm fc} < U_{\rm fc}^{\rm QCP}$. Figure 4.18 shows the magnetization

$$m = \frac{1}{N} \sum_i \langle S_i^{{\rm f}z} + S_i^{{\rm c}z} \rangle \tag{4.52}$$

versus h for $(\varepsilon_{\rm f}, U_{\rm fc}) = (-0.354, 1.458)$ (thin line) and $(-0.349, 1.442)$ (bold line), indicating that the metamagnetism defined by the diverging magnetic susceptibility

$$\chi = \frac{\partial m}{\partial h} = \infty \tag{4.53}$$

emerges at $h = h_{\rm m} = 0.01$ and 0.02, respectively. To clarify the origin, we have determined the FOVT line as well as the QCP under the magnetic field. The result is shown in Fig. 4.19. It is found that the FOVT line extends to the MV regime and the location of the QCP is shifted to the smaller $U_{\rm fc}$ and $|\varepsilon_{\rm f}|$ direction, when h is applied. This low-h behavior of the FOVT line agrees with the low-temperature limit of $T_{\rm v}$ discussed above.

We show the m-h curve at $U_{\rm fc} = 1.42$ for $\varepsilon_{\rm f}$ ranging from -0.32 to -0.36 in the inset of Fig. 4.18. $T_{\rm K}$ at $h = 0$ is estimated as 0.0353, 0.0873, 0.1346, 0.1611, and 0.1823 for $\varepsilon_{\rm f} = -0.36, -0.35, -0.34, -0.33$ and -0.32, respectively (see crosses in Fig. 4.19). From these results, the mechanism is understood as follows: At $h = 0$, $T_{\rm K}$ is originally large for $\varepsilon_{\rm f} = -0.32$ and -0.33, since the system is in the MV regime. However, by applying h, the QCP approaches, which makes $T_{\rm K}$ reduced, since the system is forced to be closer to the Kondo regime by h. At the magnetic field $h = h_{\rm m}$ where the QCP is reached, the metamagnetism occurs with a singularity $\delta m \sim \delta h^{1/3}$ [65] as shown in Fig. 4.18. On the other hand, for $\varepsilon_{\rm f} = -0.35$ and -0.36, the QCP is not approached, so that no metamagnetism appears.

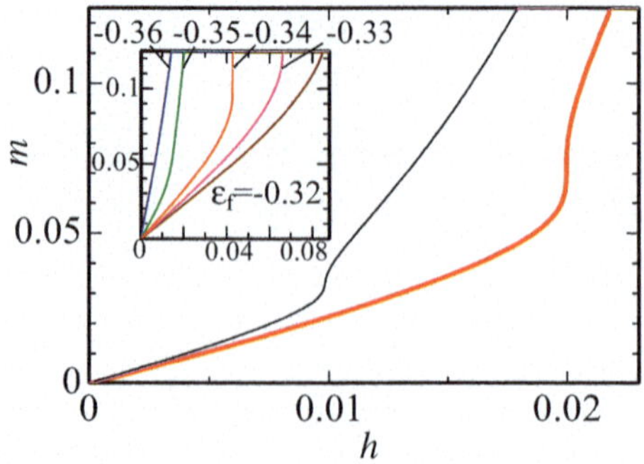

Fig. 4.18 The magnetization versus magnetic field for $(\varepsilon_{\rm f}, U_{\rm fc}) = (-0.354, 1.458)$ (black thin line) and $(-0.349, 1.442)$ (red bold line). Inset: m-h curve for $\varepsilon_{\rm f}$ ranging from -0.36 to -0.32 at $U_{\rm fc} = 1.42$. In both cases $D = 1$, $V = 0.5$ at $n = 7/8$ [4]. Reprinted from Ref. [4]. Copyright 2008 American Physical Society

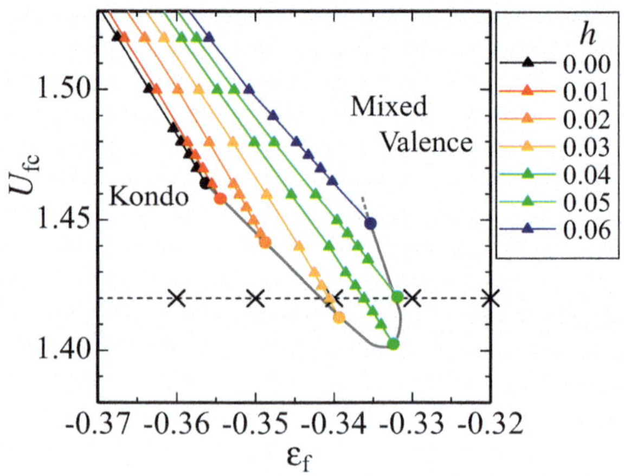

Fig. 4.19 Ground-state phase diagram in the plane of $U_{\rm fc}$ and $\varepsilon_{\rm f}$ for $D = 1$, $V = 0.5$ at $n = 7/8$. The FOVT line with a QCP for $h = 0.00$ (black), 0.01 (red) 0.02 (orange), 0.03 (bright yellow), 0.04 (yellowish green), 0.05 (green) and 0.06 (blue). The shaded line connects the QCP under h, which is a guide for the eyes. The dashed line with crosses represents $U_{\rm fc} = 1.42$ (see text) [4]. Reprinted from Ref. [4]. Copyright 2008 American Physical Society

An intriguing result is found in Fig. 4.19, which exhibits a non-monotonic h dependence of the QCP: As h increases furthermore, the QCP shows an upturn around $h = 0.04$, which is comparable to $T_{\rm K}$ at the QCP for $h = 0$. The upturn of the QCP has been also confirmed for a constant density of states $N_0(\varepsilon) = 1/(2D)$, suggesting that this behavior appears irrespective of details of the band structure.

The non-monotonic behavior can be understood from the structure of the valence susceptibility $\chi_{\rm v}$, which is given essentially by the RPA as discussed in Ref. [66]. Namely, it is given as $\chi_{\rm v}(q) \approx \chi_{\rm fc}^{(0)}(q)/[1 - U_{\rm fc}\chi_{\rm fc}^{(0)}(q)]$, where $\chi_{\rm fc}^{(0)}$ is the bubble diagram composed of f and conduction electrons. In the Kondo regime ($h \lesssim T_{\rm K}$), where f-electrons have a dominant spectral weight at around $\varepsilon \sim -\varepsilon_{\rm f}$ with width $\Delta \simeq \pi V^2 N(\varepsilon_{\rm F})$, $\chi_{\rm fc}^{(0)}$ is estimated as $\chi_{\rm fc}^{(0)} \approx 1/|\varepsilon_{\rm f}|$ and is shown to be an increasing function of h. Therefore, $U_{\rm fc}^{\rm QCP}$ is decreasing as h is applied until it reaches around $h \sim T_{\rm K}$, and $|\varepsilon_{\rm f}^{\rm QCP}| \approx U_{\rm fc}^{\rm QCP}$ is also decreasing. For $h \gtrsim T_{\rm K}$, mass enhancement ($\sim 1/z$) is quickly suppressed and the MV regime is approached. Then, $\chi_{\rm fc}^{(0)}$ is given as $\chi_{\rm fc}^{(0)} \approx 1/\Delta < 1/|\varepsilon_{\rm f}|$ with a help of shift of the f-level toward the Fermi level, i.e., $\varepsilon_{\rm f} \rightarrow \varepsilon_{\rm f} + U_{\rm fc}\delta\langle n_{\rm c}\rangle$ ($\delta\langle n_{\rm c}\rangle$ being the change of the number of conduction electrons per site due to entering the MV regime), so that $U_{\rm fc}^{\rm QCP}$ is larger than that of $U_{\rm fc}^{\rm QCP}(h \sim T_{\rm K})$.

We find that $h_{\rm m}$ corresponds to the difference of $T_{\rm K}$ at the QCP between $h = 0$ and $h \neq 0$, i.e., $h_{\rm m} \sim \Delta T_{\rm K}^{\rm QCP} = T_{\rm K}^{\rm QCP}(h \neq 0) - T_{\rm K}^{\rm QCP}(h = 0)$. This indicates the emergence of a new energy scale distinct from $T_{\rm K}$, which is characterized by the closeness to the valence QCP. Furthermore, the proximity of the valence QCP to the intermediate valence-crossover regime under h yields emergence of the metamagnetism as well as a jump in m even without showing the temperature-driven FOVT at $h = 0$.

To examine the mechanism more precisely, we have applied the density-matrix-renormalization group method to Eq. (4.44) in one spatial dimension (1D). Since valence fluctuations are basically ascribed to the atomic origin, the fundamental properties are expected to be relevant even in 1D [3]. We show here the results for $\varepsilon_k = -2\cos(k)$, $V = 1$, $U = 10^4$ at $n = 7/8$ on the lattice with $N = 40$ sites for conduction electrons illustrated in the inset of Fig. 4.20a. For $h = 0$, $\bar{n}_{\rm f}$ versus $\varepsilon_{\rm f}$ for $U_{\rm fc} = 0.0$, 1.0 and 2.0 is shown in the inset of Fig. 4.20a. As $U_{\rm fc}$ increases, the change in $\bar{n}_{\rm f}$ becomes sharp. We show here the magnetization m in the MV state of which $\bar{n}_{\rm f}$ at $h = 0$ is indicated by an arrow in the inset of Fig. 4.20a. A plateau appears at $m = 1 - n = 1/8$, which is expected to disappear if we take more realistic choice

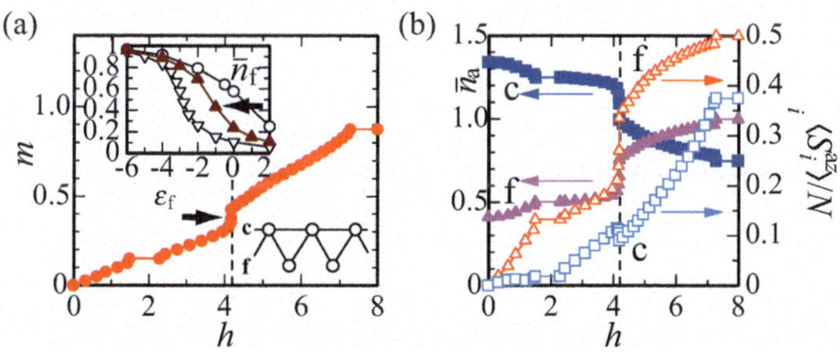

Fig. 4.20 Magnetization process for $\varepsilon_k = -2\cos(k)$, $V = 1$, $U = 10^4$, $\varepsilon_{\rm f} = -1$ and $U_{\rm fc} = 1$ at $n = 7/8$: **a** m-h curve (filled circle). An arrow indicates the metamagnetic transition. Inset: $\varepsilon_{\rm f}$ dependence of $\bar{n}_{\rm f}$ extrapolated to the bulk limit for $U_{\rm fc} = 0$ (open circle), $U_{\rm fc} = 1$ (filled triangle) and $U_{\rm fc} = 2$ (open triangle). An arrow indicates $\varepsilon_{\rm f} = -1$. Lattice structure used in the calculation. **b** $\bar{n}_{\rm f}$ (filled triangle) and $\bar{n}_{\rm c}$ (filled square). $m_{\rm f}$ (open triangle) and $m_{\rm c}$ (open square). In **a** and **b** dashed lines represent $h = h_{\rm m}$ [4]. Reprinted from Ref. [4]. Copyright 2008 American Physical Society

of parameters, e.g., the momentum dependence of V and ε_{f}. The main result is that metamagnetism emerges as indicated by an arrow. The increase of $\bar{n}_{\mathrm{f}}$ simultaneous with the decrease of $\bar{n}_{\mathrm{c}}$ at $h = h_{\mathrm{m}}$ shown in Fig. 4.20b reveals that this is caused by the field-induced extension of the QCP to the MV regime. Namely, these results indicate that the above mean-field conclusion is not altered even after taking account of quantum fluctuations and electron correlations.

To explore further the nature of this metamagnetism, we have calculated $m_{\mathrm{f}} = \sum_i \langle S_i^{\mathrm{f}z} \rangle / N$ and $m_{\mathrm{c}} = \sum_i \langle S_i^{\mathrm{c}z} \rangle / N$ and we find that m_{c} decreases at $h = h_{\mathrm{m}}$, while m_{f} increases as shown in Fig. 4.20b. Since the Kondo cloud is still formed even at $h = h_{\mathrm{m}}$, i.e., $\langle \mathbf{S}_i^{\mathrm{f}} \cdot \mathbf{S}_i^{\mathrm{c}} \rangle < 0$, a remarkable decrease of $\langle S_i^{\mathrm{c}z} \rangle$ is ascribed to *the field-induced Kondo effect*, which is a consequence of the energy benefit by both the Kondo effect and the Zeeman effect. Although this mechanism itself has been known to occur in the Kondo regime [67], the present finding is that such a mechanism works in the MV regime as a driving force of the field-induced valence QCP.

4.6.3 *Comparison with Experiments*

As discussed in Sect. 1.1.2, $\mathrm{YbInCu_4}$ exhibits the first-order valence transition. The sister compounds $\mathrm{YbXCu_4}$ (X = Ag, Au, Cd, Mg, Tl, and Zn) were synthesized and the physical properties were observed experimentally [68]. Interestingly, metamagnetic behavior was observed around 40 T in $\mathrm{YbAgCu_4}$ [68]. It is noted that the metamagnetism observed in $\mathrm{YbAgCu_4}$ is explained by the field-induced valence QCP. The sharp contrast between $\mathrm{YbAgCu_4}$ and $\mathrm{YbCdCu_4}$, both of which are characterized by the nearly the same Kondo temperature, is also explained theoretically on the basis of the extended Anderson lattice model [6]. The comparison of $\mathrm{YbAgCu_4}$ and $\mathrm{YbCdCu_4}$ is discussed in Sect. 4.6.3.1. Intriguingly, the field-induced valence crossover was observed in $\mathrm{YbAuCu_4}$. This behavior is also explained theoretically [6], which is discussed in Sect. 4.6.3.2.

4.6.3.1 $\mathrm{YbAgCu_4}$ and $\mathrm{YbCdCu_4}$

The heavy fermion metal $\mathrm{YbAgCu_4}$ and $\mathrm{YbCdCu_4}$ have the paramagnetic-metal ground states. The analysis of T dependence of uniform magnetic susceptibilities of $\mathrm{YbAgCu_4}$ and $\mathrm{YbCdCu_4}$ concluded that both have nearly the same Kondo temperatures, about 200 K [69]. However, only in $\mathrm{YbAgCu_4}$ a peak structure appears in the magnetic susceptibility around $T \sim 40$ K, while in $\mathrm{YbCdCu_4}$ a conventional Pauli paramagnetism appears [19]. Such a contrast has been also detected in the magnetic field response: $\mathrm{YbAgCu_4}$ shows a metamagnetic increase of the magnetization around $h \sim 40$ T, while $\mathrm{YbCdCu_4}$ does not [69]. These observations suggest that there exists a distinct energy scale from the Kondo temperature.

The theoretical results presented in Sect. 4.6.2 explain this sharp contrast. Figure 4.21a shows the schematic contour plot of the $4f$-hole number per site, n_{f}, which can be also regarded as the contour plot of the Kondo temperature [70],

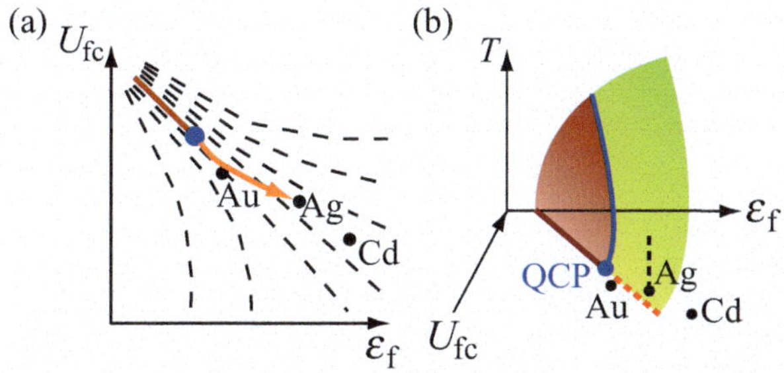

Fig. 4.21 **a** Schematic contour plot of n_f, i.e., the Kondo temperature T_K in the ε_f-U_{fc} plane for the model (4.44) at $h = 0$. The FOVT line (solid line) terminates at the QCP (filled circle). The QCP reaches $YbAgCu_4$ by applying h about 40 T represented by an arrow. $YbCdCu_4$ is located far away from the QCP. The valence-crossover surface reaches $YbAuCu_4$ by applying h about 1.3 T (see text). **b** Schematic T-ε_f-U_{fc} phase diagram at $h = 0$. $YbAgCu_4$ touches the valence-crossover surface at about $T = 40$ K. $YbCdCu_4$ is located too far from the QCP so that it needs too large T interval to reach the valence-crossover surface [6]. Reprinted from Ref. [6]. Copyright 2009 The Physical Society of Japan

$$T_K \propto \frac{1 - n_f}{1 - n_f/2}. \tag{4.54}$$

In the small (large) ε_f and U_{fc} regime, the Kondo (mixed valence) state with $n_f = 1$ ($n_f < 1$), i.e., small (large) T_K is realized. Since $YbAgCu_4$ and $YbCdCu_4$ have nearly the same T_K, both are considered to be located in the same contour area (see Fig. 4.21). If $YbAgCu_4$ is closer to the QCP with a distance about $h \sim 40$ T than $YbCdCu_4$ which is less close to the QCP, the metamagnetic behavior in $YbAgCu_4$ at $h \sim 40$ T can be naturally understood.

This picture also gives a natural explanation for the origin of the peak structure observed in the uniform susceptibility at $T \sim 40$ K in $YbAgCu_4$. Figure 4.21b illustrates a schematic phase diagram of the T-ε_f-U_{fc} space. As shown in the inset of Fig. 4.18, the metamagnetism with diverging magnetic susceptibility emerges at the QCP as well as the critical points represented by the thick solid line with a circle in Fig. 4.21b. An important result is that even at the valence-crossover surface extended from the FOVT surface in Fig. 4.21b, the magnetic susceptibility χ_s as well as the valence susceptibility χ_v is enhanced [6]. If $YbAgCu_4$ touches the valence-crossover surface at about $T_v^* \sim 40$ K, the observed peak in $\chi_s(T)$ at $T \sim 40$ K is naturally explained. The reason of no conspicuous peak in $\chi_s(T)$ in $YbCdCu_4$ can be understood as too large temperature interval to touch the valence-crossover surface (see Fig. 4.21b). Indeed, a volume expansion was observed below $T \sim 40$ K in $YbAgCu_4$ [71], which is directly related to the sharp valence crossover from $Yb^{+2.89}$ ($T > 40$ K) to $Yb^{+2.87}$ ($T < 40$ K) [68].

Indeed, a sharp valence increase was observed around the metamagnetic field 40 T in $YbAgCu_4$ as shown in Fig. 4.22 [72]. This provides the experimental evidence of the field-induced valence QCP.

Fig. 4.22 **a** Magnetic field dependence of the Yb-valence at 4.8 K in $YbAgCu_4$. **b** Magnetization (M) and magnetic differential susceptibility (dM/dH) at 4.2 K plotted as a function of the magnetic field [72]. Reprinted with permission from Ref. [72]. Copyright 2012 The Physical Society of Japan

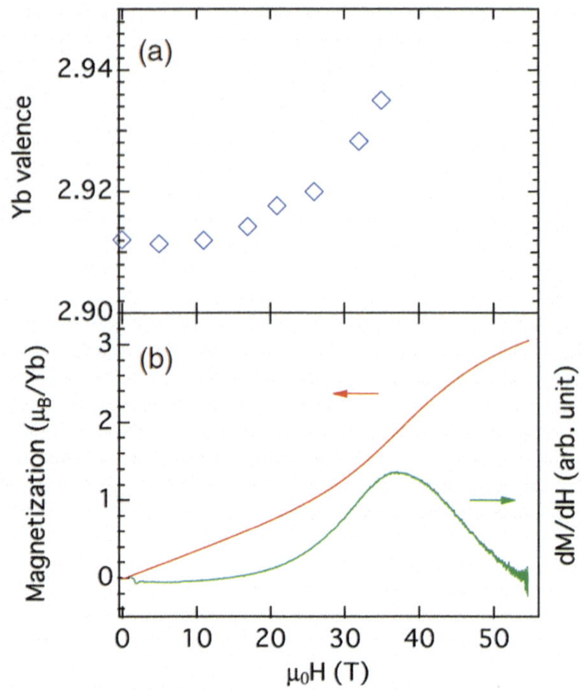

4.6.3.2 $YbAuCu_4$

The ^{63}Cu NQR measurement in $YbAuCu_4$ has revealed that the characteristic temperature T_v^* emerges in the T-h phase diagram as illustrated in Fig. 4.23a [73]. Under the magnetic field for $h > 1$ T, as T decreases, ^{63}Cu NQR frequency ν_Q decreases sharply at $T_v^*(h)$. This indicates that the electric-field gradient at the Cu site changes sharply at $T_v^*(h)$, implying that the charge distribution changes drastically around the Cu site. This strongly suggests that the valence of Yb changes at $T_v^*(h)$. As h increases and across $h_v^* \sim 1.3$ T, the T^2 coefficient A of the resistivity sharply decreases, and the residual resistivity has a cusp-like peak structure at h_v^* [73]. These observations also suggest that the Yb-valence-crossover temperature T_v^* is induced by applying the magnetic field.

This can be naturally understood if $YbAuCu_4$ is located just below the locus of the QCP as illustrated in Fig. 4.21a. As shown in Fig. 4.19, for $h < T_K^{QCP}(h = 0)$, the valence-crossover line extends to the Kondo regime (dashed lines for $h = 0.02$ and 0.03 cross that for $h = 0$). Then, by applying h, the valence-crossover surface approaches and reaches the point indicated as Au in Fig. 4.21a at $h = 1.3$ T, which gives h_v^* in Fig. 4.23. As h further increases, the QCP moves as an arrow in Fig. 4.21a. Hence, the valence-crossover temperature $T_v^*(h)$ increases as illustrated in Fig. 4.23. In the large h limit, the Yb^{3+} state is expected to be realized, since the Zeeman energy gain is mostly earned by the $n_f = 1$ state. Hence, $T_v^*(h)$ is expected to start to decrease

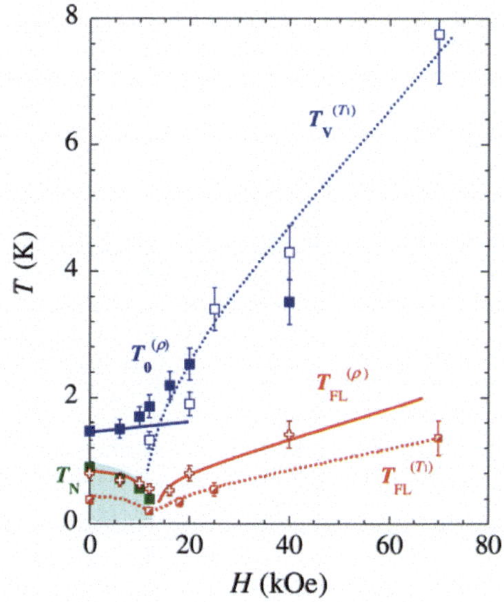

Fig. 4.23 The temperature T-magnetic field H phase diagram of $YbAuCu_4$. The Neel temperature T_N is suppressed and the Yb-valence-crossover temperature T_v is induced for $H \geq 13$ kOe. For $T < T_{FL}^{\rho}$ (red solid line) the resistivity shows the Fermi-liquid behavior $\rho \sim T^2$ [73]. Reprinted with permission from Ref. [73]. Copyright 2008 IOP Publishing

for larger h, and finally to touch absolute zero. However, as shown in Fig. 1.4c, the valence crossover is smeared out by thermal fluctuations as temperature increases. Hence, the low-temperature part of the valence-crossover line arising from the critical point at negative temperature is considered to be detected as $T_v^*(h)$ in Fig. 4.23 by the NQR measurement.

4.6.4 Consequence of Field Induced Valence QCP

We have shown that the field-induced QCP of the valence transition gives rise to metamagnetism, which offers essentially new mechanism. An important point is that at the field-induced valence QCP, uniform magnetic fluctuation diverges simultaneously with the divergence of the charge-transfer fluctuation. This indicates that simultaneous divergence of the valence susceptibility and the magnetic susceptibility occurs at the valence QCP. This finding leads to the new type of quantum criticality in the magnetic susceptibility and the other physical quantities caused by critical valence fluctuations arising from the valence QCP, which is discussed in Sect. 6.2.

References

1. W.E. Pickett, A.J. Freeman, D.D. Koelling, Phys. Rev. B **23**, 1266 (1981)
2. Y. Onishi, K. Miyake, J. Phys. Soc. Jpn. **69**, 3955 (2000)
3. S. Watanabe, M. Imada, K. Miyake, J. Phys. Soc. Jpn. **75**, 043710 (2006)

4. S. Watanabe, A. Tsuruta, K. Miyake, J. Flouquet, Phys. Rev. Lett. **100**, 236401 (2008)
5. Y. Saiga, T. Sugibayashi, D. Hirashima, J. Phys. Soc. Jpn. **77**, 114710 (2008)
6. S. Watanabe, A. Tsuruta, K. Miyake, J. Flouquet, J. Phys. Soc. Jpn. **78**, 104706 (2009)
7. S. Watanabe, K. Miyake, Phys. Status Solidi B **247**, 490 (2010)
8. S. Watanabe, K. Miyake, J. Phys. Soc. Jpn. **79**, 033707 (2010)
9. S. Watanabe, K. Miyake, J. Phys.: Condens. Matter **24**, 294208 (2012)
10. S. Watanabe, K. Miyake, J. Phys.: Condens. Matter **23**, 094217 (2011)
11. Y. Kojima, A. Koga, JPS Conf. Proc. **1**, 012106 (2014)
12. L.M. Falicov, J.C. Kimball, Phys. Rev. Lett. **22**, 997 (1969)
13. C.M. Varma, Rev. Mod. Phys. **48**, 219 (1976)
14. C.M. Varma, S. Schmitt-Rink, E. Abrahams, Solid State Commun. **62**, 681 (1987)
15. J.M. Luttinger, Phys. Rev. **119**, 1153 (1960); *ibid.* **121**, 942 (1961)
16. C.M. Varma, S. Schmitt-Rink, E. Abrahams, Solid State Commun. **62**, 681 (1987)
17. C.M. Varma, P.B. Littlewood, S. Schmitt-Rink, E. Abrahams, A.E. Ruckenstein, Phys. Rev. Lett. **63**, 1996 (1989)
18. I.E. Perakis, C.M. Varma, A.E. Ruckenstein, Phys. Rev. Lett. **70**, 3467 (1993)
19. C.M. Varma, Z. Nussinov, W. van Saarloos, Phys. Rep. **361**, 267 (2002)
20. J.G. Sereni, Physica B **215**, 273 (1995)
21. P.W. Anderson, W.F. Brinkman, Phys. Rev. Lett. **30**, 1108 (1973)
22. S. Nakjima, Prog. Theor. Phys. **50**, 1101 (1973)
23. F. Steglich, J. Aarts , C.D. Bredl, W. Lieke D. Meschede, W. Franz, H. Schäfer, Phys. Rev. Lett. **43**, 1892 (1979)
24. K. Miyake, S. Schmitt-Rink, C.M. Varma, Phys. Rev. B **34**, 6554 (1986)
25. D.J. Scalapino Jr., E. Loh, J.E. Hirsch, Phys. Rev. B **34**, 8190 (1986)
26. T. Moriya, K. Ueda, Y. Takahashi, J. Phys. Soc. Jpn. **59**, 2905 (1990)
27. P. Monthoux, D. Pines, Phys. Rev. Lett. **69**, 96 (1992)
28. T. Moriya, K. Ueda, Adv. Phys. **49**, 555 (2000)
29. N.D. Mathur, F.M. Grosche, S.R. Julian, I.R. Walker, D.M. Freye, R.K.W. Haselwimmer, G.G. Lonzarich, Nature **394**, 39 (1998)
30. F.M. Grosche, S.R. Julian, N.D. Mathur, G.G. Lonzarich, Physica B **223&224**, 50 (1996)
31. R. Movshovich, T. Graf, D. Mandrus, J.D. Thompson, J.L. Smith, Z. Fisk, Phys. Rev. B **53**, 8241 (1996)
32. P. Monthoux, G.G. Lonzarich, Phys. Rev. B **59**, 14598 (1999)
33. S. Watanabe, K. Miyake, J. Phys. Soc. Jpn. **71**, 2489 (2002)
34. B. Bellarbi, A. Benoit, D. Jaccard, J.M. Mignot, H.F. Braun, Phys. Rev. B **30**, 1182 (1984)
35. F. Thomas, C. Ayache, I.A. Fomin, J. Thomasson, C. Geibel, J. Phys.: Condens. Matter **8**, L51 (1996)
36. D. Jaccard, E. Vargoz, K. Alami-Yadri, H. Wilhelm, Rev. High Pressure Sci. Technol. **7**, 412 (1998)
37. D. Jaccard, H. Wilhelm, K. Alami-Yadri, E. Vargoz, Physica B **259–261**, 1 (1999)
38. T.C. Kobayashi T. Miyazu , N. Takeshita, K. Shimazu, K. Amaya, Y. Kitaoka, Y. Ōnuki, J. Phys. Soc. Jpn. **67**, 996 (1998)
39. Y. Kitaoka, H. Tou, G.-Q. Zheng, K. Ishida, K. Asayama, T.C. Kobayashi, A. Kohda, N. Takeshita, K. Amaya Y. Ōnuki, C. Geibel, C. Schank, F. Steglich, Physica B **206 and 207**, 55 (1995)
40. K. Miyake, O. Narikiyo, Y. Onishi, Physica B **259–261**, 676 (1999)
41. K. Miyake, H. Maebashi, J. Phys. Soc. Jpn. **71**, 1007 (2002)
42. H.Q. Yuan, F.M. Grosche, M. Deppe, C. Geibel, G. Sparn, F. Steglich, Science **302**, 2104 (2003)
43. A.T. Holmes, D. Jaccard, K. Miyake, Phys. Rev. B **69**, 024508 (2004)
44. P. Morel, P.W. Anderson, Phys. Rev. B **125**, 1263 (1962)
45. R. Read, D.M. Newns, J. Phys. C: Solid State Phys. **16**, 3273 (1983)
46. M. Lavagne, A.J. Millis, P.A. Lee, Phys. Rev. Lett. **58**, 266 (1987)
47. A. Houghton, N. Read, H. Won, Phys. Rev. B **37**, 3782 (1987)

48. A.J. Millis, P.A. Lee, Phys. Rev. B **35**, 3394 (1986)
49. C.E.T. Gonçalves da Silva, L.M. Falicov, Solid State Commun. **17**, 152 (1975)
50. A.C. Hewson, P.S. Riseborough, Solid State Commun. **22**, 379 (1977)
51. I. Singh, A.K. Ahuja, S.K. Joshi, Solid State Commun. **34**, 65 (1980)
52. S.R. White, Phys. Rev. Lett. **69**, 2863 (1992); S.R. White, Phys. Rev. B **48**, 10345 (1993)
53. I. Singh, A.K. Ahuja, S.K. Joshi, Solid State Commun. **34**, 65 (1980)
54. Y. Onishi, K. Miyake, Physica B **281–282**, 191 (2000)
55. S. Watanabe, M. Imada, J. Phys. Soc. Jpn. **73**, 3341 (2004)
56. H. Schulz, Int. J. Mod. Phys. B **5**, 57 (1991)
57. M. Ogata, M.U. Luchini, S. Sorella, F.F. Assaad, Phys. Rev. Lett. **66**, 2388 (1991)
58. When the charge velocity v_c is defined by $v_c \equiv \lim_{\Delta q \to 0} \Delta_c^*/\Delta q$ for $\Delta_c^* \equiv E_1(2nN, 0) - E(2nN, 0)$ with E_1 being the first-excited state, $K_\rho = \pi v_c \kappa/2$ holds. If the symmetric condition $\varepsilon_f + U/2 = 0$ and $U_{fc} = 0$ are satisfied in eq. (7.1), $\Delta_c = \Delta_c^*$; i.e., $v_c = v_c^*$ holds by charge SU(2) symmetry. It is expected that $v_c \sim v_c^*$ holds for eq. (7.1) in the Kondo regime
59. F. Drymiotis, J. Singleton, N. Harrison, J.C. Lashley, A. Bangura, C.H. Mielke, L. Balicas, Z. Fisk, A. Migliori, J.L. Smith, J. Phys.: Condens. Matter **17**, L77 (2005)
60. C.D. Immer, J.L. Sarrao, Z. Fisk, A. Lacerda, C. Mielke, J.D. Thompson, Phys. Rev. B **56**, 71 (1997)
61. Y.H. Matsuda, T. Inami, K. Ohwada, Y. Murata, H. Nojiri, Y. Murakami, H. Ohta, W. Zhang, K. Yoshimura, J. Phys. Soc. Jpn. **76**, 034702 (2007)
62. S. Basu, P.S. Riseborough, Physica B **378–380**, 686 (2006)
63. J.K. Freericks, V. Zlatić, Phys. Rev. B **58**, 322 (1998)
64. W.E. Pickett, A.J. Freeman, D.D. Koelling, Phys. Rev. B **23**, 1266 (1981)
65. A.J. Millis, A.J. Schofield, G.G. Lonzarich, S.A. Grigera, Phys. Rev. Lett. **88**, 217204 (2002)
66. K. Miyake, J. Phys.: Condens. Matter **19**, 125201 (2007)
67. S. Watanabe, J. Phys. Soc. Jpn. **69**, 2947 (2000)
68. J.L. Sarrao, C.D. Immer, Z. Fisk, C.H. Booth, E. Figueroa, J.M. Lawrence, R. Modler, A.L. Cornelius, M.F. Hundley, G.H. Kwei, J.D. Thompson, F. Bridges, Phys. Rev. B **59**, 6855 (1999)
69. J.L. Sarrao, Physica B **259–261**, 128 (1999)
70. T.M. Rice, K. Ueda, Phys. Rev. B **34**, 6420 (1986)
71. T. Koyama, M. Matsumoto, T. Tanaka, H. Ishida, T. Mito, S. Wada, J.L. Sarrao, Phys. Rev. B **66**, 014420 (2002)
72. Y.H. Matsuda, T. Nakamura, J.-L. Her, K. Kindo, S. Michimura, T. Inami, M. Mizumaki, N. Kawamura, M. Suzuki, B. Chen, H. Ohta, K. Yoshimura, J. Phys. Soc. Jpn. **81**, 015002 (2012)
73. S. Wada, A. Yamamoto, K. Ishida, J.L. Sarrao, J. Phys.: Condens. Matter **20**, 175201 (2008)

Chapter 5
Self-consistent Renormalization Theory

The self-consistent renormalization (SCR) theory has succeeded in explaining the quantum critical phenomena of the itinerant magnetism [1]. In this chapter, the SCR theory for the itinerant magnetism is discussed. When the control parameter of the material such as pressure, magnetic field, and chemical doping is changed, the continuous transition to the magnetically ordered phase can be suppressed to absolute zero, which is called the quantum critical point (QCP) (see Fig. 5.1). Near the QCP in the itinerant electron systems, enhanced magnetic fluctuations cause anomalous behavior in the physical quantities distinct from the Fermi-liquid behavior in normal metals, which is called the magnetic quantum critical phenomena.

The schematic phase diagram around the magnetic QCP is illustrated in Fig. 5.1. The horizontal axis g is the control parameter of the system such as pressure, magnetic field, and chemical doping and the QCP is located at $g = g_c$. For $g > g_c$, the phase transition takes place between the paramagnetic phase and the magnetically ordered phase as a function of temperature. This phase transition is driven by thermal fluctuations. On the other hand, for $T = 0$, the phase transition occurs by changing the control parameter g. This phase transition is driven by quantum-mechanical fluctuations, which is called the quantum phase transition.

At $T = 0$ and $g = g_c$, the magnetic fluctuations diverge quantum critically, giving rise to the non-Fermi-liquid behavior in physical quantities in the paramagnetic-metal phase near $g = g_c$. The region is denoted as the quantum critical regime in Fig. 5.1. As the temperature increases, the uniform susceptibility $\chi(T)$ exhibits the Curie-Weiss behavior as $\chi(T) \sim 1/(T + \theta)$. The region is denoted as the Curie-Weiss regime. In the paramagnetic-metal phase, as distanced from the QCP, the regime where the physical quantities exhibit the Fermi-liquid behavior appears at low temperatures. The quantum critical regime changes to the Fermi-liquid regime via crossover. These overall features are well understood on the basis of the SCR theory of spin fluctuations, which is explained in Sect. 5.3. Below the background of the SCR theory is overviewed.

S. Watanabe and K. Miyake, *Quantum Critical Phenomena of Valence Transition*,
Springer Tracts in Modern Physics 289,
https://doi.org/10.1007/978-981-99-3518-5_5

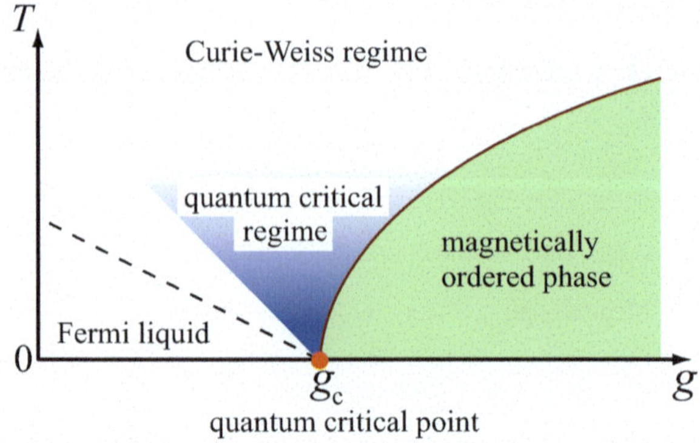

Fig. 5.1 Schematic phase diagram of magnetic quantum critical point. Vertical axis is temperature T and horizontal axis is control parameter g (pressure, magnetic field, and chemical doping)

From 1950s to 1960s, in several itinerant ferromagnetic metals such as $ZrZn_2$, the Curie-Weiss behavior was observed in the wide temperature regime higher than the ferromagnetic transition temperature T_c. The Curie temperature $T_c = 21$ K is low and the saturated magnetic moment $0.12\mu_B$ is small. The similar behavior was also observed in MnSi and Ni_3Al. In these materials, the magnetic transition temperature T_c is lower than the Fermi temperature T_F and the magnetic susceptibility follows the Curie-Weiss law, which is the same as that in the localized electron systems in spite of the small saturated moment. Namely, the Curie constant is in almost the same order of the magnitude as that in the localized electron systems with the large magnetic moment in the order of μ_B. This was a puzzling paradox. Of course, the statistical mechanics of the free electron gas tells us that the magnetic susceptibility follows the Curie law at higher temperatures than the Fermi temperature ($T \gg T_F$). However, it was the issue that in the itinerant magnets with $T_c \ll T_F$, the Curie-Weiss law holds in the wide temperature range for $T > T_c$. The reasonable explanation for this issue was given by the SCR theory developed by Moriya and coworkers.

The SCR theory was first proposed by Moriya and Kawabata in 1973 [2, 3], which succeeded in explaining the Curie-Weiss law in the magnetic susceptibility in the ferromagnetic metals [1, 4]. Furthermore, on the basis of the SCR theory, it was shown that the non-Fermi-liquid behavior in the specific heat [5], the electrical resistivity [6], and the NMR relaxation rate [7] is caused by spin fluctuations. In the antiferromagnetic metals, it was shown that the non-Fermi-liquid behavior appears [1]. As noted above, in the quantum phase transition at $T = 0$, quantum-mechanical fluctuations dominate the quantum critical phenomena. From this viewpoint, the quantum critical phenomena was studied on the basis of the renormalization group theory by Hertz in 1976 [8] and later the behavior of physical quantities near the magnetic QCP was studied by Millis [9]. The quantum criticality shown by the renormalization group theory [10] was confirmed to coincide with the results by the SCR theory [11–17].

An advantageous point of the SCR theory is that the behavior of the physical quantities is described in a unified way from the Curie-Weiss regime located in the high-temperature regime in Fig. 5.1 to the quantum critical regime located in the low-temperature regime. Later, the SCR theory was reformulated in the form of the mode-mode coupling theory of spin fluctuations and it turned out that the physical meaning was more clearly understood. Namely, this reformulation made it possible to understand that the SCR theory is regarded as the extension of the Murata-Doniach theory, which treated the mode-mode coupling of spin fluctuations in the classical framework [18], to the quantum-mechanical version. The formulation in the form of the mode-mode coupling theory made us to understand why the SCR theory can describe the quantum critical regime from the viewpoint of the renormalization group. Therefore, in Sect. 5.2, the idea of the renormalization group theory is explained before expounding the SCR theory in Sect. 5.3. Below the SCR theory is explained in the path-integral formalism based on the field theory, which seems to provide the straightforward derivation in the clearest way according to the author's viewpoint [19]. In Sect. 5.1, the action which is essential for itinerant magnetism is derived stating from the Hubbard model in the path-integral formalism. Below we employ the units where the Dirac constant and the Boltzmann constant are taken as unity $\hbar = 1$ and $k_B = 1$ for simplicity unless otherwise noted.

5.1 Magnetism in Itinerant Electron Systems with Electron Correlations

To describe the essential property of the magnetic transition in itinerant electron systems, let us consider the Hubbard model

$$H = \sum_{k\sigma} \xi_k c^{\dagger}_{k\sigma} c_{k\sigma} + U \sum_i n_{i\uparrow} n_{i\downarrow} \tag{5.1}$$

where ξ_k is the energy band relative to the Fermi energy μ defined by $\xi_k = \varepsilon_k - \mu$ and U is the onsite Coulomb interaction between the $\uparrow$ and $\downarrow$ spins. When the total electron number N_e is equal to the number of the lattice N, i.e., at the half filling, the effective model is derived for Coulomb repulsion larger than the electron transfer t as

$$H = J \sum_{\langle i,j \rangle} \boldsymbol{S}_i \cdot \boldsymbol{S}_j \tag{5.2}$$

from the strong-coupling expansion of U/t, which is called the Heisenberg model. Here $\boldsymbol{S}_i$ is the spin operator at the ith site and $\langle i, j \rangle$ denotes the pair of the i, j sites where electrons are transferred. The exchange interaction $J > 0$ implies that the antiferromagnetic interaction works between the spins of electrons. The Hubbard model in the three spatial dimension shows the magnetic transition as illustrated in the temperature-$g = U/W$ with W being the band width in Fig. 5.1 and hence the model (5.1) describes the magnetic phase transition.

The distribution function Z is expressed by using the path-integral formalism as

$$Z = \int \mathcal{D}c^\dagger \mathcal{D}c \exp\left[-\int_0^\beta d\tau L\right], \tag{5.3}$$

where $\int \mathcal{D}c^\dagger \mathcal{D}c$ is the functional integral with respect to the Grassmann number and L is the Lagrangian (see Appendix A). Here, τ is the imaginary time and β is the inverse of temperature defined by $\beta \equiv 1/T$. The Lagrangian consists of the contributions from the free electrons L_0 and the Coulomb interaction L' as

$$L = L_0 + L', \tag{5.4}$$

where L_0 and L' are given by

$$L_0 = \sum_{k\sigma} c^\dagger_{k\sigma}\left(\frac{\partial}{\partial\tau} + \xi_k\right) c_{k\sigma} \tag{5.5}$$

$$L' = U\sum_i n_{i\uparrow}n_{i\downarrow}, \tag{5.6}$$

respectively. The Coulomb interaction term can be expressed as

$$U\sum_i n_{i\uparrow}n_{i\downarrow} = \frac{U}{4}\sum_i (n_{i\uparrow} + n_{i\downarrow})^2 - \frac{U}{4}\sum_i (n_{i\uparrow} - n_{i\downarrow})^2. \tag{5.7}$$

The first term on the right-hand side (r.h.s.) describes the contributions from the charge degree of freedom and the second term describes the contribution from the spin degree of freedom. Since the magnetic order is contributed from the spin degrees of freedom, we set the second term in Eq. (5.7) as H_{s} and apply the identity called the Stratonovich-Hubbard transformation as

$$\begin{aligned}\exp\left[-\int_0^\beta d\tau H_{\mathrm{s}}(\tau)\right] &= \int \mathcal{D}\varphi \exp\left[-\frac{U}{4}\int_0^\beta d\tau \sum_i \varphi_i^2(\tau)\right] \\ &\times \exp\left[-\int_0^\beta d\tau \sum_i \frac{U}{2}\varphi_i(\tau)\left\{n_{i\uparrow}(\tau) - n_{i\downarrow}(\tau)\right\}\right].\end{aligned} \tag{5.8}$$

From the second line in the r.h.s. of Eq. (5.8), it turns out that $\varphi_i(\tau)$ is regarded as the conjugate field to the z component of the spin of electron $S_{iz} \equiv (n_{i\uparrow} - n_{i\downarrow})/2$. By performing the functional integral $\int \mathcal{D}\varphi$ in Eq. (5.8), the spin part of the Lagrangian L' is reproduced. By applying this transformation to the L' in Eq. (5.4) and by neglecting the first term in Eq. (5.7), the distribution function of the system is expressed as

$$Z = \int \mathcal{D}\varphi \mathcal{D}c^\dagger \mathcal{D}c \exp\left[-\int_0^\beta d\tau L(\varphi, c^\dagger, c)\right]. \tag{5.9}$$

By performing the Grassmann integral with respect to $c^\dagger$ and c,[1] we obtain

$$Z = \int \mathcal{D}\varphi \mathrm{e}^{-\Phi[\varphi]}, \tag{5.11}$$

where $\Phi[\varphi]$ is the action

$$\Phi[\varphi] = \frac{U}{4}\int_0^\beta d\tau \sum_i \varphi_i^2(\tau) - \mathrm{Tr}\ln\left[\frac{\partial}{\partial\tau} + \xi_k + \mathrm{sign}(\sigma)\frac{U}{2}\varphi_i(\tau)\right]. \tag{5.12}$$

Here, we have used the relation $\det A = \exp(\mathrm{Tr}\ln A)$ in the resultant determinant after performing the Grassmann integral. In Eq. (5.12), $\mathrm{sign}(\sigma)$ is introduced as $\mathrm{sign}(\uparrow) = 1$, $\mathrm{sign}(\downarrow) = -1$.

The ln term in the r.h.s. of Eq. (5.12) is expanded with respect to $U\varphi$ as

$$\begin{aligned}\Phi[\varphi] &= \frac{1}{2}\sum_{\bar{q}} \Omega_2(\bar{q})\varphi(\bar{q})\varphi(-\bar{q}) \\ &+ \sum_{\bar{q}_1,\bar{q}_2,\bar{q}_3,\bar{q}_4} \Omega_4(\bar{q}_1,\bar{q}_2,\bar{q}_3,\bar{q}_4)\varphi(\bar{q}_1)\varphi(\bar{q}_2)\varphi(\bar{q}_3)\varphi(\bar{q}_4)\delta\left(\sum_{i=1}^4 \bar{q}_i\right) \\ &+ \cdots,\end{aligned} \tag{5.13}$$

where $\bar{q}$ is defined by $\bar{q} \equiv (\boldsymbol{q}, \mathrm{i}\omega_l)$ and ω_l is boson Matsubara frequency $\omega_l = 2\pi l T$ $(l = 0, \pm1, \pm2, \cdots)$. Here, $\varphi_i(\tau)$ is defined by

$$\varphi_i(\tau) = \frac{1}{\sqrt{N\beta}}\sum_{\boldsymbol{q}}\sum_{\mathrm{i}\omega_l} \varphi(\boldsymbol{q}, \mathrm{i}\omega_l)\mathrm{e}^{\mathrm{i}(\boldsymbol{q}\cdot\boldsymbol{r}_i - \omega_l\tau)}. \tag{5.14}$$

The delta function in the fourth-order term in the r.h.s. of Eq. (5.13) represents $\sum_{i=1}^4 \boldsymbol{q}_i = \boldsymbol{0}$ and $\sum_{i=1}^4 \omega_{l_i} = 0$. In the first term in the r.h.s of Eq. (5.13), the summation for the unique $(\boldsymbol{q}, \mathrm{i}\omega_l)$ is taken. On the other hand, in the second term, the summation for different $(\boldsymbol{q}, \mathrm{i}\omega_l)$ is taken, which indicates the interaction between the spin fluctuation modes $\varphi(\boldsymbol{q}, \mathrm{i}\omega_l)$.

The coefficient of the second-order term in the r.h.s. of Eq. (5.13), i.e., the Gaussian term, is given by

$$\Omega_2(\boldsymbol{q}, \mathrm{i}\omega_l) = \frac{U}{2}\left[1 - U\chi_0(\boldsymbol{q}, \mathrm{i}\omega_l)\right], \tag{5.15}$$

[1] For Grassmann number η_i, the following equation folds:

$$\int \prod_i d\eta_i^* d\eta_i \exp\left(-\sum_{ij}\eta_i^* A_{ij}\eta_j\right) = \det A. \tag{5.10}$$

where $\chi_0(\boldsymbol{q}, \mathrm{i}\omega_l)$ is defined by

$$\begin{aligned}\chi_0(\boldsymbol{q}, \mathrm{i}\omega_l) &= -\frac{T}{N}\sum_{\boldsymbol{k}}\sum_{\mathrm{i}\varepsilon_n} G_0(\boldsymbol{k}, \mathrm{i}\varepsilon_n)G_0(\boldsymbol{k}+\boldsymbol{q}, \mathrm{i}\varepsilon_n + \mathrm{i}\omega_l) \\ &= -\frac{1}{N}\sum_{\boldsymbol{k}} \frac{f(\xi_{\boldsymbol{k}}) - f(\xi_{\boldsymbol{k}+\boldsymbol{q}})}{\xi_{\boldsymbol{k}} - \xi_{\boldsymbol{k}+\boldsymbol{q}} + \mathrm{i}\omega_l}. \end{aligned} \tag{5.16}$$

In the first line of the r.h.s. in Eq. (5.16), the Green function of free electrons is given by $G_0(\boldsymbol{k}, \varepsilon_n) = 1/(\mathrm{i}\varepsilon_n - \xi_{\boldsymbol{k}})$ with ε_n being the fermion Matsubara frequency defined by $\varepsilon_n = (2n+1)\pi T$ $(n = 0, \pm 1, \pm 2, \cdots)$.

In the uniform system, by expanding (q, ω_l) around the magnetic order ($\boldsymbol{Q} = \boldsymbol{0}$ for the ferromagnetic order or finite $\boldsymbol{Q}$ for the antiferromagnetic order) and around $\omega_l = 0$, the coefficient of the Gaussian term is expressed as

$$\Omega_2(\boldsymbol{q}, \omega_l) \approx \eta_0 + Aq^2 + C_q|\omega_l|, \tag{5.17}$$

where $C_q = C/q$ for the ferromagnetic case and $C_q = C$ for the antiferromagnetic case (see Appendix B). As for the coefficient of the fourth-order term with respect to $\varphi(\boldsymbol{q}, \omega_l)$, the higher order terms depending on q and ω_l are less important in the critical phenomena. Hence we consider $\Omega_4(\bar{q}_1, \bar{q}_2, \bar{q}_3, \bar{q}_4) \approx v_4/(\beta N)$ as the leading term. For the ferromagnetic case, v_4 is evaluated as

$$v_4 \approx \frac{U^4}{32N\beta}\sum_{\boldsymbol{k}}\sum_{\mathrm{i}\varepsilon_n}[G_0(\boldsymbol{k}, \mathrm{i}\varepsilon_n)]^4. \tag{5.18}$$

For $T \to 0$, this is evaluated as

$$v_4 \approx -\frac{U^4}{192}N''(\varepsilon_{\mathrm{F}}), \tag{5.19}$$

where $N(\varepsilon_{\mathrm{F}})$ is the density of states per spin at the Fermi level ε_{F}. From Eq. (5.19) it turns out that the curvature of the density of states at ε_{F} is reflected in the coefficient of the fourth-order term v_4.

Consequently, we obtain the action that describes the quantum critical phenomena of the magnetic transition as

$$\begin{aligned}\Phi[\varphi] = &\frac{1}{2}\sum_{\bar{q}}\left(\eta_0 + Aq^2 + C_q|\omega_l|\right)\varphi(\bar{q})\varphi(-\bar{q}) \\ &+ \frac{v_4}{\beta N}\sum_{\bar{q}_1, \bar{q}_2, \bar{q}_3, \bar{q}_4} \varphi(\bar{q}_1)\varphi(\bar{q}_2)\varphi(\bar{q}_3)\varphi(\bar{q}_4)\delta\left(\sum_{i=1}^{4}\bar{q}_i\right). \end{aligned} \tag{5.20}$$

Since Eq. (5.20) contains the Matsubara frequency ω_l, this action describes the effect of the quantum dynamics. The phase transition at finite temperature is driven by thermal fluctuations and its critical phenomena are described by the Landau-Ginzburg-Wilson (LGW) action. The action in Eq. (5.20) is regarded as the extension to the general form including quantum fluctuations. In general, magnetic correlation length ξ increases as the critical point is approached and eventually diverges at the critical point. Therefore, in the vicinity of the critical point by changing the length scale the LGW action can be isomorphic, which ensures the scale invariance. On the contrary, in the action $\Phi[\varphi]$ describing the quantum phase transition, $\boldsymbol{q}$ and ω_l appear with the non-symmetric form as shown in Eq. (5.20), which indicates that there exists anisotropy for scaling in space and time. Hertz applied the scale transformation to space and time and analyzed the quantum critical phenomena by renormalization group [8]. Later, the approach was refined by Millis, who discussed the temperature dependence of the physical quantities [9].

5.2 Perturbation Renormalization Group Approach

Let us consider the vicinity of the quantum critical point for $T \to 0$. Then, the Matsubara frequency ω_l can be regarded as continuous variable ω. By rescaling the momentum and energy in Eq. (5.20), we obtain

$$\begin{aligned}\Phi[\varphi] &= \frac{1}{2}\beta N \int_0^{q_c} \frac{d^d q}{(2\pi)^d} \int_0^{\omega_c} \frac{d\omega}{2\pi} \left(r_0 + q^2 + \bar{C}_q|\omega|\right) |\varphi(q,\omega)|^2 \\ &+ u(\beta N)^2 \int_0^{q_c} \prod_{i=1}^{3} \frac{d^d q_i}{(2\pi)^d} \int_0^{\omega_c} \prod_{i=1}^{3} \frac{d\omega_i}{2\pi} \varphi(q_1,\omega_1)\varphi(q_2,\omega_2)\varphi(q_3,\omega_3) \\ &\times \varphi(-q_1-q_2-q_3, -\omega_1-\omega_2-\omega_3), \end{aligned} \tag{5.21}$$

where $\bar{C}_q$ is given as $\bar{C}_q = 1/q$ for the ferromagnetic case and $\bar{C}_q = 1$ for the antiferromagnetic case. Namely, by adjusting the coefficients A and C in Eq. (5.20) to be 1 so that the parameters η_0 and v_4 are transformed to r_0 and u, respectively, with constant factors be multiplied, which results in Eq. (5.21). In Eq. (5.21), q_c and ω_c are cutoff of the momentum and energy, respectively.

We apply the scale transformation to the action in Eq. (5.21) in the following procedure:
(a) Integrating out high momentum and frequency parts for $q_c/s < q < q_c$ and $\omega_c/s^z < \omega < \omega_c$, respectively (Fig. 5.2), with s being a dimensionless scaling parameter ($s \geq 1$) and z the dynamical exponent.
(b) Scaling of q and ω by $q' = sq$ and $\omega' = s^z\omega$.
(c) Re-scaling of φ by $\varphi'(q', \omega') = s^a \varphi(q'/s, \omega'/s^z)$.
Then, to make the Gaussian term in Eq. (5.21) be invariant, a must satisfy $a = -(d + z + 2)/2$ with d being the spatial dimension and $z = 3$ for the ferromagnetic

case and $z = 2$ for the antiferromagnetic case. This implies that ω_l is scaled by q^z in Eq. (5.17). As for the coefficient of the fourth-order term of φ, to make the fourth-order term invariant by the scale transformation, the relation between u and u' which is the coefficient after scale transformation

$$u(s) = u' s^{4-(d+z)} \tag{5.22}$$

is required to hold. This leads to the scaling equation

$$\frac{du}{d \ln s} = [4 - (d+z)]u + O(u^2). \tag{5.23}$$

By solving this equation, it turns out that the coefficient of the u term in the r.h.s. becomes negative for $d + z > 4$ so that u decreases as the scale s increases from 1, giving rise to

$$\lim_{s \to \infty} u(s) = 0. \tag{5.24}$$

This implies that as the quantum critical point is approached by increasing the scale s, the effect of the interaction of the fourth order with respect to φ becomes small and eventually vanishing. Hence u is referred to as irrelevant parameter. For $d + z > 4$, the critical property of the system is essentially described by the Gausian term in Eq. (5.20) and the threshold spatial dimension 4 is referred to as upper critical dimension $d_\mathrm{u} = 4$. For $d + z > d_\mathrm{u} = 4$, the low-energy states of the system are described by the Gausian fixed point in terms of the renormalization group.

In the case of the antiferromagnetic order in two-spatial dimension ($d = 2, z = 2$), which gives $d + z = d_\mathrm{u} = 4$, it is necessary to consider the higher order term of $O(u^2)$ in Eq. (5.23). This case is referred to as marginal. By solving Eq. (5.23) including the contribution from $O(u^2)$ it turns out that the logarithmic correction appears in physical quantities [9].

As for the higher order terms (the sixth-order term of φ) in Eq. (5.23), the scaling equation can be also derived in the similar way above, it can be shown that the upper critical dimension is smaller than 4[2]. Hence, the higher order terms of φ for ≥ 4 become irrelevant for $d + z > 4$. These analyses of the renormalization group, the interactions between the magnetic modes $\varphi(q, \omega)$ with different q and ω are irrelevant for $d \geq 2$ and $T \to 0$, which are scaled to zero as the quantum critical point is approached.

[2] For example, for the sixth-order term of φ, the upper critical dimension is given by $d_\mathrm{u} = 3$. In general, for the $2n$th order interaction term, the upper critical dimension is shown to be $d_\mathrm{u} = 2n/(n-1)$ for $n \geq 2$.

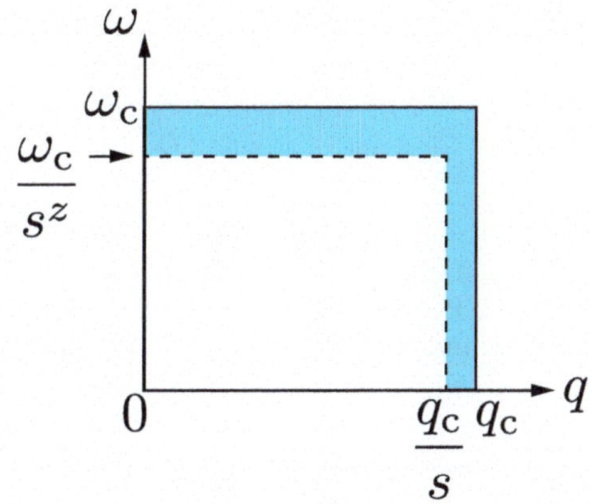

Fig. 5.2 Scale transformation in the q-ω plane

5.3 Mode-Coupling Theory for Magnetic Fluctuations

By the perturbative renormalization group analysis in the previous section, the quantum critical phenomena of ferromagnetism and antiferromagnetism for $d \geq 2$ are essentially described by the Gaussian fixed point. This implies that it is appropriate to construct the best Gaussian, i.e., the best quadratic action in terms of $\varphi(q, i\omega_l)$ taking into account the inter-mode interactions. The framework is the self-consistent renormalization (SCR) theory of spin fluctuations.

The SCR theory was developed by Moriya and coworkers [1–3]. The SCR equation was originally derived in the phenomenological manner, which has also been derived in the microscopic formulation later [1]. Here, starting from the general form of the action for interacting itinerant electrons derived in the previous section, we show the microscopic formulation of the SCR theory.

5.3.1 SCR Theory in Path-Integral Formalism

As discussed in Sect. 5.1, the action of the itinerant electrons with Coulomb interaction responsible for magnetic phase transition is expressed as

$$\Phi[\varphi] = \frac{1}{2}\sum_{\bar{q}} \Omega_2(\bar{q})\varphi(\bar{q})\varphi(-\bar{q}) + \sum_{\bar{q}_1,\bar{q}_2,\bar{q}_3,\bar{q}_4} \Omega_4(\bar{q}_1,\bar{q}_2,\bar{q}_3,\bar{q}_4)\varphi(\bar{q}_1)\varphi(\bar{q}_2)\varphi(\bar{q}_3)\varphi(\bar{q}_4)\delta\left(\sum_{i=1}^{4}\bar{q}_i\right). \quad (5.25)$$

It is noted that this action describes the isotropic spin space [20]. Here, $\bar{q}$ is the abbreviation for $\bar{q} \equiv (\boldsymbol{q}, i\omega_l)$ where $\omega_l = 2\pi lT$ with l being integer. Since long wavelength $|\boldsymbol{q}| \ll q_c$ around the magnetically ordered vector $\boldsymbol{Q}$ and low frequency $|\omega| \ll \omega_c$ regions play the dominant role in the critical phenomena with q_c and ω_c being the cutoffs for momentum and frequency, respectively, Ω_i for $i = 2, 4$ are expanded for q and ω around $(\boldsymbol{Q}, 0)$:

$$\Omega_2(\boldsymbol{q}, i\omega_l) \approx \eta_0 + Aq^2 + C_q|\omega_l|, \tag{5.26}$$

where C_q is defined as $C_q \equiv C/q^{z-2}$ with z being the dynamical exponent (e.g., $z = 3$ for ferromagnetism and $z = 2$ for antiferromagnetism) and $\Omega_4(\bar{q}_1, \bar{q}_2, \bar{q}_3, \bar{q}_4) \approx v_4/(\beta N)$ with $\beta \equiv 1/T$.

To construct the effective action for the best Gaussian, taking account of the mode-mode coupling effects up to the fourth order in $\Phi[\varphi]$, we employ Feynman's inequality [22] on the free energy:

$$F \leq F_{\text{eff}} + T\langle \Phi - \Phi_{\text{eff}} \rangle_{\text{eff}} \equiv \tilde{F}(\eta). \tag{5.27}$$

Here, the effective action Φ_{eff} is parametrized as

$$\Phi_{\text{eff}}[\varphi] = \frac{1}{2}\sum_l \sum_q \left(\eta + Aq^2 + C_q|\omega_l|\right) |\varphi(q, i\omega_l)|^2, \tag{5.28}$$

where η expresses the effect of the mode-mode coupling of spin fluctuations and parameterizes the closeness to the quantum criticality. In Eq. (5.27), $\langle \cdots \rangle_{\text{eff}}$ denotes the statistical average taken by the weight $\exp(-\Phi_{\text{eff}}[\varphi])$ and F_{eff} is given by

$$F_{\text{eff}} = -T \ln \int \mathcal{D}\varphi \exp\left(-\Phi_{\text{eff}}[\varphi]\right). \tag{5.29}$$

By optimal condition $\frac{d\tilde{F}(\eta)}{d\eta} = 0$, the self-consistent renormalization (SCR) equation for η is given by

$$\frac{\eta_0 - \eta}{2} + \frac{6v_4}{N}\langle \varphi^2 \rangle_{\text{eff}} = 0, \tag{5.30}$$

where spin fluctuation $\langle \varphi^2 \rangle_{\text{eff}}$ is defined as

$$\langle \varphi^2 \rangle_{\text{eff}} = T\sum_q \sum_l \frac{1}{\eta + Aq^2 + C_q|\omega_l|}. \tag{5.31}$$

Here, $\langle \varphi^2 \rangle_{\text{eff}}$ consists of the quantum (zero-point) fluctuation $\langle \varphi^2 \rangle_{\text{zero}}$ and thermal fluctuation $\langle \varphi^2 \rangle_{\text{th}}$ as

$$\langle \varphi^2 \rangle_{\text{eff}} = \langle \varphi^2 \rangle_{\text{zero}} + \langle \varphi^2 \rangle_{\text{th}}, \tag{5.32}$$

where $\langle\varphi^2\rangle_{\text{zero}}$ and $\langle\varphi^2\rangle_{\text{th}}$ are expressed as

$$\langle\varphi^2\rangle_{\text{zero}} = \frac{N_{\text{F}}}{\pi}\sum_q \frac{1}{C_q}\int_0^{\omega_{\text{c}}} d\omega \frac{\omega}{\Gamma_q^2+\omega^2}, \tag{5.33}$$

$$\langle\varphi^2\rangle_{\text{th}} = \frac{N_{\text{F}}}{\pi}\sum_q \frac{2}{C_q}\int_0^{\omega_{\text{c}}} d\omega \frac{1}{e^{\beta\omega}-1}\frac{\omega}{\Gamma_q^2+\omega^2}, \tag{5.34}$$

respectively. Here, Γ_q is defined by $\Gamma_q \equiv (\eta + Aq^2)/C_q$.

From Eq. (5.28), the dynamical magnetic susceptibility is given by

$$\chi_{\boldsymbol{Q}}(q,\omega) = \frac{1}{\eta + Aq^2 - \mathrm{i}C_q\omega}, \tag{5.35}$$

where $\boldsymbol{Q}$ is the wavenumber vector of the magnetically ordered phase (e.q., $\boldsymbol{Q}=\boldsymbol{0}$ for ferromagnetism and $\boldsymbol{Q}\neq\boldsymbol{0}$ for antiferromagnetism).

The free energy $\tilde{F}$ defined by Eq. (5.27) is expressed as

$$\begin{aligned}\tilde{F} = &\frac{1}{\pi}\sum_q\int_0^{\omega_{\text{c}}} d\omega \frac{\Gamma_q}{\omega^2+\Gamma_q^2}\left\{\frac{\omega}{2} + T\ln\left(1-e^{-\frac{\omega}{T}}\right)\right\}\\ &+\frac{\eta_0-\eta}{2N_{\text{F}}}\langle\varphi^2\rangle_{\text{eff}} + \frac{3v_4}{N}\langle\varphi^2\rangle_{\text{eff}}^2 - \frac{1}{\pi}\sum_q\frac{\pi\omega_{\text{c}}}{4}.\end{aligned} \tag{5.36}$$

Here, let us define dimensionless parameters for η

$$y \equiv \frac{\eta}{Aq_{\text{B}}^2} \tag{5.37}$$

and the wave number $x \equiv q/q_{\text{B}}$ with q_{B} being the wave number of the Brillouin zone. Thus, Γ_q is expressed as

$$\Gamma_q = 2\pi T_0 x^{z-2}(y+x^2), \tag{5.38}$$

where the characteristic temperature of spin fluctuation is defined as

$$T_0 \equiv \frac{Aq_{\text{B}}^2}{2\pi C_{q_{\text{B}}}}. \tag{5.39}$$

Near the QCP, quantum spin fluctuation $\langle\varphi^2\rangle_{\text{zero}}$ is calculated [16] for the cases above and just at the upper critical dimension 4, respectively, as

$$\langle\varphi^2\rangle_{\text{zero}} = Nd\frac{T_0}{2T_A}\begin{cases} C_1 - C_2 y + \cdots, & \text{for } d+z>4 \\ C_1 + y\ln y - C_2 y + \cdots, & \text{for } d+z=4 \end{cases} \tag{5.40}$$

where T_A is defined as

$$T_A \equiv \frac{Aq_{\mathrm{B}}^2}{2}. \tag{5.41}$$

The constants C_1 and C_2 are given by

$$C_1 = \int_0^{x_c} dx x^{d+z-3} \ln \left| \frac{\omega_{cT_0}^2 + x^{2z}}{x^{2z}} \right|, \tag{5.42}$$

$$C_2 = \begin{cases} 2\int_0^{x_c} dx x^{d+z-5} \frac{\omega_{cT_0}^2}{\omega_{cT_0}^2 + x^{2z}}, & \text{for } d+z>4 \\ 1 + \ln x_c^2 - \frac{1}{2}\ln \left| \frac{\omega_{cT_0}^2 + x_c^4}{\omega_{cT_0}^2} \right|, & \text{for } d+z=4 \end{cases} \tag{5.43}$$

respectively. Here, the cut-off of the wave number is set to be q_c in the q integration, which is expressed as $x_c \equiv \frac{q_c}{q_B}$ in the dimensionless scaled form and ω_{cT} is defined as $\omega_{cT} \equiv \frac{\omega_c}{2\pi T}$. The thermal spin fluctuation $\langle\varphi^2\rangle_{\mathrm{th}}$ is calculated as

$$\langle\varphi^2\rangle_{\mathrm{th}} = Nd\frac{T_0}{T_A}\int_0^{x_c} dx x^{d+z-3}\left\{\ln u - \frac{1}{2u} - \psi(u)\right\}, \tag{5.44}$$

where $\psi(u)$ is the digamma function with u defined as

$$u \equiv \frac{\Gamma_q}{2\pi T} = \frac{x^{z-2}(y+x^2)}{t}. \tag{5.45}$$

Here, t is defined as the dimensionless scaled temperature

$$t \equiv \frac{T}{T_0}. \tag{5.46}$$

By substituting Eqs. (5.40) and (5.44) into (5.32), the SCR equation (Eq. (5.30)) is written in the scaled form for $d + z > 4$ as [12–14]

$$y = y_0 + \frac{d}{2}y_1 \int_0^{x_c} dx x^{d+z-3}\left\{\ln u - \frac{1}{2u} - \psi(u)\right\}, \tag{5.47}$$

and for $d + z = 4$ as [16]

$$y = y_0 + \frac{y_1}{2}\left(y\ln y + d\int_0^{x_c} dx x\left\{\ln u - \frac{1}{2u} - \psi(u)\right\}\right), \tag{5.48}$$

where y_0 and y_1 are given by

$$y_0 = \frac{\frac{\eta_0}{Aq_B^2} + 3dv_4\frac{T_0}{T_A^2}C_1}{1 + 3dv_4\frac{T_0}{T_A^2}C_2}, \tag{5.49}$$

$$y_1 = \frac{12v_4\frac{T_0}{T_A^2}}{1 + 3dv_4\frac{T_0}{T_A^2}C_2}, \tag{5.50}$$

respectively. Here, note that y_0 is different from that obtained by substituting η_0 for η in the r.h.s. of Eq. (5.37).

The solution of the SCR equation y is proportional to the inverse susceptibility

$$y = \frac{1}{2T_A}\frac{1}{\chi_Q(0,0)}, \tag{5.51}$$

which is obtained by substituting Eqs. (5.37) into (5.35) with the use of Eq. (5.41).

5.3.2 Effect of Mode-Mode Coupling of Spin Fluctuations

The dynamical susceptibility of spin fluctuations is obtained from Eq. (5.28) as

$$\begin{aligned}\chi_Q(q, i\omega_l) &= \langle\varphi(q, i\omega_l)\varphi(-q, -i\omega_l)\rangle_{\text{eff}}, \\ &= \frac{1}{\eta + Aq^2 + C_q|\omega_l|}.\end{aligned} \tag{5.52}$$

On the other hand, if we regard Eq. (5.25) as the action derived from the Hubbard model, the dynamical susceptibility obtained by the random phase approximation (RPA) is given in the vicinity of $(q, i\omega_l) = (0, 0)$ by Eq. (5.26)

$$\chi_Q^{\text{RPA}}(q, i\omega_l) = \frac{1}{\eta_0 + Aq^2 + C_q|\omega_l|}. \tag{5.53}$$

We see that Eqs. (5.52) and (5.53) have the similar form but the first term in the denominator is different. The former has the temperature dependence $[\eta = \eta(T)]$ reflecting the effect of the mode-mode coupling of spin fluctuations, while the latter has the constant η_0 not depending on temperature. To grasp the implication intuitively, let us rewrite Eq. (5.52) as follows:

$$\begin{aligned}\chi_Q^{-1}(q, i\omega_l) &= \eta + Aq^2 + C_q|\omega_l| \\ &= \chi_Q^{\text{RPA}}(q, i\omega_l)^{-1} + \eta - \eta_0\end{aligned} \tag{5.54}$$

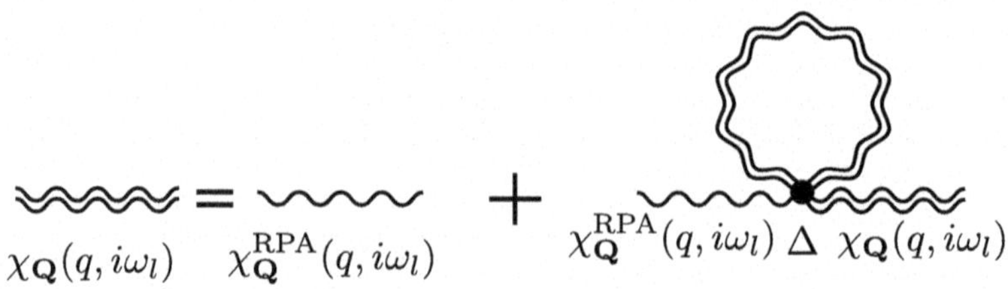

Fig. 5.3 Feynman diagram of mode-mode coupling of spin fluctuations. Critical spin fluctuation (double wavy line) consists of RPA spin fluctuation (wavy line) taking into account the mode-mode coupling (filled circle) up to the infinite order

The second and third terms in the r.h.s. of Eq. (5.54) are equal to $12v_4\langle\varphi^2\rangle_{\mathrm{eff}}/N$ according to the SCR equation (5.30). Hence, by setting this to be $-\Delta$, the dynamical susceptibility is expressed as

$$\chi_Q(q, \mathrm{i}\omega_l) = \chi_Q^{\mathrm{RPA}}(q, \mathrm{i}\omega_l) + \chi_Q^{\mathrm{RPA}}(q, \mathrm{i}\omega_l)\Delta\chi_Q(q, \mathrm{i}\omega_l). \tag{5.55}$$

This is illustrated by the Feynmann diagram in Fig. 5.3. Then it turns out that by the mode coupling in the second term of the r.h.s. of Eq. (5.55), which is illustrated as the filled circle, the effect of the mode-mode coupling of spin fluctuations is taken into account to the spin fluctuation by the RPA up to the infinite order. Therefore, the dynamical susceptibility obtained by inputting the solution $y = y(t) \propto \eta(T)$ of the SCR equation (5.30) i.e., Eq. (5.47) or (5.48) into the denominator of Eq. (5.52) takes into account the effect of the mode-mode coupling of spin fluctuations up to the infinite order.

5.3.3 *Solution of the SCR Equation and Criticality*

In the SCR equation (5.47) or (5.48), y_0 is equal to y at $t = 0$, i..e, $y_0 = y(t = 0)$. By inputting $y_0 = 0$ into the SCR equation, we obtain the solution of $y(t = 0) = 0$. Hence, the static susceptibility $\chi_Q(0, 0)$ diverges at $t = 0$. This implies that the magnetic quantum critical point is expressed as $y(t = 0) = 0$. The paramagnetic-metal phase is expressed as $y_0 > 0$ and the magnetically ordered phase is expressed as $y_0 < 0$.

Numerical solutions of the SCR equation $y(t)$ for each temperature t by inputting the initial value of y_0 are shown in Fig. 5.4a–d. Here, as a typical value, we set $y_1 = 1$ and cutoff $x_c = 1$ for the dimensionless wave number. For each class, we show the t dependence of $1/y$ in the inset. We note that $1/y$ is proportional to the magnetic susceptibility $\chi_Q(0, 0)$ (see Eq. (5.51)). For $y_0 > 0$, the paramagnetic-metal phase is realized for all the temperature ranges. For $y_0 = 0$, the quantum critical point is realized at $t = 0$. As y_0 increases from 0, $y(t \to 0)$ increases from 0. This implies that the susceptibility $\chi_Q(0, 0)$, which diverges at the quantum critical point, is

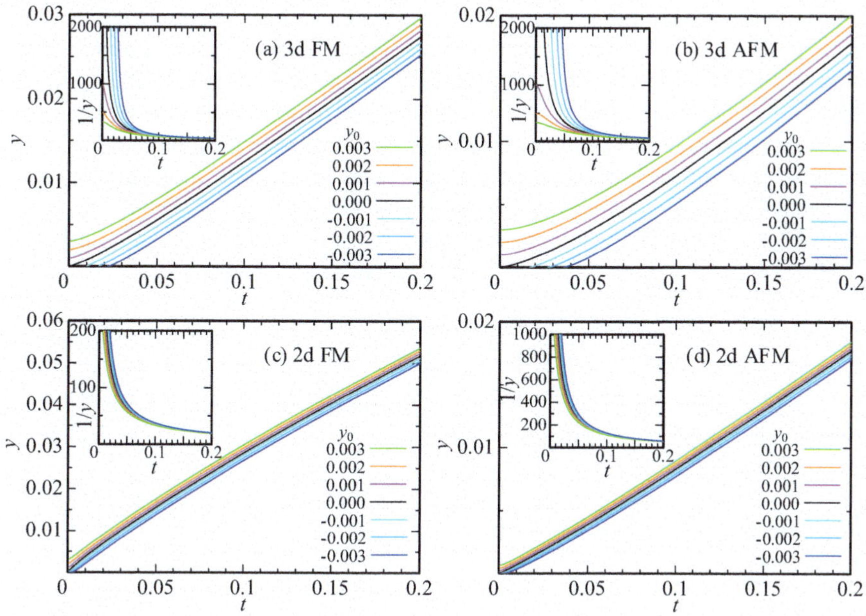

Fig. 5.4 Scaled temperature dependence of y for **a** 3d FM, **b** 3d AFM, **c** 2d FM, and **d** 2d AFM. The inset shows the scaled temperature dependence of $1/y \propto \chi_Q(0, 0)$ [19]. Reprinted from Ref. [19]. Copyright 2019 American Physical Society

suppressed to be finite as the system enters into the paramagnetic-metal phase. This corresponds to the crossover to the $g < g_c$ regime from $g = g_c$ in Fig. 5.1. In the case of $y_0 < 0$, the magnetic order appears for $t < t_c \equiv T_c/T_0$, while the paramagnetic-metal phase appears for $t > t_c$. As shown in Fig. 5.4a, b, in three-dimensional system, the magnetic transition temperature t_c increases continuously from 0 as y_0 decreases from 0.

This corresponds to the fact that the magnetic transition temperature $T_c(g)$ increases continuously from 0 as g changes from $g = g_c$ to $g > g_c$ in Fig. 5.1.

On the other hand, as shown in Fig. 5.4c, d, in two-dimensional system, $y(t = 0) = 0$ is realized only at $t = 0$. This implies that the magnetic transition occurs at $T_c = 0$, which satisfies the Mermin-Wagner theorem [24]. The starting action of the present SCR theory (Eq. (5.25)) was derived from the Hubbard model (5.1). As explained in Sect. 5.1, the Heisenberg model is derived from the Hubbard model in the strong-coupling expansion. The Mermin-Wagner theorem shows that there exists no magnetic transition for $T > 0$ in the Heisenberg model in the two-spatial dimension [24]. The maximum spatial dimension where the ordered state cannot remain because of the enhanced fluctuations is defined as the lower critical dimension denoted as d_l. In the present case, $d_l = 2$ for $T > 0$.

As seen for $t \gtrsim 0.05$ in Fig. 5.4a, the Curie-Weiss behavior $\chi_Q(0, 0) \propto y^{-1} \sim t^{-1}$ appears in the high-temperature regime $t \gtrsim 0.05$ in the paramagnetic-metal phase.

In general, the characteristic temperature of spin fluctuations T_0 is comparable to the effective Fermi temperature of correlated electron systems T_F^*. An important point is that in the low-temperature region smaller than one order of magnitude of T_F^*, the Curie-Weiss behavior which usually appears in the localized spin system is realized. This is ascribed to the fact that the amplitude of the thermal spin fluctuations $\langle\varphi^2\rangle_{\rm th}$ is proportional to temperature T in Eq. (5.44). This resolves the long-standing issue in the metallic magnetism in terms of spin fluctuations [1].

Next, let us focus on the quantum criticality for $t \sim 0$. In $d = 3$, the x integral in Eq. (5.47) converges for $y \to 0$ and then the solution is obtained as

$$y \propto t^{1+\frac{1}{z}} \tag{5.56}$$

at the QCP with $y_0 = 0.0$, as shown in Appendix C.1. This yields $y \sim t^{4/3}$ for the three-dimensional (3d) ferromagnetic QCP ($z = 3$) and $y \sim t^{3/2}$ for the 3d antiferromagnetic QCP ($z = 2$).

In $d = 2$, the x integral in Eq. (5.47) shows logarithmic divergence for $y \to 0$. At the ferromagnetic quantum critical point for $z = 3$, the solution of Eq. (5.47) is obtained as

$$y = -\frac{y_1}{12} t \ln t \tag{5.57}$$

(see Appendix C.2). At the antiferromagnetic quantum critical point for $z = 2$, the solution of Eq. (5.48) is obtained as [23]

$$y = -t\frac{\ln(-\ln t)}{2\ln t} \tag{5.58}$$

(see Appendix C.3). The criticality y ($\propto \eta$) for each class is summarized in Table 5.1.

It is noted that y is related to the magnetic correlation length as $y \propto \xi^{-2}$. The quantum criticality shown in Eqs. (5.56), (5.57), and (5.58) coincides with the results of $\xi = \xi(T)$ for each class in $d = 3$ and $d = 2$ shown by the renormalization group theory [9]. In Ref. [9], "SCR procedure does not yield the log corrections

Table 5.1 Quantum criticality at the magnetic QCP for each class specified by $z = 3$ (FM) and $z = 2$ (AFM) in $d = 3$ and 2 [1]. Electrical resistivity $\rho(T)$, specific-heat coefficient C/T, uniform susceptibility $\chi(T)$, and NMR relaxation rate $(T_1T)^{-1}$. For χ, $\to$ C.W. denotes the crossover to the Curie-Weiss behavior. Note that $\eta \propto y$ holds (see Eq. (5.37)) [19]

Class	η	ρ	C/T	χ	$(T_1T)^{-1}$	References
3d FM	$T^{4/3}$	$T^{5/3}$	$-\ln T$	$T^{-4/3} \to$ C.W.	$T^{-4/3}$	[6, 14]
3d AFM	$T^{3/2}$	$T^{3/2}$	const $-T^{1/2}$	const $-T^{1/4} \to$ C. W.	$T^{-3/4}$	[13, 25]
2d FM	$-T\ln T$	$T^{4/3}$	$T^{-1/3}$	$-1/(T\ln T) \to$ C.W.	$-1/(T\ln T)^{3/2}$	[12]
2d AFM	$\frac{-T\ln(-\ln T)}{\ln T}$	T	$-\ln T$	–	$-\ln T/T$	[16, 23]

in $d = 2$ for $z = 3$ and $z = 2$". is stated. However, this statement is *not* correct. The SCR results coincide with those derived by the renormalization group theory including the log corrections in $d = 2$ for $z = 3$ [12] and $z = 2$. See, for details, Appendixs C.2 and C.3, respectively.

Ferromagnetic spin fluctuations or antiferromagnetic spin fluctuations diverging at the quantum critical point cause the non-Fermi-liquid behavior in physical quantities, as shown in Table 5.1.

5.3.3.1 Electrical Resistivity

Electrical resistivity due to the scattering by spin fluctuations is calculated on the basis of the Boltzmann equation, which gives

$$\rho(T) \sim T^{\frac{d+2(z-2)}{z}} \tag{5.59}$$

at low temperatures [6, 11–13].

5.3.3.2 Uniform Magnetic Susceptibility

The uniform magnetic susceptibility $\chi = \chi_0(0, 0)$ in the vicinity of the ferromagnetic quantum critical point is given by Eq. (5.35). On the other hand, the uniform magnetic susceptibility in the vicinity of the antiferromagnetic quantum critical point is obtained by evaluating the effect of spin fluctuations in the magnetic field dependence of the free energy [25].

5.3.3.3 NMR Spin-Lattice Relaxation Rate

The contribution of spin fluctuations to the spin-lattice relaxation rate in the NMR measurement is given by

$$\frac{1}{T_1} = T\frac{k_B\gamma_n^2}{\mu_B^2}\sum_q |\bar{A}_q|^2 \frac{\mathrm{Im}\chi_Q(q, \omega_0)}{\omega_0}, \tag{5.60}$$

where γ_n is gyromagnetic ratio of nuclear spin and $\bar{A}_q$ is the hyperfine-coupling constant [26]. Here, μ_B is the Bohr magneton and ω_0 is the nuclear resonance frequency. By substituting the dynamical susceptibility Eqs. (5.35) into (5.60) and neglecting the q dependence of $\bar{A}_q$ since the hyperfine coupling is local, we obtain

$$\frac{1}{T_1 T} \propto \begin{cases} y^{-1} & (d = 3) \\ y^{-3/2} & (d = 2) \end{cases}$$

in the vicinity of the ferromagnetic quantum critical point. We also obtain

$$\frac{1}{T_1 T} \propto \begin{cases} y^{-1/2} & (d = 3) \\ y^{-1} & (d = 2) \end{cases}$$

in the vicinity of the antiferromagnetic quantum critical point [11, 12, 15]. These results indicate that the criticality is directly reflected in the spin-lattice relaxation rate.

5.3.4 *Entropy and Specific Heat*

In the original SCR theory, the specific heat was calculated with the zero-point spin fluctuation being neglected [12–14]. Taking into account the zero-point spin fluctuation [17, 30] as well as the stationary condition of the free energy adequately [30], the specific heat was calculated. This has shown that the dominant contribution to the quantum criticality comes from the thermal spin fluctuation and the critical indices [12–14] endorsed by the renormalization group theory [9, 10] do not change.

However, in the calculation of the thermal-expansion coefficient and the Grüneisen parameter in the SCR theory, the zero-point spin fluctuation as well as the stationary condition of the free energy should be taken into account correctly, which has not been addressed in Refs. [27, 28]. Takahashi considered these effects in the extended SCR theory by introducing the conservation law of the total spin fluctuation amplitude and discussed the magneto-volume effect [29].

In Ref. [19], the thermal-expansion coefficient α and the Grüneisen parameter Γ in the original SCR theory are derived in the thermodynamically consistent way. By taking into account zero-point spin fluctuation as well as the stationary condition of the free energy correctly, the specific heat C_V near the ferromagnetic (FM) QCP and the antiferromagnetic (AFM) QCP in three spatial dimensions ($d = 3$) and two-spatial dimension ($d = 2$) are reexamined. Then, the thermal-expansion coefficient α for each class is derived starting from the entropy, which is proven to be equivalent to that obtained from the explicit form of the free energy with the use of the stationary condition in the SCR theory [19]. On the basis of these correctly calculated C_V and α, Grüneisen parameter Γ is obtained. By performing analytical and numerical calculations of C_V, α, and Γ near the magnetic QCP, their quantum critical properties are clarified [19]. In this section, the entropy S and the specific heat C_V are discussed. The thermal-expansion coefficient α and Grüneisen parameter Γ are discussed in Sects. 5.3.5 and 5.3.6, respectively.

The entropy $S = -\left(\frac{\partial \tilde{F}}{\partial T}\right)_V$ is obtained by differentiating the free energy $\tilde{F}$ in Eq. (5.36) with respect to the temperature. Noting that the terms with $\left(\frac{\partial \eta}{\partial T}\right)_V$ and also $\left(\frac{\partial \langle \varphi^2 \rangle_{\rm eff}}{\partial T}\right)_V$ vanish with the use of the SCR equation (Eq. (5.30)), the entropy is derived as [30]

$$S = -Nd \int_0^{x_c} dx x^{d-1} \left\{ \ln\sqrt{2\pi} - u + \left(u - \frac{1}{2}\right) \ln u - \ln\Gamma(u) \right\}$$
$$+ Nd \int_0^{x_c} dx x^{d-1} u \left\{ \ln u - \frac{1}{2u} - \psi(u) \right\}, \tag{5.61}$$

where $\Gamma(u)$ is the Gamma function.

The specific heat under a constant volume is obtained by differentiating the entropy S in Eq. (5.61) with respect to the temperature [17, 30] as

$$C_V = T \left(\frac{\partial S}{\partial T} \right)_V,$$
$$= C_a - C_b, \tag{5.62}$$

where C_a and C_b are given by

$$C_a = -Nd \int_0^{x_c} dx x^{d-1} u^2 \left\{ \frac{1}{u} + \frac{1}{2u^2} - \psi'(u) \right\}, \tag{5.63}$$

$$C_b = \tilde{C}_b \left(\frac{\partial y}{\partial t} \right)_V, \tag{5.64}$$

respectively. Here, $\psi'(u)$ is the trigamma function and $\tilde{C}_b$ is given by

$$\tilde{C}_b = -Nd \int_0^{x_c} dx x^{d+z-3} u \left\{ \frac{1}{u} + \frac{1}{2u^2} - \psi'(u) \right\}. \tag{5.65}$$

As for C_a, the x integral in Eq. (5.63) shows no divergence from $x = 0$ even for $y \to 0$ irrespective of spatial dimensions [30]. Hence, C_a for $t \ll 1$ at the QCP has no explicit y dependence.

As for $\tilde{C}_b$, in $d = 3$, the x integral in Eq. (5.65) shows no divergence from $x = 0$ for $y \to 0$ and is evaluated as [30]

$$\frac{\tilde{C}_b}{N} \sim t^{1+\frac{1}{z}}, \tag{5.66}$$

for $t \to 0$ at the QCP. Namely, $\tilde{C}_b$ shows the same temperature dependence as y, i.e., $\tilde{C}_b \sim y$, as seen in Eq. (5.56). In $d = 2$, the x integral in Eq. (5.65) shows logarithmic divergence arising from $x = 0$ for $y \to 0$. Hence, $\tilde{C}_b$ has the $\ln y$ dependence at the QCP.

The temperature dependence of the specific heat at the magnetic QCP for each class is summarized in the following subsections. The numerical calculation of Eq. (5.62) is also performed. To calculate $\left(\frac{\partial y}{\partial t}\right)_V$, by differentiating the SCR equation (Eqs. (5.47) and (5.48)) with respect to the scaled temperature t, we have

$$\left(\frac{\partial y}{\partial t}\right)_V = \begin{cases} \dfrac{\frac{y_1}{2t}\tilde{C}_b\frac{1}{N}}{1-\frac{dy_1}{2t}M} & \text{for } d+z>4, \\ \dfrac{\frac{y_1}{2t}\tilde{C}_b\frac{1}{N}}{1-\frac{y_1}{2}(\ln y+1)-\frac{dy_1}{2t}M} & \text{for } d+z=4, \end{cases} \tag{5.67}$$

where M is given by

$$M = \int_0^{x_c} dx x^{d+2z-5}\left\{\frac{1}{u}+\frac{1}{2u^2}-\psi'(u)\right\}. \tag{5.68}$$

To calculate the t dependence of the specific heat just at the QCP, we set $y_0 = 0.0$ in the SCR equation (Eqs. (5.47) and (5.48)) with setting as $y_1 = 1.0$ and $x_c = 1.0$. By solving the SCR equation, we obtain the solution $y(t)$. Then, by inputting $y(t)$ to Eq. (5.67), we obtain $\left(\frac{\partial y}{\partial t}\right)_V$. Finally, by substituting $y(t)$ and $\left(\frac{\partial y}{\partial t}\right)_V$ into Eq. (5.62) for each class, we obtain $C_V(t)$, which is shown in Fig. 5.5a–d, respectively.

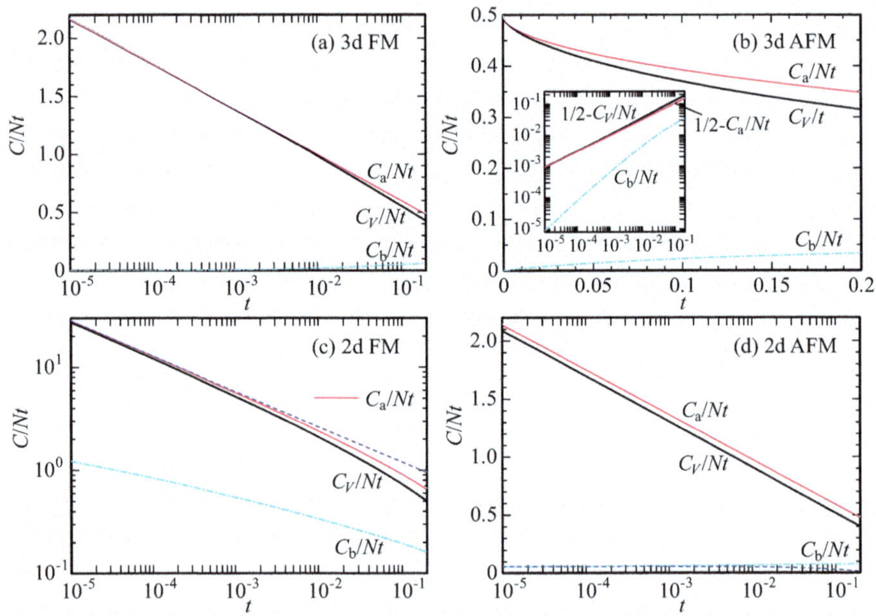

Fig. 5.5 Specific-heat coefficient vs. scaled temperature just at the QCP. C_V/Nt (bold solid line), C_a/Nt (thin line), and C_b/Nt (dash-dotted line) are calculated numerically in Eqs. (5.62), (5.63), and (5.64), respectively, for $y_0 = 0.0$ and $y_1 = 1.0$. **a** 3d FM QCP. **b** 3d AFM QCP. The inset shows log-log plot of $1/2 - C_V/Nt$ (thick solid line), $1/2 - C_a/Nt$ (thin solid line), and C_b/Nt (dash-dotted line). **c** 2d FM QCP. The dashed line represents the least-square fit of C_a/Nt for $10^{-5} \le t \le 10^{-4}$ with $at^{-1/3}$. **d** 2d AFM QCP. The dashed line represents the least-square fit of C_b/t for $10^{-5} \le t \le 10^{-4}$ with $-a\frac{\{\ln(-\ln t)\}^2}{-\ln t}$ [19]. Reprinted from Ref. [19].

5.3.4.1 3d Ferromagnetic Case

For $t \ll 1$, $C_{\rm a}$ is evaluated as [5, 14, 30]

$$\frac{C_{\rm a}}{N} \approx -\frac{t}{6}\ln t. \tag{5.69}$$

For $t \ll 1$, $\tilde{C}_{\rm b}$ is evaluated as [30]

$$\frac{\tilde{C}_{\rm b}}{N} \sim t^{\frac{4}{3}}. \tag{5.70}$$

Since $C_{\rm b}$ behaves as $C_{\rm b} \sim t^{5/3}$ in Eq. (5.64) with $\partial y/\partial t \sim t^{1/3}$ for $t \ll 1$, the specific heat in Eq. (5.62) behaves as

$$\frac{C_V}{N} \approx \frac{C_{\rm a}}{N} \sim -t\ln t \tag{5.71}$$

for $t \ll 1$, where the dominant contribution comes from $C_{\rm a}$, as seen in Fig. 5.5a. This reproduces the criticality shown by the past SCR theory, which is summarized in Table 5.1 (see C/T for 3d FM). It is noted that the same temperature dependence as Eq. (5.71) was also derived from the renormalization group theory [10].

5.3.4.2 3d Antiferromagnetic Case

For $t \ll 1$, $C_{\rm a}$ is evaluated as [13, 30]

$$\frac{C_{\rm a}}{N} \approx \frac{1}{2}t\left(x_{\rm c} - \frac{15}{2}a^*_{\frac{5}{2}}t^{\frac{1}{2}}\right), \tag{5.72}$$

where $a^*_{\frac{5}{2}}$ is a constant given by $a^*_{\frac{5}{2}} = \int_0^{\infty} du u^{3/2}\left\{\frac{1}{2u^2} - \ln u + \frac{1}{2u} + \psi(u)\right\}$. For $t \ll 1$, $\tilde{C}_{\rm b}$ is evaluated as [30]

$$\frac{\tilde{C}_{\rm b}}{N} \sim t^{\frac{3}{2}}. \tag{5.73}$$

Since $C_{\rm b}$ behaves as $C_{\rm b} \sim t^2$ in Eq. (5.64) with $\partial y/\partial t \sim t^{1/2}$ for $t \ll 1$, the specific heat in Eq. (5.62) behaves as

$$\frac{C_V}{N} \approx \frac{C_{\rm a}}{N} \sim \frac{1}{2}t\left(x_{\rm c} - \frac{15}{2}a^*_{\frac{5}{2}}t^{\frac{1}{2}}\right) \tag{5.74}$$

for $t \ll 1$, where the dominant contribution comes from $C_{\rm a}$, as seen in Fig. 5.5b. This reproduces the criticality shown by the past SCR theory (see C/T for 3d AFM in

Table 5.1). It is noted that the same temperature dependence as Eq. (5.74) was also derived from the renormalization group theory [10].

5.3.4.3 2d Ferromagnetic Case

For $t \ll 1$, $C_{\rm a}$ is evaluated as [12, 30]

$$\frac{C_{\rm a}}{N} \approx \frac{10}{9} a_{\frac{5}{3}} t^{\frac{2}{3}}, \tag{5.75}$$

where a_ν is a constant given by $a_\nu \equiv \pi\zeta(\nu)\Gamma(\nu)/[(2\pi)^\nu \sin(\nu\pi/2)]$ and $a_{5/3} = 0.5629\cdots$. For $y \ll t^{\frac{2}{3}} \ll 1$, we have

$$\frac{\tilde{C}_{\rm b}}{N} \approx \frac{t}{2} \ln\left[\frac{1}{y}\left(\frac{t}{6}\right)^{\frac{2}{3}}\right]. \tag{5.76}$$

Since $y \approx -\frac{y_1}{12} t \ln t$ at the QCP is obtained from the SCR equation (Eq. (5.47)) for $d = 2$ and $z = 3$ by setting $y_0 = 0.0$ [12], we have

$$\frac{\tilde{C}_{\rm b}}{N} \approx -\frac{t}{6} \ln t. \tag{5.77}$$

Since the coefficient $\frac{10}{9} a_{\frac{5}{3}}$ in Eq. (5.75) is the quantity of $O(1)$, which is much larger than that in $\tilde{C}_{\rm b} \left(\frac{\partial y}{\partial t}\right)_V \sim \frac{y_1}{72} t \ln t (\ln t + 1)$, the specific heat in Eq. (5.62) is dominated by $C_{\rm a}$ as

$$\frac{C_V}{N} \approx \frac{C_{\rm a}}{N} \sim t^{\frac{2}{3}} \tag{5.78}$$

for $t \ll 1$. This can be confirmed by numerical calculation of Eq. (5.62), as shown in Fig. 5.5c. This reproduces the criticality shown by the past SCR theory (see C/T for 2d FM in Table 5.1). It is noted that the same temperature dependence as Eq. (5.78) was also derived from the renormalization group theory [32].

5.3.4.4 2d Antiferromagnetic Case

For $t \ll 1$, $C_{\rm a}$ is evaluated as [30]

$$\frac{C_{\rm a}}{N} \approx -\frac{t}{6} \ln t. \tag{5.79}$$

For $y \ll t \ll 1$, $\tilde{C}_{\rm b}$ is evaluated as

$$\frac{\tilde{C}_{\rm b}}{N} \approx -\frac{t}{2}\ln\left(6\frac{y}{t}\right). \tag{5.80}$$

Since $y \approx -t\frac{\ln(-\ln t)}{2\ln t}$ at the QCP is obtained from the SCR equation (Eq. (5.48)) for $d = 2$ and $z = 2$ by setting $y_0 = 0.0$ [23], we have

$$\frac{\tilde{C}_{\rm b}}{N} \sim t\ln\left(-\ln t\right). \tag{5.81}$$

Since $C_{\rm b}$ behaves as $C_{\rm b} \sim -t\frac{\{\ln(-\ln t)\}^2}{\ln t}$ in Eq. (5.64) with $\partial y/\partial t \sim -\ln(-\ln t)/\ln t$ for $0 < t \ll 1$ (see the dashed line in Fig. 5.5c), the specific heat in Eq. (5.62) behaves as

$$\frac{C_V}{N} \approx \frac{C_{\rm a}}{N} \sim -t\ln t \tag{5.82}$$

for $t \ll 1$, where the dominant contribution comes from $C_{\rm a}$, as seen in Fig. 5.5d. This reproduces the criticality shown by the past SCR theory (see C/T for 2d AFM in Table 5.1). It is noted that the same temperature dependence as Eq. (5.82) was also derived from the renormalization group theory [32].

5.3.5 Thermal-Expansion Coefficient

The thermal-expansion coefficient is expressed as

$$\alpha = -\frac{1}{V}\left(\frac{\partial S}{\partial P}\right)_T. \tag{5.83}$$

The detailed derivation is explained in Appendix D.1. Then, α dominated by spin fluctuations near the magnetic QCP is derived from the entropy S in Eq. (5.61). By differentiating the entropy S with respect to the pressure P under a constant temperature, we obtain

$$\left(\frac{\partial S}{\partial P}\right)_T = -\frac{\tilde{C}_{\rm b}}{t}\left(\frac{\partial y}{\partial P}\right)_T - \frac{C_{\rm a}}{T_0}\left(\frac{\partial T_0}{\partial P}\right)_T, \tag{5.84}$$

where $\tilde{C}_{\rm b}$ and $C_{\rm a}$ appeared in the formula of the specific heat, which are given by Eqs. (5.65) and (5.63), respectively. Then, the thermal-expansion coefficient is obtained as

$$\alpha = -\frac{1}{V}\left(\frac{\partial S}{\partial P}\right)_T \tag{5.85}$$

$$= \alpha_{\rm a} + \alpha_{\rm b}, \tag{5.86}$$

where α_{a} and α_{b} are defined by

$$\alpha_{\mathrm{a}} \equiv \frac{1}{V}\frac{C_{\mathrm{a}}}{T_0}\left(\frac{\partial T_0}{\partial P}\right)_T, \tag{5.87}$$

$$\alpha_{\mathrm{b}} \equiv \frac{1}{V}\frac{\tilde{C}_{\mathrm{b}}}{t}\left(\frac{\partial y}{\partial P}\right)_T, \tag{5.88}$$

respectively. Here, $(\partial y/\partial P)_T$ can be calculated by differentiating the SCR equation (Eqs. (5.47) and (5.48)) with respect to the pressure under a constant temperature as

$$\left(\frac{\partial y}{\partial P}\right)_T = \begin{cases} \dfrac{\left(\frac{\partial y_0}{\partial P}\right)_T + \left(\frac{\partial y_1}{\partial P}\right)_T \frac{d}{2}L - \frac{1}{T_0}\left(\frac{\partial T_0}{\partial P}\right)_T \tilde{C}_{\mathrm{b}}\frac{y_1}{2}\frac{1}{N}}{1 - \frac{dy_1}{2t}M} & (d+z>4), \\[2ex] \dfrac{\left(\frac{\partial y_0}{\partial P}\right)_T + \left(\frac{\partial y_1}{\partial P}\right)_T \left(\frac{d}{2}L + \frac{1}{2}y\ln y\right) - \frac{1}{T_0}\left(\frac{\partial T_0}{\partial P}\right)_T \tilde{C}_{\mathrm{b}}\frac{y_1}{2}\frac{1}{N}}{1 - \frac{y_1}{2}(\ln y + 1) - \frac{dy_1}{2t}M} & (d+z=4), \end{cases} \tag{5.89}$$

where L is defined as

$$L \equiv \int_0^{x_{\mathrm{c}}} dx x^{d+z-3}\left\{\ln u - \frac{1}{2u} - \psi(u)\right\}. \tag{5.90}$$

The procedure for calculating $\alpha(t)$ is as follows: First, we solve the SCR equation (Eq. (5.47) or (5.48)) by inputting $y_0 = 0$, which corresponds to the QCP, with setting $y_1 = 1$ and $x_{\mathrm{c}} = 1$. Then, by using this solution $y(t)$, $C_{\mathrm{a}}(t)$ and $\tilde{C}_{\mathrm{b}}(t)$ are calculated as in Eqs. (5.63) and (5.65), respectively. By inputting the solution $y(t)$ into Eq. (5.89) with setting $(\partial y_0/\partial P)_T = 1$, $(\partial y_1/\partial P)_T = 1$, and $(\partial T_0/\partial P)_T/T_0 = 1$ as representative values (the reason for this parameterization is explained below), $(\partial y/\partial P)_T$ is obtained by calculating the right-hand side of Eq. (5.89). Finally, we obtain $\alpha_{\mathrm{a}}(t)$ in Eq. (5.87) and $\alpha_{\mathrm{b}}(t)$ in Eq. (5.88), resulting in $\alpha(t)$ in Eq. (5.86).

Here, we note the unit and the parametrization of α. When we input the value of $(\partial T_0/\partial P)_T/T_0$ in the unit of GPa^{-1} and the molar volume as V in the unit of Å^3 into Eq. (5.87), α_{a} can be expressed in the unit of K^{-1} as

$$\alpha_{\mathrm{a}} = \underline{\frac{0.0138}{V}\frac{1}{T_0}\left(\frac{\partial T_0}{\partial P}\right)_T} \times \frac{C_{\mathrm{a}}}{N}. \tag{5.91}$$

When we input the value of $(\partial y/\partial P)_T$ in the unit of GPa^{-1} and V in the unit of Å^3 into Eq. (5.88), α_{b} can be expressed in the unit of K^{-1} as

$$\alpha_{\mathrm{b}} = \underline{\frac{0.0138}{V}\left(\frac{\partial y}{\partial P}\right)_T} \times \frac{\tilde{C}_{\mathrm{b}}}{Nt}. \tag{5.92}$$

Hence, multiplying the numerical values of the underlined terms in Eqs. (5.91) and (5.92) for each material to the following results of α_{a} and α_{b}, respectively, shown in Fig. 5.6a–d, one can make a direct comparison with experiments.

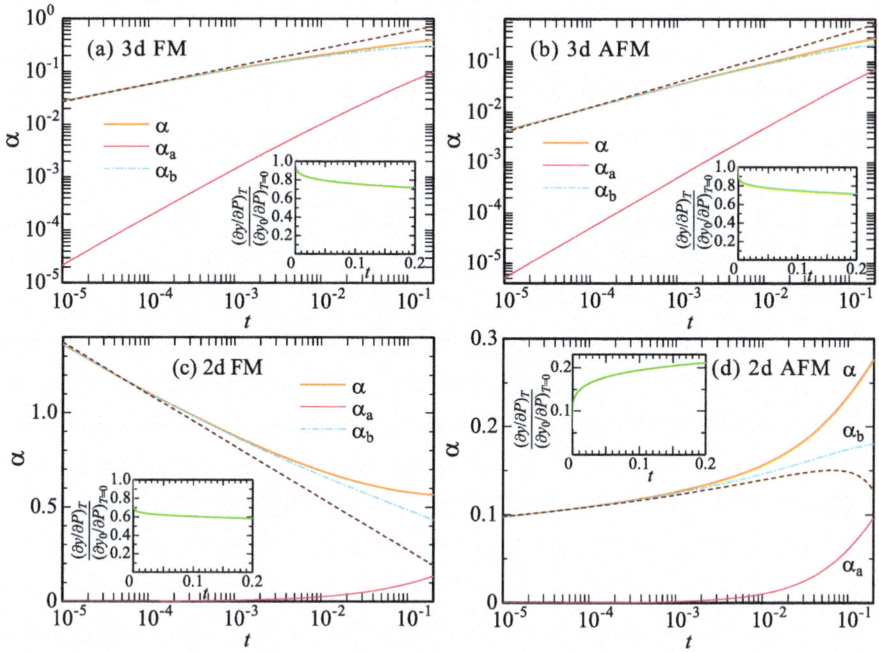

Fig. 5.6 (Color online) Thermal-expansion coefficient α vs scaled temperature t just at the QCP. The thick solid line is for α, the thin solid line is for α_a, and the dashed-dotted line is for α_b. The dashed line represents the least-squares fit of α with $f_i(t)$ (i = a, b, c) for $10^{-5} \leq t \leq 10^{-4}$ and $f_d(t)$ for $10^{-7} \leq t \leq 10^{-4}$. **a** 3d FM QCP: $f_a(t) = at^{1/3}$. **b** 3d AFM QCP: $f_b(t) = at^{1/2}$. **c** 2d FM QCP: $f_c(t) = a \ln t$. **d** 2d AFM QCP: $f_d(t) = -a\frac{\ln(-\ln t)}{\ln\left(-\frac{t}{\ln t}\right)}$. The inset shows the t dependence of $\left(\frac{\partial y}{\partial P}\right)_T / \left(\frac{\partial y_0}{\partial P}\right)_{T=0}$ [19]. Reprinted from Ref. [19]. Copyright 2019 American Physical Society

The results of the numerical calculations of $\alpha(t)$ for the input parameters mentioned above for each universality class are shown in Fig. 5.6a–d. Note that, in each class, α_b becomes dominant as t decreases, while α_a gives a minor contribution.

For $d + z > 4$, α behaves as

$$\alpha \propto \frac{\tilde{C}_b}{t} \tag{5.93}$$

for $t \ll 1$, where $(\partial y/\partial P)_T$ is regarded as temperature independent, which is verified by the almost t-independent behavior shown in the inset of Fig. 5.6a–c. The critical behavior expressed as Eq. (5.93) actually appears for $t \ll 1$ as shown by the dashed line in Fig. 5.6a–c, which is summarized in the first column of Table 5.3. The thermal-expansion coefficient in the 2d FM case diverges $\alpha(t) \to \infty$ for $t \to 0$, whereas $\alpha(t) \to 0$ for $t \to 0$ in the 3d FM and 3d AFM cases. These t dependences coincide with those shown by the renormalization group theory [32]. However, this asymptotic criticality appears only at sufficiently low temperatures for $t \lesssim 10^{-3}$ in the 3d FM

Table 5.2 Quantum criticality at the magnetic QCP for each class specified by $z = 3$ (FM) and $z = 2$ (AFM) in spatial dimension $d = 3$ and 2 [1, 31]. Temperature dependences of $C_{\rm a}$, $\tilde{C}_{\rm b}$, and C_V at the QCP for $t \ll 1$ [5, 12–14, 30]

Class	y	$C_{\rm a}$	$\tilde{C}_{\rm b}$	C_V
3d FM	$t^{\frac{4}{3}}$	$-t \ln t$	$t^{\frac{4}{3}}$	$-t \ln t$
3d AFM	$t^{\frac{3}{2}}$	$\text{const.} - t^{1/2}$	$t^{\frac{3}{2}}$	$\text{const.} - t^{1/2}$
2d FM	$-t \ln t$	$t^{\frac{2}{3}}$	$-t \ln t$	$t^{\frac{2}{3}}$
2d AFM	$-\frac{t \ln(-\ln t)}{\ln t}$	$-t \ln t$	$t \ln(-\ln t)$	$-t \ln t$

Table 5.3 Temperature dependences of α and Γ at the QCP of FM ($z = 3$) and AFM ($z = 2$) in spatial dimensions ($d = 3, 2$) [19]

Class	α	Γ
3d FM	$t^{1/3}$	$-\frac{t^{-2/3}}{\ln t}$
3d AFM	$t^{1/2}$	$\frac{t^{-1/2}}{\text{const.} - t^{1/2}}$
2d FM	$-\ln t$	$-t^{-2/3} \ln t$
2d AFM	$-\frac{\ln(-\ln t)}{\ln(-\frac{t}{\ln t})}$	$\frac{1}{t \ln t} \frac{\ln(-\ln t)}{\ln(-\frac{t}{\ln t})}$

case (Fig. 5.6a) and the 3d AFM case (Fig. 5.6b), and for $t \lesssim 10^{-4}$ in the 2d FM case (Fig. 5.6c). This is due to the presence of the weakly temperature-dependent $(\partial y/\partial P)_T$ in $\alpha_{\rm b}$, as shown in the inset of Fig. 5.6a–c.

For the marginal case $d + z = 4$, i.e., the 2d AFM case, $\alpha(t)$ for $t \ll 1$ is evaluated as

$$\alpha \propto \frac{\tilde{C}_{\rm b}}{t} \left(\frac{\partial y}{\partial P} \right)_T \sim -\frac{\ln(-\ln t)}{\ln\left(-\frac{t}{\ln t}\right)}. \tag{5.94}$$

This can be seen in Fig. 5.6d, where α is well fit by the dashed line expressed as Eq. (5.94). Here, we found that $(\partial y/\partial P)_T$ has the temperature dependence even for $t \ll 1$ as $(\partial y/\partial P)_T \approx (\partial y_0/\partial P)_{T=0}(-b_4)/\ln(-t/\ln t)$ with b_4 being a positive constant, which can be confirmed in the inset of Fig. 5.6d. This is due to the logarithmic correction term in Eq. (5.48). Namely, $(\partial y/\partial P)_T$ affects the criticality in Eq. (5.94), which was not reported in the past renormalization group studies [32, 33]. Hence, the t dependence of $\tilde{C}_{\rm b}/t$ showing divergence as $\sim \ln(-\ln t)$ for $t \to 0$ (see the third column in Table 5.2), in agreement with the renormalization group theory [32], is counteracted by $(\partial y/\partial P)_T \to 0$ for $t \to 0$.

5.3.6 Grüneisen Parameter

The Grüneisen parameter Γ is defined by

$$\Gamma \equiv \frac{\alpha V}{C_V \kappa_T}, \tag{5.95}$$

where κ_T is the isothermal compressibility (see Appendix D.2). Since α is expressed as $\alpha_a + \alpha_b$ in Eq. (5.86), Γ is expressed as

$$\Gamma = \Gamma_a + \Gamma_b, \tag{5.96}$$

where Γ_i (i = a, b) is defined by

$$\Gamma_i \equiv \frac{\alpha_i V}{C_V \kappa_T}. \tag{5.97}$$

At low temperatures, C_V is governed by C_a as $C_V = C_a - C_b \approx C_a$. By using Eqs. (5.87) and (5.88), Γ (Eq. (5.96)) is expressed for $t \ll 1$ as

$$\begin{aligned}\Gamma &\approx \frac{1}{\kappa_T}\frac{1}{T_0}\left(\frac{\partial T_0}{\partial P}\right)_T + \frac{\tilde{C}_b}{C_a}\frac{1}{t}\frac{1}{\kappa_T}\left(\frac{\partial y}{\partial P}\right)_T, \\ &= -\frac{V}{T_0}\left(\frac{\partial T_0}{\partial V}\right)_T - \frac{\tilde{C}_b}{C_a}\frac{V}{t}\left(\frac{\partial y}{\partial V}\right)_T,\end{aligned} \tag{5.98}$$

where the definition of the isothermal compressibility Eq. (D.4) in Appendix D.2 has been used to derive the second line. One can see that the first and second terms of Eq. (5.98) correspond to Γ_a and Γ_b, respectively.

Here, we note that Eq. (5.98) is consistent with Γ derived under an adiabatic process. By differentiating both sides of Eq. (5.61) with respect to the volume under a constant entropy and using the expression $\Gamma = -\frac{V}{T}\left(\frac{\partial T}{\partial V}\right)_S$, which is equivalent to Eq. (5.95), we obtain

$$\Gamma = -\frac{V}{T_0}\left(\frac{\partial T_0}{\partial V}\right)_S - \frac{\tilde{C}_b}{C_a}\frac{V}{t}\left(\frac{\partial y}{\partial V}\right)_S. \tag{5.99}$$

We see that the first and second terms correspond to those in Eq. (5.98), respectively.

As for the first term in Eq. (5.98), $\Gamma_a(T=0) = -\frac{V}{T_0}\left(\frac{\partial T_0}{\partial V}\right)_{T=0}$ is the volume derivative of the characteristic temperature of spin fluctuation. In heavy electron systems, $|\Gamma_a|$ typically has an enhanced value with $O(10)$ being in the same order of the Grüneisen parameter in the Fermi-liquid regime Γ_{FL}. Here, Γ_{FL} is defined as $\Gamma_{FL} \equiv -\frac{V}{T_K}\left(\frac{\partial T_K}{\partial V}\right)_{T=0}$, where T_K is the characteristic temperature of heavy electrons called Kondo temperature. Since Γ_{FL} typically has an enhanced value [34, 38] of

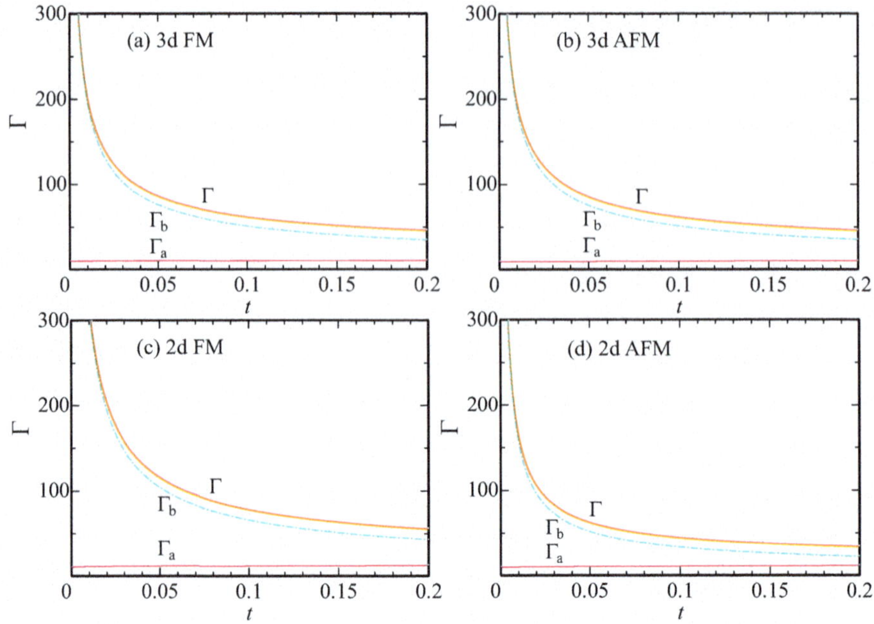

Fig. 5.7 Temperature dependence of the Grüneisen parameter $\Gamma(t)$ (bold line), $\Gamma_{\rm a}(t)$ (thin line), and $\Gamma_{\rm b}(t)$ (thin dash-dotted line) at each quantum critical point for **a** 3d FM, **b** 3d AFM, **c** 2d FM, and **d** 2d AFM [19]. Reprinted from Ref. [19]. Copyright 2019 American Physical Society

$O(10)$ and T_0 is shown to be proportional to $T_{\rm K}$, $|\Gamma_{\rm a}|$ is also enhanced although $|\Gamma_{\rm a}(t)|$ is almost t independent.

To analyze the t dependence of Γ at the QCP, we performed the numerical calculation of Eq. (5.95). We calculate α by the procedure in Sect. 5.3.5. As for C_V, we calculate $C_{\rm a}$ and $C_{\rm b}$ in Eqs. (5.63) and (5.64), respectively, where $(\partial y/\partial t)_V$ is obtained by calculating Eq. (5.67). The input parameters set is the same as that set for Fig. 5.6. As for the isothermal compressibility, we confirmed that κ_T does not show divergence even at the QCP for each class but has a finite value in general and hence we input $\kappa_T = 0.1$ as a typical value for heavy electron systems. This is because $\frac{1}{T_0}\left(\frac{\partial T_0}{\partial P}\right)_T = \Gamma_{\rm a}(T=0)\kappa_T = 1$ was used in Sect. 5.3.5 and κ_T is set so as to reproduce $\Gamma_{\rm a}(T=0) = 10$.

The results of the numerical calculations for each class are shown in Fig. 5.7a–d. Since $\Gamma_{\rm a}$ has a minor t dependence here, we show the t dependences of Γ and $\Gamma_{\rm b}$. Reflecting the fact that α is dominated by $\alpha_{\rm b}$ for low t (see Fig. 5.6), Γ is mainly contributed from $\Gamma_{\rm b}$. Then, as t decreases, Γ increases and finally diverges for $t \to 0$ in each class because of the factor $1/t$ in the last term of Eq. (5.98). Our analysis has revealed that the divergence of the Grüneisen parameter arises from the term with the volume derivative of the inverse susceptibility of spin fluctuation [see the last term of Eq. (5.98)].

For $d + z > 4$, Γ behaves as

$$\Gamma \propto -\frac{\tilde{C}_{\mathrm{b}}}{C_{\mathrm{a}}}\frac{1}{t} \tag{5.100}$$

for $t \ll 1$ because $(\partial y/\partial P)_T$ is almost t independent as mentioned in Sect. 5.3.5. The critical t dependence for each class, which can be known analytically from Table 5.2, is actually confirmed by the numerical result well fitted by the dashed line in Fig. 5.7a–c. The criticality of Γ is summarized in the second column of Table 5.3, which is in agreement with the renormalization group result [32]. Note, however, that the criticality appears at a sufficiently low-t regime for $t \lesssim 10^{-3}$, indicating that the temperature dependence of $(\partial y/\partial P)_T$ affects the intermediate-t region.

For $d + z = 4$, i.e., the 2d AFM case, Γ behaves as

$$\Gamma \propto -\frac{\tilde{C}_{\mathrm{b}}}{C_{\mathrm{a}}}\frac{V}{t}\left(\frac{\partial y}{\partial V}\right)_T \sim \frac{1}{t \ln t}\frac{\ln(-\ln t)}{\ln\left(-\frac{t}{\ln t}\right)} \tag{5.101}$$

for $t \ll 1$. The critical t dependence comes from the multiplication of $\tilde{C}_{\mathrm{b}}/(C_{\mathrm{a}}t)$ (see Table 5.2) and the prefactor $(\partial y/\partial P)_T \approx 1/\ln(-t/\ln t)$, which can be confirmed numerically by the dashed line in Fig. 5.7d. The t dependence of $\tilde{C}_{\mathrm{b}}/(C_{\mathrm{a}}t)$ for $t \ll 1$ is in agreement with the renormalization group theory [32], except for $\ln\left(-\frac{t}{\ln t}\right)$ in the denominator, which arises from the prefactor $(\partial y/\partial P)_T$ in α_{b} as discussed in Sect. 5.3.5. Even after the inclusion of the t dependence of $(\partial y/\partial P)_T$, Γ diverges for $t \to 0$ since the factor $1/t$ in $\tilde{C}_{\mathrm{b}}/(C_{\mathrm{a}}t)$ overcomes the logarithmic correction.

Next, the Grüneisen parameter near the QCP for each class is discussed on the basis of the SCR theory. We calculate $\Gamma(t)$ in the paramagnetic phase by solving the SCR equation Eq. (5.47) for $d + z > 4$ and Eq. (5.48) for $d + z = 4$ in the paramagnetic region ($y_0 > 0$) and also in the region where the magnetic order takes place ($y_0 < 0$). The input parameters other than y_0 are the same as those in Sect. 5.3.6. The results are shown in Fig. 5.8a–d (see corresponding Fig. 5.4a–d, respectively).

At the QCP specified by $y_0 = 0$, $\Gamma(t)$ shows the divergence for $t \to 0$ in each class. As getting away from the QCP in the paramagnetic region, $\Gamma(t)$ for $t \to 0$ becomes finite, whose value decreases as y_0 increases from 0, as shown in Fig. 5.8a–d.

On the other hand, as y_0 decreases from 0, the magnetic order occurs at finite temperature $t = t_c \equiv T_c/T$ for 3d FM (Fig. 5.8a) and 3d AFM (Fig. 5.8b). In the high-t regime for $t \gtrsim 0.7$, $\Gamma(t)$ shows the Curie-Weiss behavior in each class as mentioned above. As t decreases, $\Gamma(t)$ increases and turns to decrease, which finally converges into a finite value $\Gamma(t) \to \Gamma_{\mathrm{a}}(t = t_{\mathrm{c}}) = -\frac{V}{T_0}\left(\frac{\partial T_0}{\partial V}\right)_{T=T_{\mathrm{c}}} = 10.0$ for $t \to t_{\mathrm{c}+}$. This is because $\tilde{C}_{\mathrm{b}} \to 0$ is realized for $y \to 0$ as t approaches t_{c} from the high-t side, making $\alpha_{\mathrm{b}} \propto \frac{\tilde{C}_{\mathrm{b}}(t_{\mathrm{c}})}{t_{\mathrm{c}}} \to 0$ in Eq. (5.88). Then, by Eq. (5.88), it turns out that the Grüneisen parameter at t_{c} is expressed as $\Gamma(t) \to \Gamma_{\mathrm{a}}(t)$ for $t \to t_{\mathrm{c}+}$.

For 2d FM and 2d AFM, the magnetic order occurs only at $t = t_{\mathrm{c}} = 0$. The Grüneisen parameters for $y_0 < 0$ in this case also show $\Gamma(t) \to \Gamma_{\mathrm{a}}(t) = -\frac{V}{T_0}\left(\frac{\partial T_0}{\partial V}\right)_{T=0} = 10.0$ for $t \to 0_+$, as seen in Fig. 5.8c, d, respectively.

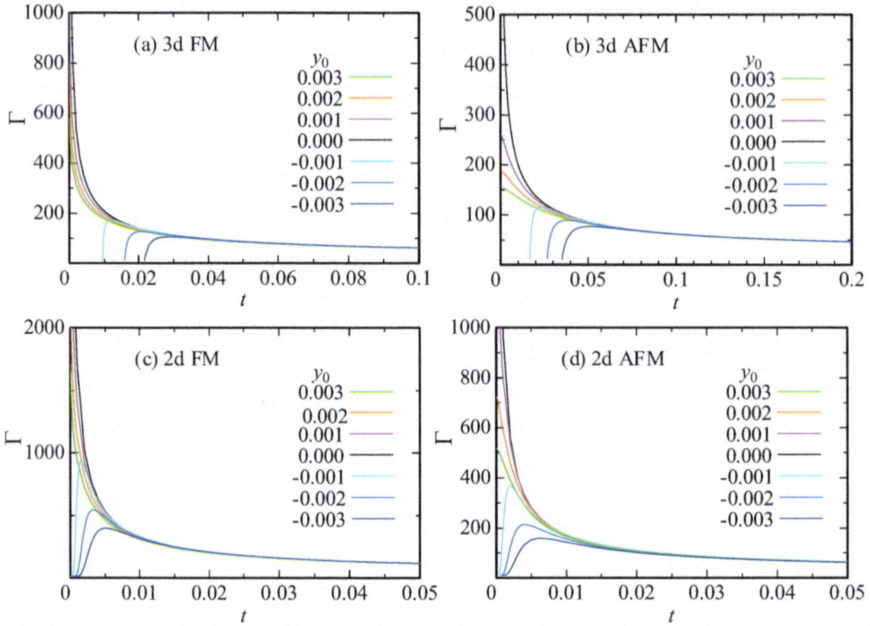

Fig. 5.8 Scaled temperature dependence of the Grüneisen parameter for **a** 3d FM, **b** 3d AFM, **c** 2d FM, and **d** 2d AFM [19]. Reprinted from Ref. [19]. Copyright 2019 American Physical Society

5.3.7 Comparison with Experiments

The thermal-expansion coefficient is generally expressed as

$$\alpha = \alpha_e + \alpha_{ph} + \alpha_{mag}, \tag{5.102}$$

where α_e and α_{ph} are contributions from itinerant electrons and acoustic phonons, respectively. At low temperatures, α_e behaves as $\alpha_e = aT$, as shown by the free electron model, and α_{ph} is given by $\alpha_{ph} = \beta T^3$ [35]. In Eq. (5.102), α_{mag} arises from spin fluctuations, which become profound near the continuous magnetic transition point, as discussed in Sect. 5.3.5 (note that α_{mag} was denoted as α in Sect. 5.3.5).

The Grüneisen parameter is generally expressed as

$$\Gamma = \Gamma_e + \Gamma_{ph} + \Gamma_{mag}, \tag{5.103}$$

where Γ_i is defined by Eq. (5.97) with i=e, ph, and mag, corresponding to each term in Eq. (5.102). For sufficiently lower temperatures than the Fermi temperature, Γ_e is given as a constant [35]. In heavy electron systems, the characteristic temperature of the quasiparticles is the effective Fermi temperature, which is referred to as the

Kondo temperature T_K in the lattice system. Then, the characteristic temperature T^* in the Fermi liquid (see Eq. (D.9) in Appendix D.2) is set to be T_K, which leads to [34, 36, 38]

$$\Gamma_e = -\frac{V}{T_K}\left(\frac{\partial T_K}{\partial V}\right)_S. \tag{5.104}$$

To grasp the main property, let us take the view from the strong limit of onsite Coulomb repulsion of f-electrons. By inputting $T_K = De^{-\frac{1}{N_{cF}J}}$ to Eq. (5.104), where D and N_{cF} are the half band width and the density of states at the Fermi level of conduction electrons per "spin", respectively, and J is the effective Kondo exchange coupling ($J > 0$) in the lattice system [39, 40], we obtain for $N_{cF}J \ll 1$

$$\Gamma_e \approx -\frac{1}{N_{cF}J}\left\{\frac{V}{J}\left(\frac{\partial J}{\partial V}\right)_S + c\right\}, \tag{5.105}$$

where c is a constant of $O(1)$ (e.g., $c = 2/3$ for free conduction electrons in $d = 3$). In heavy electron systems, $(JN_{cF})^{-1}$ typically has a magnitude of $O(10)$. Thus we see that the factor $(N_{cF}J)^{-1}$ gives rise to the enhancement of $|\Gamma_e|$, which is often observed in the heavy electron metals with about 10–100 times larger values than those of ordinary metals [34, 36, 41, 42].

In the Fermi-liquid regime of the heavy fermion system, the coefficient A in the resistivity $\rho(T) = \rho_0 + AT^2$ is related to the specific-heat coefficient γ_e in the specific heat $C_{FL}(T) = \gamma_e T$ as

$$A \propto \gamma_e^2 \propto \frac{1}{T_K^2}, \tag{5.106}$$

as discussed in Sect. 3 [43–45].

In the heavy fermion metal Ce_7Ni_3, the antiferromagnetic transition takes place at $T_N = 1.8$ K and T_N is suppressed under pressure, i.e., $T_N \to 0$ at $P_c = 0.39$ GPa, as shown in Fig. 5.9a [50]. By further applying pressure to Ce_7Ni_3, the Fermi-liquid regime appears at low temperatures. Umeo et al. estimated the pressure dependence of T_K by applying Eq. (5.106) to the resistivity and the specific-heat coefficient observed for $P \geq 0.7$ GPa, as shown in Fig. 5.9b [50]. By evaluating the volume dependence of T_K in Eq. (5.106), the Grüneisen parameter is estimated as $\Gamma_e \approx 100$ at $P = 0.7$ GPa [50], which is greatly enhanced.

When the system approaches the continuous magnetic transition point by varying parameters, e.g., by applying pressure or magnetic field, or chemical doping, Γ_{mag} arising from spin fluctuations becomes predominant in Eq. (5.103), as discussed in Sect. 5.3.6 (note that Γ_{mag} was denoted as Γ in Sect. 5.3.6). In the following subsections, let us discuss the pressure tuning to the magnetic QCP in the Ce- and Yb-based heavy electron systems.

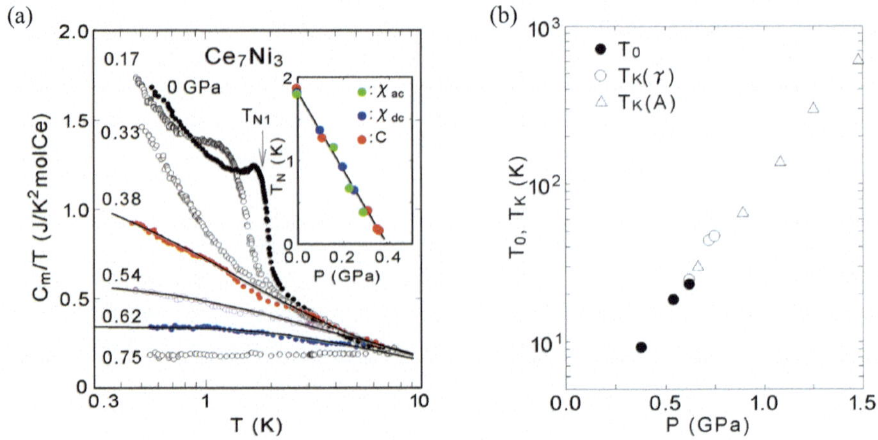

Fig. 5.9 **a** Temperature dependence of the specific-heat coefficient of the 4f-electrons in Ce_7Ni_3 under pressure. The solid line is the result of fitting by the SCR theory for the three-dimensional antiferromagnet. The inset shows the temperature dependence of the antiferromagnetic transition temperature T_N [37]. **b** The characteristic temperature of spin fluctuations T_0 (filled circle), the Kondo temperature T_K estimated from the specific-heat coefficient (open circle), and T_K estimated from the resistivity (open triangle) in Ce_7Ni_3 [37]. Reprinted with permission from Ref. [37]. Copyright 2007 Maruzen Publishing Co., Ltd.

5.3.7.1 Pressure Effects on Ce-based Systems

In the Ce-based heavy electron systems, by applying pressure, the hybridization between f and conduction electrons $|V_{\rm fc}|$ increases and the f-level $\varepsilon_{\rm f}$ increases in general. Hence, the Kondo coupling

$$J \sim \frac{V_{\rm fc}^2}{\varepsilon_{\rm F} - \varepsilon_{\rm f}} \tag{5.107}$$

increases, giving rise to increase in the Kondo temperature $T_{\rm K}$. This yields $\left(\frac{\partial J}{\partial V}\right)_S < 0$, which leads to $\Gamma_{\rm e} > 0$ in Eq. (5.105) [46].

Namely, the system becomes more itinerant under pressure, which makes the characteristic temperature of spin fluctuation T_0 increase. Direct evaluation of T_0 in Eq. (5.39) gives

$$T_0 = \frac{\tilde{A} v_{\rm F} \tilde{q}_{\rm B}}{\pi^2 n^{2/3}} \tag{5.108}$$

in $d = 3$, where $v_{\rm F}$ is the Fermi velocity and n is the filling defined by $n \equiv \frac{N_{\rm e}}{2N}$. Here, $\tilde{q}_{\rm B}$ is given by $\tilde{q}_{\rm B} = q_{\rm B}$ for $z = 3$ and $\tilde{q}_{\rm B} = Q$ for $z = 2$, and $\tilde{A}$ is a dimensionless constant defined by the q^2 coefficient around the ordered vector $\boldsymbol{Q}$ in the irreducible susceptibility at $\omega = 0$ (e.g., $\tilde{A} = \frac{1}{12}$ for the free electron model [8]). Since $v_{\rm F} \sim T_{\rm K}$ holds, i.e., T_0 is proportional to $T_{\rm K}$

$$T_0 \propto T_K, \tag{5.109}$$

applying pressure induces increase in v_F reflecting the pressure-induced expansion of the effective band width of the quasiparticles. This effect contributes to increase in T_0 under pressure, i.e., $\left(\frac{\partial T_0}{\partial P}\right) > 0$.

Indeed, in Ce_7Ni_3, it was observed that T_0 increases as pressure increases, as shown in Fig. 5.9b. Umeo et al. estimated the pressure dependence of T_0 by fitting the SCR theory for the three-dimensional antiferromagnet [13] to the temperature dependence of the specific-heat under pressure, as shown in the solid lines in Fig. 5.9b [50]. The result shows that $T_0(P)$ varies smoothly with $T_K(P)$. This smooth variation of T_0 and T_K observed under pressure is also understandable from Eq. (5.109), since $T_0(P)$ can get close to $T_K(P)$ according to the parameters of $\tilde{A}$ and n. The plot of $T_0(P)$ as well as $T_K(P)$ determined from the measurements of the specific heat and resistivity in Ce_7Ni_3 under pressure [50] enables us to estimate $\frac{1}{T_K}\left(\frac{\partial T_K}{\partial P}\right) = 4.0$ GPa^{-1}, which is comparable to $\frac{1}{T_0}\left(\frac{\partial T_0}{\partial P}\right)$ [36]. The bulk modulus is observed as $\kappa_T^{-1} = 24.6$ GPa at room temperature. The Grüneisen parameter is estimated to be $\Gamma_a \approx 110$ at $P \approx 0.38$ GPa [36].

Figure 5.10 with the right-pointing P axis illustrates the T-P phase diagram of the Ce-based heavy electron systems. As P increases, the magnetic transition temperature T_c is suppressed to be absolute zero at the QCP denoted by P_c. At $P = P_c$, the magnetic susceptibility $\chi_Q(0, 0)$ diverges with $y = 0$ (see Eq. (5.35)]. Since y increases as P increases from P_c as shown by the dashed line, $\left(\frac{\partial y}{\partial P}\right)_{T=0} > 0$ holds (For the AFM QCP ($z = 2$) in $d = 2$, y starts to appear with zero slope $\left(\frac{\partial y}{\partial P}\right)_{T=0} = 0$ at P_c [19]. Then, from Eq. (5.88) and resultant Eq. (5.86), the positive thermal-expansion coefficient appears $\alpha_{mag} > 0$ for $P > P_c$ at low temperatures.

This is understandable from the P dependence of the entropy S. When the QCP is approached from the paramagnetic side for $P > P_c$, $S/T \approx C_V/T$ (see Table 5.1) increases toward P_c at the infinitesimal temperature. This gives $\left(\frac{\partial S}{\partial P}\right)_T < 0$ for $P > P_c$, leading to $\alpha_{mag} > 0$ by Eq. (5.83) and hence $\Gamma_{mag} > 0$.

On the other hand, when P further decreases from P_c, the continuous transition to the magnetically ordered phase makes S/T decrease continuously for $P < P_c$ at the infinitesimal T. This gives $\left(\frac{\partial S}{\partial P}\right)_T > 0$ for $P < P_c$, leading to $\alpha_{mag} < 0$ by Eq. (5.83) and hence $\Gamma_{mag} < 0$. Namely, the sign change of α_{mag} and Γ_{mag} occurs at P_c as a consequence of the entropy accumulation near the QCP [32, 33]. It is noted that the sign change of α_{mag} was shown in Refs. [1, 51] and the sign change of Γ_{mag} as well was discussed in Ref. [33].

5.3.7.2 Pressure Effects on Yb-Based Systems

On the other hand, in the Yb-based heavy electron systems, the electronic state with $4f^{13}$ configuration for Yb^{3+} is understood on the basis of the hole picture. Hence, the f-hole level ε_f decreases as pressure increases in the Yb-based systems. In case this effect outweighs the increase in the f-c hybridization, the Kondo coupling

$$J \sim \frac{V_\mathrm{fc}^2}{\varepsilon_\mathrm{F} - \varepsilon_\mathrm{f}} \tag{5.110}$$

decreases, giving rise to decrease in the Kondo temperature T_K under pressure. This yields $\left(\frac{\partial J}{\partial V}\right)_S > 0$, which leads to $\Gamma_\mathrm{e} < 0$ in Eq. (5.105) and hence the negative volume expansion $\alpha_\mathrm{e} < 0$ in Eq. (5.102).

Namely, pressure induces the system where f-electrons becomes more localized. When T_K decreases under pressure, decrease in the f-hole level is more effective than increase in the f-c hybridization, which makes v_F decrease. This effect contributes to the decrease in T_0 under pressure, i.e., $\left(\frac{\partial T_0}{\partial P}\right) < 0$. This yields negative thermal-expansion coefficient and Grüneisen parameter $\alpha_\mathrm{a} < 0$ in Eq. (5.87) and $\Gamma_\mathrm{a} < 0$ in Eq. (5.97).

Figure 5.10 with the left-pointing P axis illustrates the T-P phase diagram of the Yb-based heavy electron systems, where by applying pressure to the paramagnetic-metal phase, the magnetic transition occurs at T_c starting from the QCP denoted by P_c. As P approaches P_c, the magnetic susceptibility $\chi_\mathbf{Q}(0,0) \propto y^{-1}$ increases and diverges at P_c for $T = 0$. Hence, $\left(\frac{\partial y}{\partial P}\right)_T < 0$ holds for $P < P_\mathrm{c}$, as shown by the dashed line. Thus, from Eq. (5.88), $\alpha_\mathrm{b} < 0$ appears, which results in the negative thermal-expansion coefficient $\alpha_\mathrm{mag} < 0$ and hence $\Gamma_\mathrm{mag} < 0$ for $P < P_\mathrm{c}$ at low temperatures.

On the other hand, when P further increases from P_c, the continuous transition to the magnetically ordered phase makes S/T decrease continuously for $P > P_\mathrm{c}$ at the infinitesimal T. This gives $\left(\frac{\partial S}{\partial P}\right)_T < 0$ for $P > P_\mathrm{c}$, leading to $\alpha_\mathrm{mag} > 0$ by Eq. (5.83) and hence $\Gamma_\mathrm{mag} > 0$. Namely, sign change of α_mag and Γ_mag occurs at P_c due to the entropy accumulation near the QCP.

5.3.7.3 Observation of α and Γ Near the Magnetic QCP

To detect the thermal-expansion coefficient α and the Grüneisen parameter Γ near the QCP, experimental efforts have been devoted [28, 53–55]. So far, a few data have been reported to exhibit the quantum criticality shown in Tables 5.1 and 5.3 in the stoichiometric compounds. To access the QCP, chemical doping has often been performed for high accuracy measurement of α at ambient pressure. However, the chemical doping more or less brings about effects of disorder, which sometimes masks true critical behaviors expected in clean systems. In this subsection, keeping this aspect in mind, experimental data to be compared with the criticality in Tables 5.1 and 5.3 are discussed.

In $\mathrm{CeNi_2Ge_2}$, the specific heat $C_{4\mathrm{f}}/T \sim \gamma_0 - a_C T^{1/2}$ and resistivity $\rho \sim T^{n_\rho}$ ($1.2 \le n_\rho \le 1.5$) are observed in the low T region at ambient pressure as shown in Fig. 5.11, suggesting close proximity to the 3d AFM QCP (see Table 5.1) [56]. The

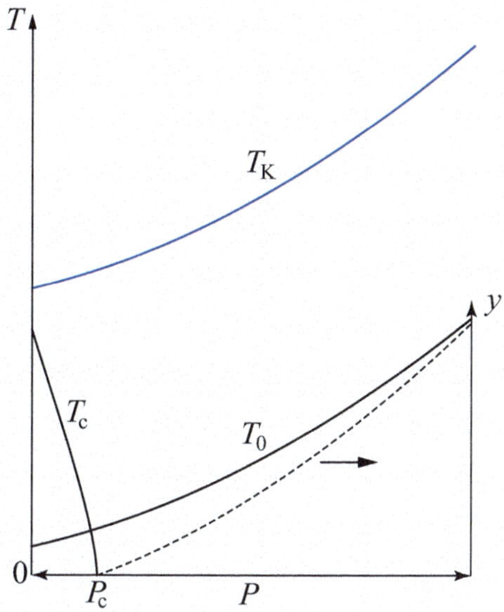

Fig. 5.10 (Color online) Schematic temperature-pressure phase diagram of Ce-based heavy electron systems (right-pointing P axis) and Yb-based heavy electron systems (left-pointing P axis) in $d = 3$. The Kondo temperature T_K and the characteristic temperature of spin fluctuation T_0 are given by the solid lines (left axis). Note that vertical scales for T_K and T_0 can be different in general. Depending on the parameters in each material, it is possible that $T_0(P)$ gets close to $T_K(P)$ (see text). The magnetic transition temperature T_c (solid line, left axis) is suppressed to 0 at the QCP denoted by P_c. The dashed line gives $y = 1/[2T_A\chi_Q(0,0)]$ (right axis) for $P > P_c$. Note that in $d = 2$ for $z = 2$ the dashed line starts to appear from P_c with a vanishing slope $\left(\frac{\partial y}{\partial P}\right)_{T=0} = 0$ [19]. Note that it is possible that a crossing of $T_K(P)$ and $T_c(P)$ occurs depending on the material parameters. If the system has XY or Heisenberg symmetry, no magnetic transition occurs for finite T in $d = 2$

measured thermal-expansion coefficient $\alpha = a_\alpha\sqrt{T} + b_\alpha T$ is in accordance with $\alpha_{\text{mag}} \sim T^{1/2}$ in Eq. (5.93) (see Fig. 5.12) and $\alpha_e \sim T$ in Eq. (5.102). As shown in the inset in Fig. 5.11, the Grüneisen parameter $\Gamma \approx 57$ at $T = 5$ K is already enhanced reflecting the contribution from Γ_a in Eq. (5.96) and the heavy electron background as noted around Eq. (5.105). Further enhancement of Γ for lowering T is observed as $\Gamma \approx 98 \pm 10$ at $T \approx 0.1$ K, which suggests the contribution from Γ_b as shown in Eq. (5.100) (see Fig. 5.7b).

As for the sign change of the thermal-expansion coefficient, $\alpha_{\text{mag}} < 0$ in the AFM phase for $x < x_c$ and $\alpha_{\text{mag}} > 0$ in the paramagnetic phase for $x > x_c$ were observed in $\text{CeIn}_{3-x}\text{Sn}_x$ with $x_c = 0.67 \pm 0.03$ [57] and in $\text{CeRhIn}_{5-x}\text{Sn}_x$ with $x_c = 0.48$ [58].

Since the systematic study of $T_0(P)$ and $T_K(P)$ has already been performed in Ce_7Ni_3 [36, 50], the measurements of $\alpha(T)$ and $\Gamma(T)$ at the QCP specified by $P_c = 0.39$ GPa and their analyses based on Eqs. (5.86) and (5.96) are an interesting subject for future studies.

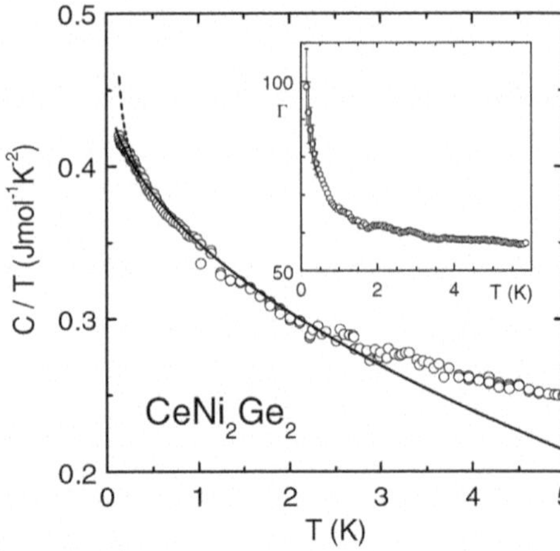

Fig. 5.11 Temperature dependence of the specific-heat coefficient in $CeNi_2Ge_2$. The inset shows temperature dependence of the Grüneisen parameter [56]. Reprinted with permission from Ref. [56]. Copyright 2003 American Physical Society

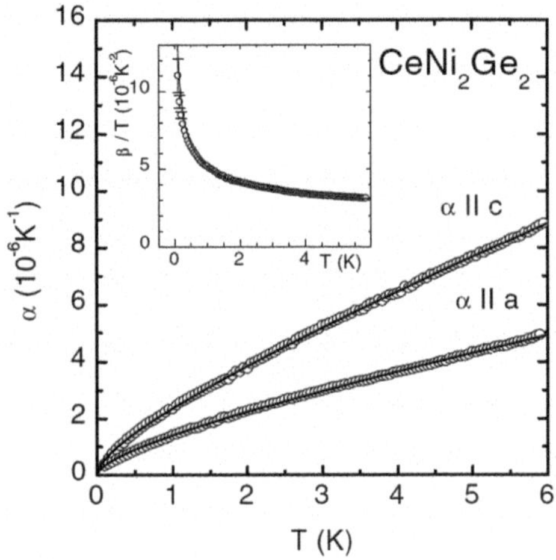

Fig. 5.12 Temperature dependence of the thermal-expansion coefficient in $CeNi_2Ge_2$. The inset shows temperature dependence of β/T where β is defined by $\beta \equiv 2\alpha_a + \alpha_c$ in $CeNi_2Ge_2$ [56]. Reprinted with permission from Ref. [56]. Copyright 2003 American Physical Society

Furthermore, experimental observation of the quantum criticality shown in Tables 5.1 and 5.3 for each class is also greatly desired. Near the FM QCP, $\left(\frac{\partial y}{\partial P}\right)_T$ can be directly observed by measuring the pressure dependence of the uniform susceptibility since $\chi_0(0,0)^{-1} \propto y$ holds. Near the AFM QCP, by measuring the pressure dependence of the NMR relaxation rate $(T_1T)^{-1}$ or resistivity $\rho(T)$ at low temperatures, $\left(\frac{\partial y}{\partial P}\right)_T$ can be extracted [1, 31]. Observation of $T_0(P)$ as well as $T_K(P)$ and evaluation of $\frac{1}{T_0}\left(\frac{\partial T_0}{\partial P}\right)_T$ as done in Ce_7Ni_3 and $\left(\frac{\partial y}{\partial P}\right)_T$ enables us to make the complete analysis of $\alpha(T)$ and $\Gamma(T)$ at the QCP on the basis of Eqs. (5.86) and (5.96). Such measurements are much to be desired.

References

1. T. Moriya, *Spin Fluctuations in Itinerant Electron Magnetism* (Springer, Berlin, 1985)
2. T. Moriya, A. Kawabata, J. Phys. Soc. Jpn. **34**, 639 (1973)
3. T. Moriya, A. Kawabata, J. Phys. Soc. Jpn. **35**, 669 (1973)
4. G.G. Lonzarich, L. Taillefer, J. Phys. C **18**, 4339 (1985)
5. K. Makoshi, T. Moriya, J. Phys. Soc. Jpn. **38**, 10 (1975)
6. K. Ueda, T. Moriya, J. Phys. Soc. Jpn. **39**, 605 (1975)
7. T. Moriya, J. Magn. Magn. Matter. **100**, 261 (1991)
8. J.A. Hertz, Phys. Rev. B **14**, 1165 (1976)
9. A.J. Millis, Phys. Rev. B **48**, 7183 (1993)
10. U. Zülicke, A.J. Millis, Phys. Rev. B **51**, 8996 (1995)
11. T. Moriya, Y. Takahashi, K. Ueda, J. Phys. Soc. Jpn. **59**, 2905 (1990)
12. M. Hatatani, T. Moriya, J. Phys. Soc. Jpn. **64**, 3434 (1995)
13. T. Moriya, T. Takimoto, J. Phys. Soc. Jpn. **64**, 960 (1995)
14. A. Ishigaki, T. Moriya, J. Phys. Soc. Jpn. **65**, 376 (1996)
15. A. Ishigaki, T. Moriya, J. Phys. Soc. Jpn. **65**, 3402 (1996)
16. A. Ishigaki, T. Moriya, J. Phys. Soc. Jpn. **67**, 3924 (1998)
17. A. Ishigaki, T. Moriya, J. Phys. Soc. Jpn. **68**, 3673 (1999)
18. K.K. Murata, S. Doniach, Phys. Rev. Lett. **29**, 285 (1972)
19. S. Watanabe, K. Miyake, Phys. Rev. B **99**, 035108 (2019)
20. We employ Eq. (5.25) which essentially describes isotropic spin space, i.e., Heisenberg symmetry. It is noted that the action with the factor 3 multiplied to the r.h.s. of Eq. (5.25) corresponds to the case for Heisenberg symmetry discussed in Ref. 17 and Refs. 21,30
21. It is straightforward to extend the present results to the case of magnetic fluctuations of XY or Heisenberg type, where the factor of 2 or 3 is multiplied to the free energy in Eq. (5.36), respectively (Note that the solution of the SCR equation correctly satisfies the Mermin-Wagner theorem for the Heisenberg symmetry as shown in Fig. 5.4). Hence, the factor is also to be multiplied to the entropy in Eq. (5.61) and the specific heat in Eqs. (5.63) and (5.65). In real materials, there exists more or less anisotropy of the spin space, leading to anisotropic spin fluctuation. Therefore, we plot α as well as C_V for a single component of the spin fluctuation with the factor of 1 in a series of Figures for convenience of comparison with experiments. As for the Grüneisen parameter Γ, the results shown in Fig. 5.7 and Fig. 5.8 are not affected by this factor except for κ_T, since the factors in α and C_V are cancelled each other in Eq. (5.97)
22. R.P. Feynman, *Statistical Mechanics* (Addison-Wesley, Reading, MA, 1990), Sect. 3.4
23. Although the criticality in $d = 2$ for $z = 2$, $y \sim -\frac{t}{\ln t}$ was reported in Ref. 16, more precisely it can be shown that $y \sim -t\frac{\ln(-\ln t)}{\ln t}$ is the solution of the SCR equation, which coincides with the result of the renormalization group discussed in Ref. 9 (see Appendix C.3)
24. N.D. Mermin, H. Wagner, Phys. Rev. **17**, 1133 (1966)
25. M. Hatatani, O. Narikiyo, K. Miyake, J. Phys. Soc. Jpn. **67**, 4002 (1998); M. Hatatani, Ph.D. thesis, Graduate School of Engineering Science, Osaka University (2000)
26. T. Moriya, J. Phys. Soc. Jpn. **18**, 516 (1963)
27. T. Moriya, K. Usami, Solid State Commun. **34**, 95 (1980)
28. S. Kambe, J. Flouquet, P. Lejay, P. Haen, A. de Visser, J. Phys.: Cond. Matt. **9**, 4917 (1997)
29. Y. Takahashi, H. Nakano, J. Phys.: Condens. Matter **18**, 521 (2006)
30. Y. Takahashi, J. Phys.: Condens. Matter **11**, 6439 (1999)
31. T. Moriya, K. Ueda, Rep. Prog. Phys. **66**, 1299 (2003)
32. L. Zhu, M. Garst, A. Rosch, Q. Si, Phys. Rev. Lett. **91**, 066404 (2003)
33. M. Garst, A. Rosch, Phys. Rev. B **72**, 205129 (2005)
34. J. Flouquet, Prog. Low. Temp. Phys. **15**, 149 (2005)
35. N.W. Ashcroft, N.D. Mermin, *Solid State Physics* (Saunders College, 1976)
36. K. Umeo, H. Kadomatsu, T. Takabatake, Phys. Rev. B **54**, 1194 (1996)
37. K. Umeo, in *Handbook of High Pressure Technique* (Japanese) ed. by N. Mori, K. Murata, Y. Uwatoko, H. Takahashi (Maruzen, 2007), p. 354

38. A.V. Goltsev, M.M. Abd-Elmeguid, J. Phys.: Cond. Mat. **17**, 5813 (2005)
39. Y. Ōno, T. Matsuura, Y. Kuroda, Physica C **159**, 878 (1989)
40. N. Shibata, T. Nishino, K. Ueda, C. Ishii, Phys. Rev. B **53**, R8828 (1996)
41. A. de Visser, J.N.M. Franse, J. Flouquet, Physica B **161**, 324 (1989)
42. J.D. Thompson, J.M. Lawrence, Z. Fisk, J. Low Temp. Phys. **95**, 59 (1994)
43. K. Miyake, M. Matsuura, C. Varma, Solid State Commun. **71**, 1149 (1989)
44. N.K. Sato, A. Sumiyama, S. Kunii, H. Nagano, T. Kasuya, J. Phys. Soc. Jpn. **54**, 1923 (1985)
45. K. Kadowaki, S.B. Woods, Solid State Commun. **58**, 507 (1986)
46. It is generally expected that the first term in Eq. (5.105) dominates over the second term since the volume dependence of the Kondo exchange coupling is much stronger than that of the density of states of conduction electrons. If we adopt the tight-binding picture with the distance dependence of the hybridization between the f and conduction electrons $V_{\rm fc} \sim 1/r^{\ell+\ell'+1}$ with the azimthal quantum numbers $\ell = 3$, ℓ', respectively [47-49], we estimate $\frac{V}{J}\left(\frac{\partial J}{\partial V}\right) \sim -\frac{1}{d}(2\ell' + 6 + O(\frac{\varepsilon_{\rm f}}{\varepsilon_{\rm F}-\varepsilon_{\rm f}}))$ in the d-dimensional system. The absolute value is larger than $c \equiv \frac{V}{N_{\rm F}}\left(\frac{\partial N_{\rm F}}{\partial V}\right) \sim \frac{2}{d}$ estimated from the nearly-free-electron picture, since $\varepsilon_{\rm f}$ is not so close to $\varepsilon_{\rm F}$ in typical heavy-electron systems. This reflects the fact that the locality of the 4f wavefunction reacts severely to the volume change of the system
47. O.K. Andersen, O. Jepsen, Physica B **91**, 317 (1977)
48. O.K. Andersen, W. Klose, H. Nohl, Phys. Rev. B **17**, 1209 (1978)
49. W.A. Harrison, *Electronic Structure and the Properties of Solids* (W. H. Freeman and Co., San Francisco, CA, 1980)
50. K. Umeo, H. Kadomatsu, T. Takabatake, J. Phys.: Cond. Matt. **8**, 9743 (1996)
51. It should be noted that sign change of $\alpha_{\rm mag}$ in the ferromagnetically ordered phase ($\alpha_{\rm mag}<0$) and the paramagnetic phase ($\alpha_{\rm mag}>0$) was shown by the Moriya-Usami theory [27] to explain the observation in $\rm ZrZn_2$ [53]. See Fig. 5.4 in p. 89 in Sect. 5.2 in Ref. 1
52. S. Ogawa, N. Kasai, J. Phys. Soc. Jpn. **27**, 789 (1969)
53. R. Küchler, P. Gegenwart, C. Geibel, F. Steglich, Sci. Technol. Adv. Mater. **8**, 428 (2007). (and references therein)
54. A. Steppke, R. Küchler, S. Lausberg, E. Lengyel, L. Steinke, R. Borth, T. Luhmann, C. Krellner, M. Nicklas, C. Geibel, F. Steglich, M. Brando, Science **339**, 933 (2013)
55. P. Gegenwart, Rep. Prog. Phys. **79**, 114502 (2016). (and references therein)
56. R. Küchler, N. Oeschler, P. Gegenwart, T. Cichorek, K. Neumaier, O. Tegus, C. Geibel, J.A. Mydosh, F. Steglich, L. Zhu, Q. Si, Phys. Rev. Lett. **91**, 066405 (2003)
57. R. Küchler, P. Gegenwart, J. Custers, O. Stockert, N. Caroca-Canales, C. Geibel, J.G. Sereni, F. Steglich, Phys. Rev. Lett. **96**, 256403 (2006)
58. J.G. Donath, F. Steglich, E.D. Bauer, F. Ronning, J.L. Sarrao, P. Gegenwart, Europhys. Lett. **87**, 57011 (2009)

Chapter 6
Quantum Criticality of Valence Transition—Experiments and Theory

In this chapter, the quantum criticality of the valence transition is discussed from both aspects of the experiments and the theory. In Sect. 6.1, the systematic anomalies associated with the valence instability observed in several heavy-fermion systems are overviewed. The theory of critical valence fluctuations is explained in Sects. 6.2, 6.3, and 6.4. In Sect. 6.2, the action which is essential for the valence transition is derived from the extended Anderson lattice model. In Sect. 6.3, the perturbative renormalization group method is applied to the action. In Sect. 6.4, the mode-mode coupling theory of critical valence fluctuations is explained.

A new type of scaling in the magnetic susceptibility, which is expressed as a single scaling function of the ratio of the temperature T and magnetic field B, was observed in the heavy-fermion metal β-$YbAlB_4$. The T/B scaling in β-$YbAlB_4$ is explained by the theory of critical valence fluctuations in Sect. 6.5.

The common quantum criticality as well as the T/B scaling was observed in the quasicrystal $Yb_{15}Al_{34}Au_{51}$. Interestingly, the quantum criticality persists even under pressure at least up to $P = 1.6$ GPa, which is in sharp contrast to the quantum critical phenomena ever known in periodic crystals. In Sect. 6.6, the quantum critical phenomena in the quasicrystal $Yb_{15}Al_{34}Au_{51}$ are discussed from both aspects of the experiments and the theory of critical valence fluctuations.

6.1 Systematic Anomalies Associated with Critical Valence Transition

Quantum critical phenomena have been one of the central issues in condensed matter physics. The nature of the quantum critical point (QCP) emerging when the magnetically ordered temperature is suppressed to absolute zero has been intensively studied and the role of spin fluctuations in the critical phenomena has been well understood [1–3].

S. Watanabe and K. Miyake, *Quantum Critical Phenomena of Valence Transition*,
Springer Tracts in Modern Physics 289,
https://doi.org/10.1007/978-981-99-3518-5_6

However, anomalous critical phenomena, which do not follow the conventional spin-fluctuation theories [1–3] have been discovered in paramagnetic metal phases in $YbRh_2Si_2$ [4, 5] and $YbRh_2(Si_{0.95}Ge_{0.05})_2$ [6, 7]. In Table 6.1, the temperature dependences of the uniform magnetic susceptibility χ, the specific-heat coefficient C/T, the resistivity subtracted by the residual resistivity $\rho - \rho_0$, and the NMR spin-lattice relaxation rate $(T_1T)^{-1}$ at low temperatures are shown. In the Fermi-liquid metal, χ, C/T, and $(T_1T)^{-1}$ show the temperature-independent behavior and the resistivity shows $\rho - \rho_0 \propto T^2$ behavior. As the representative case for the magnetic quantum critical phenomena (see Table 1.1), the three-dimensional (3D) antiferromagnetic (AF) case is listed in Table 6.1. The low-T behavior in the physical quantities observed in $YbRh_2Si_2$ does not follow either of the Fermi liquid nor any class of the magnetic criticality listed in Table 1.1.

The most striking anomaly is that low-temperature uniform spin susceptibility exhibits divergent behavior $\chi(T) \sim T^{-\zeta}$ with the anomalous critical exponent $\zeta = 0.6$ in $YbRh_2(Si_{0.95}Ge_{0.05})_2$ in spite of no sign of a ferromagnetic phase nearby. In these materials, the specific-heat constant exhibits the logarithmic divergence $\gamma = C/T \sim -\ln T$ [4, 7], giving rise to a large Wilson ratio, i.e., a dimensionless ratio of χ to γ, $R_W = 17.5$ in $YbRh_2(Si_{0.95}Ge_{0.05})_2$ [6] exceeding the conventional strong-coupling value $R_W = 2$. The linear-T dependence of low-T resistivity emerges in a wide-T range in $YbRh_2Si_2$ [4], $YbRh_2(Si_{0.95}Ge_{0.05})_2$ [7]. The NMR spin-lattice relaxation rate shows $(T_1T)^{-1} \sim T^{-0.5}$ [8].

Interestingly, the inhomogeneity of electronic states, one of which is the Fermi-liquid state and the other is the non-Fermi liquid, grows under the magnetic field applied to the c-axis direction in $YbRh_2Si_2$ [9]. This observation supports the field-induced first-order valence transition explained in Sect. 4.6. Since the phase separation of the state with a relatively larger valence and the state with the relatively lower valence, which is accompanied by the first-order valence transition, is expected to occur in real materials, the field-induced inhomogeneity is compatible with the theory of the valence fluctuations.

So far, to explain $YbRh_2Si_2$ and $YbRh_2(Si_{0.95}Ge_{0.05})_2$, theoretical efforts have been made [10–15]. In particular, a scenario asserting that f electrons undergo a localized to itinerant transition was extensively discussed [10, 11]. However, $YbRh_2Si_2$ shows a large specific-heat constant as $\sim 1.7\,\mathrm{J/(molK^2)}$ even inside of the antiferromagnetic (AF) phase [7]. This fact indicates that heavy quasiparticles are responsible for the AF state. Indeed, a band-structure calculation showed evidence contrary to the scenario by demonstrating that a tiny valence change of Yb can explain the Hall-coefficient measurement in $YbRh_2Si_2$ [16], which seems to be a basis of the above scenario [17]. At low temperature $T = 10\,\mathrm{K}$, the large Fermi surface was observed by the angle-resolved photoemission measurement [18], which also provides the clear evidence of the c-f hybridization.

In β-$YbAlB_4$, the similar quantum criticality was observed [19, 20] as shown in Table 6.1. This material is the first superconductor of the Yb-based heavy fermion systems with the transition temperature $T_c = 80\,\mathrm{mK}$. In the paramagnetic metal phase for $T > T_c$, the uniform magnetic susceptibility shows diverging enhancement with the power smaller than 1 similar to $YbRh_2Si_2$, $\chi \sim T^{-0.5}$. The specific-heat coeffi-

Table 6.1 Temperature dependences of uniform susceptibility χ, electrical specific heat coefficient C/T, electrical resistivity ρ, and NMR spin-lattice relaxation rate $(T_1T)^{-1}$ at low temperatures in correlated metals. In the quantum valence criticality, $0.5 \le \zeta \le 0.7$. In α-$YbAl_{0.086}Fe_{0.014}B_4$, the dynamical muon relaxation rate T_1^{-1} shows the quantum valence criticality for $T < 0.2$ K [24]. The reason is explained in Ref. [25] (see text)

Theory/Material	χ	C/T	$\rho - \rho_0$	$(T_1T)^{-1}$
Fermi liquid	const.	const.	T^2	const.
3D AF	const.$-T^{1/4}$	const.$-T^{1/2}$	$T^{3/2}$	$T^{-3/4}$
Quantum valence criticality	$T^{-\zeta}$	$-\log T$	T	$T^{-\zeta}$
$YbRh_2Si_2$	$T^{-0.6}$	$-\log T$	T	$T^{-0.5}$
β-$YbAlB_4$	$T^{-0.5}$	$-\log T$	$T^{1.5} \to T$	–
α-$YbAl_{0.086}Fe_{0.014}B_4$	$T^{-0.5}$	$-\log T$	$T^{1.6} \to T$	$T^{-\zeta}$
$Yb_{15}Al_{34}Au_{51}$	$T^{-0.5}$	$-\log T$	T	$T^{-0.5}$

cient exhibits the logarithmic increase $C/T \sim -\log T$, giving rise to the large Wilson ratio $R_W = 6.5$. The T-linear resistivity appears at low temperatures at least below 3 K down to a few hundred mK. The valence of Yb was detected as $Yb^{+2.75}$ (0.75 4f hole per Yb) at $T = 20$ K in β-$YbAlB_4$, indicating strong valence fluctuations [21]. By applying pressure to β-$YbAlB_4$, the magnetic order occurs at $P = 2.5$ GPa [22]. Since the physical quantities for $P \lesssim 2.5$ GPa behave differently from those around $P = 0$, the origin of the non-Fermi liquid behavior is considered to arise from the non-magnetic origin.

The sister compound α-$YbAlB_4$, which is a locally isostructural polymorph of β-$YbAlB_4$, shows the Fermi-liquid behavior at low temperatures. However, by replacing Al with Fe 1.4 %, the same quantum criticality as that observed in β-$YbAlB_4$ appears as shown in Table 6.1 [23]. At just $x = 0.014$, a sharp change in the Yb valence as well as the volume change was detected (see Fig. 6.9) [23]. This is the direct evidence of the quantum valence criticality, as explained in Sect. 6.5.2.

In α-$YbAl_{0.086}Fe_{0.014}B_4$, the temperature dependence of the dynamical muon spin relaxation rate, i.e, T_1^{-1} revealed that the quantum valence criticality appears below 0.1 K (see Table 6.1) [24]. This can be understood as the cooperative phenomena of the quantum valence criticality and the muon-induced local Kondo effect, as follows [25]. The muon with the positive charge attracts conduction electrons, which form the local Kondo singlet. Then, as temperature decreases, the dynamical relaxation rate, i.e., T_1^{-1} increases and shows the Koringa relation, i.e., temperature-independent behavior below the muon-induced local Kondo temperature, which is estimated to be $T_K \sim 0.2$ K. Then, for $T \lesssim T_K$, the bulk quantum valence criticality appears.

Interestingly, the common criticality was discovered in the quasicrystal $Yb_{15}Al_{34}Au_{51}$ at ambient pressure and zero magnetic field [26, 27]. As temperature decreases, the uniform magnetic susceptibility diverges as $\chi \sim T^{-0.5}$ and the specific-heat coefficient shows logarithmic increase, giving rise to a large Wilson ratio. The NMR relaxation rate $(T_1T)^{-1}$ shows the similar temperature dependence to the susceptibility and approximately T-linear resistivity appears (see Table 6.1). The quasicrystal has no periodicity and there exists no magnetic order in the temperature-pressure phase diagram [28]. The experimental evidence of the quantum valence criticality was also observed in the quasicrystal $Yb_{15}Al_{34}Au_{51}$ [29] as explained in Sect. 6.6.6.

The quantum criticality listed in Table 6.1 seems to show the new universality class irrespective of the periodic or aperiodic lattice structures. As the key origin, the theory of critical valence fluctuations was proposed in ref. [30], which was later endorsed by the direct measurements of the sharp Yb-valence change with the quantum valence criticality in the periodic crystal α-$YbAl_{0.086}Fe_{0.014}B_4$ [23] and the quasicrystal $Yb_{15}Al_{34}Au_{51}$ [26].

6.2 The Action Derived from the Extended Anderson Lattice Model

Let us construct the mode-coupling theory for critical valence fluctuations starting from the extended Anderson lattice model, which describes the essential part of the Ce- and Yb-based systems in the standard notation:

$$H = H_{\rm c} + H_{\rm f} + H_{\rm hyb} + H_{U_{\rm fc}}, \tag{6.1}$$

where $H_{\rm c} = \sum_{k\sigma} \varepsilon_k c^{\dagger}_{k\sigma} c_{k\sigma}$, $H_{\rm f} = \varepsilon_{\rm f} \sum_{i\sigma} n^{\rm f}_{i\sigma} + U_{\rm ff} \sum_{i=1}^{N} n^{\rm f}_{i\uparrow} n^{\rm f}_{i\downarrow}$, $H_{\rm hyb} = \sum_{k\sigma} V_k \left(f^{\dagger}_{k\sigma} c_{k\sigma} + c^{\dagger}_{k\sigma} f_{k\sigma} \right)$, and $H_{U_{\rm fc}} = U_{\rm fc} \sum_{i=1}^{N} n^{\rm f}_i n^{\rm c}_i$. The $U_{\rm fc}$ term is the Coulomb repulsion between 4f and conduction electrons (holes) in Ce (Yb) systems, which is considered to play an important role in the valence transition [31].

To consider correlation effects by $U_{\rm ff}$, we employ the slave-boson large-N expansion framework [32]. The Hamiltonian Eq. (6.1) is generalized to the case with N-fold degeneracy from $\sigma = \uparrow, \downarrow$, and the slave-boson operator b_i is introduced to eliminate the doubly occupied state for $U_{\rm ff} \to \infty$ under the constraint $\sum_m n^{\rm f}_{im} + N b^{\dagger}_i b_i = 1$. The Lagrangian is written as $\mathcal{L} = \mathcal{L}_0 + \mathcal{L}'$:

$$\begin{aligned}
\mathcal{L}_0 &= \sum_{km} c^{\dagger}_{km} \left(\frac{\partial}{\partial \tau} + \bar{\varepsilon}_k \right) c_{km} + \sum_{kk'm} f^{\dagger}_{km} \left(\frac{\partial}{\partial \tau} + \bar{\varepsilon}^{\rm f}_{k-k'} \right) f_{k'm} \\
&+ \frac{V}{\sqrt{N_{\rm s}}} \sum_{kk'm} \left(c^{\dagger}_{km} f_{k'm} b^{\dagger}_{k-k'} + {\rm h.c.} \right) + \frac{N}{N_{\rm s}} \sum_{kk'} b^{\dagger}_k \lambda_{k-k'} b_{k'}, \\
\mathcal{L}' &= -\frac{U_{\rm fc}}{2} \sum_{im} \left(n^{\rm c}_{im} + n^{\rm f}_{im} \right) + \frac{U_{\rm fc}}{N} \sum_{imm'} n^{\rm f}_{im} n^{\rm c}_{im'},
\end{aligned}$$

where λ_k is the Lagrange multiplier to impose the constraint, and $\bar{\varepsilon}_k \equiv \varepsilon_k + \frac{U_{fc}}{2}$ and $\bar{\varepsilon}^{f}_{k-k'} \equiv \left(\varepsilon_f + \frac{U_{fc}}{2}\right)\delta_{kk'} + \frac{1}{\sqrt{N_s}}\lambda_{k-k'}$. We here separate $\mathcal{L}$ as $\mathcal{L}_0$ and $\mathcal{L}'$ to perform the expansion with respect to the U_{fc} term after taking account of the local correlation of the U_{ff} term.

For $\exp(-S_0)$ with the action $S_0 = \int_0^\beta d\tau \mathcal{L}_0(\tau)$, the saddle point solution is obtained via the stationary condition $\delta S_0 = 0$ by approximating spatially uniform and time-independent ones, i.e., $\lambda_q = \lambda\delta_q$ and $b_q = b\delta_q$. The solution is obtained by solving mean-field equations $\partial S_0/\partial\lambda = 0$ and $\partial S_0/\partial b = 0$ self-consistently.

For $S' = \int_0^\beta d\tau \mathcal{L}'(\tau)$, we introduce the identity applied by a Stratonovich-Hubbard transformation

$$
\begin{aligned}
e^{-S'} = \int \mathcal{D}\varphi \exp[\sum_{im} \int_0^\beta d\tau\{-\frac{U_{fc}}{2}\varphi_{im}(\tau)^2 \\
+\frac{U_{fc}}{\sqrt{N}}(c_{im} f^\dagger_{im} + f_{im} c^\dagger_{im})\}\varphi_{im}(\tau))]. \quad (6.2)
\end{aligned}
$$

The partition function is expressed as

$$
Z = \int \mathcal{D}(cc^\dagger ff^\dagger\varphi)\exp(-S) \quad (6.3)
$$

with $S = S_0 + S'$. By performing Grassmann number integrals for $cc^\dagger$ and $ff^\dagger$, we obtain $Z = \int \mathcal{D}\varphi \exp(-S[\varphi])$ with

$$
\begin{aligned}
S[\varphi] = \sum_m \sum_{\bar{q}} \varphi_m(\bar{q})\varphi_m(-\bar{q}) - \mathrm{Trln}[-\hat{G}_0^{-1} + \hat{V}] - \frac{\beta N}{\sqrt{N_s}}\lambda|b|^2 \\
+ \beta N q_0 \sqrt{N_s}\lambda. \quad (6.4)
\end{aligned}
$$

Here, the abbreviation $\bar{q} \equiv (\boldsymbol{q}, i\omega_l)$ with $\omega_l = 2l\pi T$ is used, and $\hat{G}_0$ and $\hat{V}$ are defined as

$$
\hat{G}_0^{-1} \equiv \begin{bmatrix} i\varepsilon_n - \bar{\varepsilon}_k & -\frac{Vb^*}{\sqrt{N_s}} \\ -\frac{Vb}{\sqrt{N_s}} & i\varepsilon_n - \bar{\varepsilon}^f_0 \end{bmatrix}\delta_{kk'},\ \hat{V} \equiv \begin{bmatrix} 0 & \tilde{\varphi}_{m\bar{k}\bar{k}'} \\ \tilde{\varphi}_{m\bar{k}\bar{k}'} & 0 \end{bmatrix}
$$

with $\tilde{\varphi}_{m\bar{k}\bar{k}'} \equiv \frac{U_{fc}}{\sqrt{\beta N N_s}}\varphi_m(\bar{k} - \bar{k}')$, respectively. Here, $\bar{k} \equiv (\boldsymbol{k}, i\varepsilon_n)$ with $\varepsilon_n = (2n+1)\pi T$, and each matrix element of $\hat{G}_0$ is defined as $(\hat{G}_0)_{11} \equiv G_0^{cc}$, $(\hat{G}_0)_{12} \equiv G_0^{cf}$, $(\hat{G}_0)_{21} \equiv G_0^{fc}$, and $(\hat{G}_0)_{22} \equiv G_0^{ff}$. By using $\mathrm{Trln}[-\hat{G}_0^{-1} + \hat{V}] = \mathrm{Trln}[-\hat{G}_0^{-1}] - \sum_{n=1}^{\infty} \frac{1}{n}\mathrm{Tr}(\hat{G}_0\hat{V})^n$, we obtain

$$
\begin{aligned}
S[\varphi] = \sum_{m} \Bigg[& \frac{1}{2} \sum_{\bar{q}} \Omega_2(\bar{q}) \varphi_m(\bar{q}) \varphi_m(-\bar{q}) \\
+ & \sum_{\bar{q}_1, \bar{q}_2, \bar{q}_3} \Omega_3(\bar{q}_1, \bar{q}_2, \bar{q}_3) \varphi_m(\bar{q}_1) \varphi_m(\bar{q}_2) \varphi_m(\bar{q}_3) \delta \left(\sum_{i=1}^{3} \bar{q}_i \right) \\
+ & \sum_{\bar{q}_1, \bar{q}_2, \bar{q}_3, \bar{q}_4} \Omega_4(\bar{q}_1, \bar{q}_2, \bar{q}_3, \bar{q}_4) \\
\times & \; \varphi_m(\bar{q}_1) \varphi_m(\bar{q}_2) \varphi_m(\bar{q}_3) \varphi_m(\bar{q}_4) \delta \left(\sum_{i=1}^{4} \bar{q}_i \right) + \cdots \Bigg]. \qquad (6.5)
\end{aligned}
$$

Here, constant terms independent of $\varphi_m(\bar{q})$ are omitted in Eq. (6.5) since they merely shift the origin of the free energy of the system. The coefficient of the quadratic term is given by

$$
\Omega_2(\boldsymbol{q}, i\omega_l) = U_{\text{fc}} \left[1 - \frac{2U_{\text{fc}}}{N} \left\{ \chi_0^{\text{ffcc}}(\boldsymbol{q}, i\omega_l) + \chi_0^{\text{cfcf}}(\boldsymbol{q}, i\omega_l) \right\} \right], \qquad (6.6)
$$

where $\chi_0^{\alpha\beta\gamma\delta}(\boldsymbol{q}, i\omega_l) \equiv -\frac{T}{N_s} \sum_{\boldsymbol{k},n} G_0^{\alpha\beta}(\boldsymbol{k}+\boldsymbol{q}, i\varepsilon_n + i\omega_l) G_0^{\gamma\delta}(\boldsymbol{k}, i\varepsilon_n)$. Since long wavelength $|\boldsymbol{q}| \ll q_c$ around $\boldsymbol{q} = \boldsymbol{0}$ and low-frequency $|\omega| \ll \omega_c$ regions play dominant roles in critical phenomena with q_c and ω_c being cutoffs for momentum and frequency in the order of inverse of the lattice constant and the effective Fermi energy, respectively, Ω_i for $i = 2$, 3, and 4 are expanded for q and ω around $(\boldsymbol{0}, 0)$:

$$
\Omega_2(\boldsymbol{q}, i\omega_l) \approx \eta + Aq^2 + C\frac{|\omega_l|}{q} \qquad (6.7)
$$

where $\eta = U_{\text{fc}}[1 - \frac{2U_{\text{fc}}}{N}\{\chi_0^{\text{ffcc}}(\boldsymbol{0}, 0) + \chi_0^{\text{cfcf}}(\boldsymbol{0}, 0)\}]$, $\Omega_3(q_1, q_2, q_3) \approx v_3/\sqrt{\beta N_s}$, and $\Omega_4(q_1, q_2, q_3, q_4) \approx v_4/(\beta N_s)$.

6.3 Perturbation Renormalization Group Approach

Different from ordinary critical phenomena of spin fluctuations [1–3], there appears a cubic term in Eq. (6.5) in general for the valence fluctuation case [33]. Let us here apply the Hertz's renormalization-group procedure [2] to $S[\varphi]$:

(a) Integrating out high-momentum and -frequency parts for $q_c/s < q < q_c$ and $\omega_c/s^z < \omega < \omega_c$, respectively, with s being a dimensionless scaling parameter ($s \geq 1$) and z the dynamical exponent.

(b) Scaling of q and ω by $q' = sq$ and $\omega' = s^z\omega$.

(c) Re-scaling of φ by $\varphi'(\boldsymbol{q}', \omega') = s^a \varphi(\boldsymbol{q}'/s, \omega'/s)$.

Then, we determined that to make the Gaussian term in Eq. (6.5) scale-invariant, a must satisfy $a = -(d+z+2)/2$ with d spatial dimension and the dynamical exponent $z = 3$. The renormalization-group equations for coupling constants v_j are derived as

$$\frac{dv_3}{ds} = [6 - (d+z)]\, v_3 + O(v_3^2), \tag{6.8}$$

$$\frac{dv_4}{ds} = [4 - (d+z)]\, v_4 + O(v_4^2), \tag{6.9}$$

for cubic and quadratic terms, respectively. By solving these equations, it is shown that higher order terms than the Gaussian term are irrelevant

$$\lim_{s\to\infty} v_j(s) = 0 \ \text{ for } \ j \geq 3 \tag{6.10}$$

for $d + z > 6$. For the case of $d = 3$ and $z = 3$, it is shown that the cubic term is marginally irrelevant [31]. It is also noted that as shown in Fig. 1.4, the valence crossover line is parallel to the temperature axis in general. Hence, as we approach the QCP on cooling, the cubic term does not appear. Hence, the universality class of the criticality of valence fluctuations belongs to the Gaussian fixed point. This implies that critical valence fluctuations are qualitatively described by the RPA framework with respect to $U_{\rm fc}$. The coefficient of the Gaussian term in Eq. (6.5) is nothing but the inverse of the valence susceptibility $\Omega_2(\boldsymbol{q}, i\omega_l) \equiv \chi_{\rm v}(\boldsymbol{q}, i\omega_l)^{-1}$. Since evaluation of $\chi_0^{\rm ffcc}(\boldsymbol{q}, i\omega_l)$ and $\chi_0^{\rm cfcf}(\boldsymbol{q}, i\omega_l)$ using the saddle point solution for $\exp(-S_0)$ concludes $\chi_0^{\rm ffcc} \gg \chi_0^{\rm cfcf}$ (see Fig. 6.2 and text below), it turns out that $\chi_{\rm v}$ is expressed by the RPA form

$$\begin{aligned}\chi_{\rm v}(\boldsymbol{q}, i\omega_l) &= \int_0^\beta d\tau \langle T_\tau n_{\rm f}(\boldsymbol{q},\tau) n_{\rm f}(-q, 0)\rangle e^{i\omega_l \tau} \\ &\approx U_{\rm fc}^{-1}[1 - \frac{2U_{\rm fc}}{N}\chi_0^{\rm ffcc}(\boldsymbol{q}, i\omega_l)]^{-1},\end{aligned} \tag{6.11}$$

as shown in Fig. 6.1.

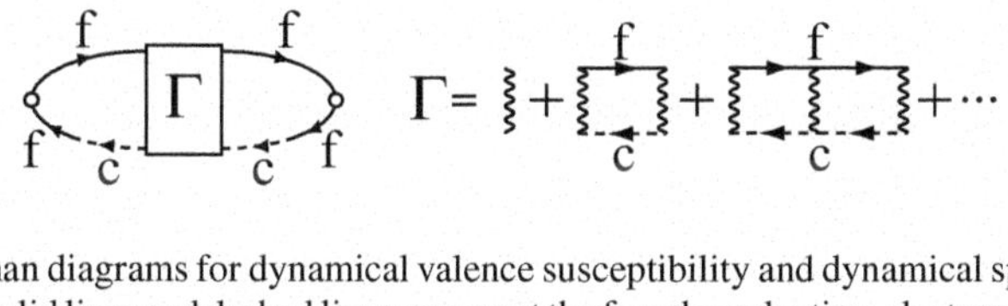

Fig. 6.1 Feynman diagrams for dynamical valence susceptibility and dynamical spin susceptibility for f electrons. Solid lines and dashed lines represent the f- and conduction-electron Green functions, $G_0^{\rm ff}$ and $G_0^{\rm cc}$, respectively. Wiggly lines represent $U_{\rm fc}$

An important consequence of this result is that dynamical f-spin susceptibility

$$\chi_{\rm f}^{+-}(\boldsymbol{q}, i\omega_l) \equiv \int_0^\beta d\tau \langle T_\tau S_{\rm f}^+(\boldsymbol{q}, \tau) S_{\rm f}^-(-\boldsymbol{q}, 0)\rangle e^{i\omega_l \tau} \qquad (6.12)$$

has a common structure with $\chi_{\rm v}$ in the RPA framework as shown in Fig. 6.1. At the quantum critical point of the valence transition, namely, the QCP, the valence susceptibility $\chi_{\rm v}(\mathbf{0}, 0)$ diverges. The common structure indicates that $\chi_{\rm f}^{+-}(\mathbf{0}, 0)$ also diverges at the QCP. The uniform spin susceptibility is given by $\chi \approx \chi_{\rm s}^{\rm f} \approx \frac{3}{2}\mu_{\rm B}^2 g_{\rm f}^2 \chi_{\rm f}^{+-}(\mathbf{0}, 0)$ with $\chi_{\rm s}^{\rm f}$ uniform f-spin susceptibility, $\mu_{\rm B}$ the Bohr magneton, and $g_{\rm f}$ Lande's g factor for f electrons. This gives a qualitative explanation for the fact that uniform spin susceptibility diverges at the QCP of valence transition under a magnetic field, which was shown by the slave-boson mean-field theory applied to Eq. (6.1) [34]. Numerical calculations for Eq. (6.1) in $d = 1$ by the DMRG [34] and in $d = \infty$ by the DMFT [35] also showed the simultaneous divergence of $\chi_{\rm v}$ and uniform spin susceptibility under the magnetic field, reinforcing the above argument based on RPA.

The other important point of the present theory is that the "unperturbed" term $\mathcal{L}_0$, i.e., $\hat{G}_0$, already contains the local correlation effect by $U_{\rm ff}$. This effect plays a key role in critical phenomena in Ce- and Yb-based systems, which will be shown below to be the origin of the unconventional criticality. The local correlation effect emerges as dispersionless, almost q-independent $\chi_0^{\rm ffcc}(\mathbf{q}, 0)$ and $\chi_0^{\rm cfcf}(\mathbf{q}, 0)$ in Eq. (6.6), as shown in Fig. 6.2a. Here, the saddle point solution for $\exp(-S_0)$ is employed for a typical parameter set of heavy-electron systems: $D = 1$, $V = 0.5$, and $U_{\rm ff} = \infty$ at total filling $n = 7/8$ with $\varepsilon_{\boldsymbol{k}} = \boldsymbol{k}^2/(2m_0) - D$ and $n \equiv \bar{n}_{\rm f} + \bar{n}_{\rm c}$, where $\bar{n}_{\rm f}$ and $\bar{n}_{\rm c}$ are the number of f electrons and conduction electrons per "spin" and site, respectively. The bare mass m_0 is chosen such that the integration from $-D$ to D of the density of states of conduction electrons per "spin" is equal to 1. This local nature is reflected in the inverse of valence susceptibility in Eq. (6.7) as an extremely small coefficient A. We note here that this flat-q result is obtained not only for deep $\varepsilon_{\rm f}$ with $\bar{n}_{\rm f} = 1/2$

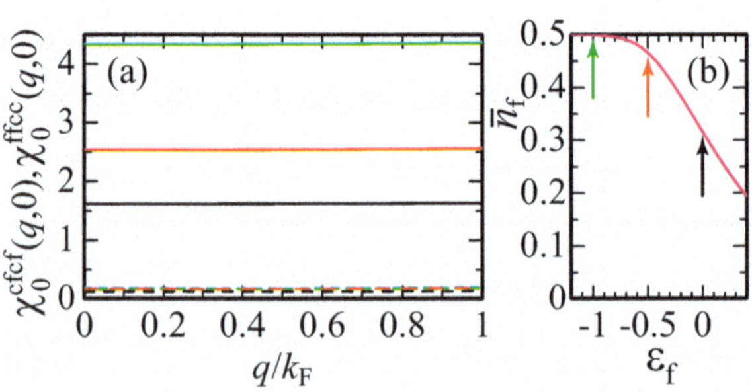

Fig. 6.2 **a** q dependence of $\chi_0^{\rm ffcc}(q, 0)$ (solid line) and $\chi_0^{\rm cfcf}(q, 0)$ (dashed line) calculated by using saddle point solution of $\exp(-S_0)$ for $\varepsilon_{\rm f} = -1$ (green), -0.5 (red), and 0.0 (black). **b** $\bar{n}_{\rm f}$ versus $\varepsilon_{\rm f}$. **a** and **b** are results for $D = 1$, $V = 0.5$, and $U_{\rm ff} = \infty$ at $n = 7/8$ [30]. Reprinted from Ref. [30]. Copyright 2010 American Physical Society

in the Kondo regime but also for shallow $\varepsilon_{\rm f}$ with $\bar{n}_{\rm f} < 1/2$ in the valence-fluctuating regime (see Fig. 6.2b). Here, we note that the c-f hybridization is always finite.

6.4 Mode-Coupling Theory for Critical Valence Fluctuations

To clarify how this local nature causes unconventional criticality, we construct a self-consistent renormalization (SCR) theory for valance fluctuations. Although higher order terms v_j $(j \geq 3)$ in $S[\varphi]$ are irrelevant as shown in Eq. (6.10), the effect of their mode couplings affects low-T physical quantities significantly as is well known in spin-fluctuation theories [1–3]. To construct the action using the best Gaussian taking account of the mode-coupling effects up to the fourth-order $(j \leq 4)$ in $S[\varphi]$, we employ Feynman's inequality on the free energy:

$$F \leq F_{\rm eff} + T\langle S - S_{\rm eff}\rangle_{\rm eff} \equiv \tilde{F}(\eta). \tag{6.13}$$

Here, the action is given by

$$S_{\rm eff}[\varphi] = \frac{1}{2}\sum_{\boldsymbol{q},l}(\eta + Aq^2 + C_q|\omega_l|)|\varphi(\boldsymbol{q}, i\omega_l)|^2 \tag{6.14}$$

where Q_q is defined by $C_q \equiv C/\max\{q, l_{\rm i}^{-1}\}$ with $l_{\rm i}$ being the mean free path by impurity scattering [36]. Here we consider the case that the system is clean, i.e., $C_q = C/q$. In Eq. (6.13), $\langle\cdots\rangle_{\rm eff}$ denotes the statistical average taken by the weight $\exp(-S_{\rm eff}[\varphi])$ and $F_{\rm eff}$ is given by

$$F_{\rm eff} = -T\ln\int\mathcal{D}\varphi\exp(-S_{\rm eff}[\varphi])\,. \tag{6.15}$$

In Eq. (6.13), the valence susceptibility $\chi_{\rm v}(\boldsymbol{q}, i\omega_l)$ is defined as

$$\chi_{\rm v}(\boldsymbol{q}, i\omega_l)^{-1} = \eta + Aq^2 + C_q|\omega_l|, \tag{6.16}$$

where η expresses the effect of the mode-mode coupling of critical valence fluctuations and parametrizes the closeness to the QCP. The free energy $\tilde{F}$ defined by Eq. (6.13) is expressed as [37]

$$\begin{aligned}\tilde{F} &= \frac{1}{\pi}\sum_{q}\int_0^{\omega_{\rm c}} d\omega\frac{\Gamma_q}{\omega^2+\Gamma_q^2}\left\{\frac{\omega}{2} + T\ln\left(1 - {\rm e}^{-\frac{\omega}{T}}\right)\right\}\\ &+ \frac{\eta_0 - \eta}{2}\langle\varphi^2\rangle_{\rm eff} + \frac{3v_4}{N}\langle\varphi^2\rangle_{\rm eff}^2 - \frac{1}{\pi}\sum_{q}\frac{\pi\omega_{\rm c}}{4},\end{aligned} \tag{6.17}$$

where Γ_q is defined by $\Gamma_q \equiv (\eta + Aq^2)/C_q$ and valence fluctuation $\langle \varphi^2 \rangle_{\rm eff}$ is defined as

$$\langle \varphi^2 \rangle_{\rm eff} = T \sum_q \sum_l \frac{1}{\eta + Aq^2 + C_q|\omega_l|}. \tag{6.18}$$

Here, $\langle \varphi^2 \rangle_{\rm eff}$ consists of the quantum (zero-point) fluctuation $\langle \varphi^2 \rangle_{\rm zero}$ and thermal fluctuation $\langle \varphi^2 \rangle_{\rm th}$ as

$$\langle \varphi^2 \rangle_{\rm eff} = \langle \varphi^2 \rangle_{\rm zero} + \langle \varphi^2 \rangle_{\rm th}, \tag{6.19}$$

where $\langle \varphi^2 \rangle_{\rm zero}$ and $\langle \varphi^2 \rangle_{\rm th}$ are expressed as

$$\langle \varphi^2 \rangle_{\rm zero} = \frac{1}{\pi} \sum_q \frac{1}{C_q} \int_0^{\omega_c} d\omega \frac{\omega}{\Gamma_q^2 + \omega^2}, \tag{6.20}$$

$$\langle \varphi^2 \rangle_{\rm th} = \frac{1}{\pi} \sum_q \frac{2}{C_q} \int_0^{\omega_c} d\omega \frac{1}{e^{\beta\omega} - 1} \frac{\omega}{\Gamma_q^2 + \omega^2}, \tag{6.21}$$

respectively.

Near the QCP of the valence transition, quantum valence fluctuation $\langle \varphi^2 \rangle_{\rm zero}$ in Eq. (6.20) is calculated as

$$\langle \varphi^2 \rangle_{\rm zero} = N \frac{3T_0}{2T_A} C_1 - C_2 y + \cdots, \tag{6.22}$$

where the characteristic temperature of valence fluctuation is defined as

$$T_0 \equiv \frac{Aq_{\rm B}^2}{2\pi C_{q_{\rm B}}} \tag{6.23}$$

with $q_{\rm B}$ being the wave number of the Brillouin Zone. In Eq. (6.22), T_A is defined as

$$T_A \equiv \frac{Aq_{\rm B}^2}{2}. \tag{6.24}$$

and y is defined as

$$y \equiv \frac{\eta}{Aq_{\rm B}^2}. \tag{6.25}$$

The constants C_1 and C_2 in Eq. (6.22) are given by

$$C_1 = \int_0^{x_c} dx x^3 \ln \left| \frac{\omega_{cT_0}^2 + x^6}{x^6} \right|, \tag{6.26}$$

$$C_2 = 2 \int_0^{x_c} dx x \frac{\omega_{cT_0}^2}{\omega_{cT_0}^2 + x^6}, \tag{6.27}$$

respectively, where x is the dimensionless wave number defined as $x \equiv q/q_B$. Here, the cut off is expressed as $x_c \equiv q_c/q_B$ in the dimensionless scaled form and ω_{cT} is defined as $\omega_{cT} \equiv \omega_c/(2\pi T)$.

The thermal valence fluctuation $\langle\varphi^2\rangle_{th}$ in Eq. (6.21) is calculated as

$$\langle\varphi^2\rangle_{th} = 3N \frac{T_0}{T_A} \int_0^{x_c} dx x^3 \left\{ \ln u - \frac{1}{2u} - \psi(u) \right\}, \tag{6.28}$$

where $\psi(u)$ is the digamma function with u being defined as

$$u \equiv \frac{\Gamma_q}{2\pi T} = \frac{x(y + x^2)}{t}. \tag{6.29}$$

Here, t is defined as the dimensionless scaled temperature

$$t \equiv \frac{T}{T_0}. \tag{6.30}$$

Under the optimal condition $\frac{d\tilde{F}(\eta)}{d\eta} = 0$, the SCR equation in the $Aq_B^2 \lesssim \eta$ regime with q_B being the wavenumber of the Brillouin zone is obtained as

$$y = y_0 + \frac{3}{2} y_1 t \left[\frac{x_c^3}{6y} - \frac{1}{2y} \int_0^{x_c} dx \frac{x^3}{x + \frac{t}{6y}} \right], \tag{6.31}$$

where y_0 and y_1 are dimensionless constants [30].

When $y \gg t$, $y \propto t^{2/3}$ is obtained from Eq. (6.31) at the QCP with $y_0 = 0$. This indicates that the valence susceptibility shows unconventional criticality $\chi_v(\mathbf{0}, 0) = \eta^{-1} \propto t^{-2/3}$. Figure 6.3a shows numerical solutions of Eq. (6.31). Note here that the coefficient A is quite small as shown above, giving rise to quite small $T_0(= Aq_B^3/(2\pi C) \ll T_F)$ so that the region of $t \equiv T/T_0$ shown in Fig. 6.3 corresponds to that of $T \ll T_F \sim O(D)$. Hence, a wide range of $t = T/T_0$ is shown in the plot. The least-square fit of the data for $5 \le t \le 100$ gives $y \propto t^{0.551}$. Since the Gaussian fixed point ensures the simultaneous divergence of valence susceptibility and uniform f-spin susceptibility as discussed above, χ_s^f shows divergent behavior $\sim t^{-\zeta}$ with $0.5 \lesssim \zeta \lesssim 0.7$ depending on the temperature range in agreement with experiments in $YbRh_2(Si_{0.95}Ge_{0.05})_2$ and β-$YbAlB_4$.

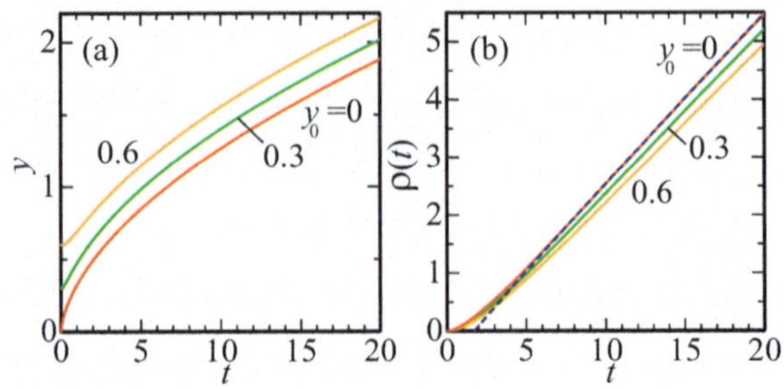

Fig. 6.3 (color online) **a** Numerical solutions of Eq. (6.31) for $y_0 = 0.0$ (at QCP), 0.3, and 0.6 at $y_1 = 1$ and $x_c = 1$. **b** Electrical resistivity $\rho(T)$ calculated by using $y(t)$ in **a**. Dashed line represents the linear-t fit [30]. Reprinted from Ref. [30]. Copyright 2010 American Physical Society

We should note here that in the $T \to 0$ limit, although it may be experimentally difficult to access such a low temperature overcoming the smallness of A, the SCR equation follows the conventional $z = 3$ type [1–3] in the clean system, giving rise to $y \propto t^{4/3}$. Then, at the QCP ($y_0 = 0$), as t decreases, a crossover from $y \propto t^{2/3}$ to $y \propto t^{4/3}$ occurs. In reality, however, because of the smallness of A, the low-T range is extremely elongated by the relation $t = T/T_0$ with $T_0 \propto A$, which makes it possible that unconventional criticality dominates over the experimentally accessible low-temperature region.

6.4.1 NMR Relaxation Rate

The spin-lattice relaxation rate is given by

$$\frac{1}{T_1 T} = 2\left(\frac{\gamma_n^2}{g_f \mu_B}\right)^2 \sum_q |D_q|^2 \frac{\operatorname{Im}\chi^R(\boldsymbol{q}, \omega_0)}{\omega_0}, \tag{6.32}$$

where $\chi^R(\boldsymbol{q}, \omega_0)$ is the retarded dynamical f-spin susceptibility with the nuclear resonance frequency ω_0, γ_n is the gyromagnetic ratio of nuclear spin, g_f is a Lande's g factor for f electrons, and D_q is the hyperfine-coupling constant. Then, we have $(T_1 T)^{-1} \approx y^{-1}$ near the QCP. Namely, the uniform magnetic susceptibility and spin-lattice relaxation rate are shown to have the same temperature dependence: $\chi(t) \sim t^{-2/3}$ and $(T_1 T)^{-1} \sim t^{-2/3}$ for $y \gg t$. When T is decreased down to $T \sim T_0$, y in Eq. (6.31) is evaluated as $y \sim t^{0.5}$ by the least-square fit of the numerical solution of Eq. (6.31). Hence, depending on the flatness of critical valence fluctuation mode and measured temperature range, $\chi(T) \sim t^{-\alpha}$ and $(T_1 T)^{-1} \sim t^{-\alpha}$ with $0.5 \lesssim \alpha \lesssim 0.7$ are observed.

We note that the NMR or NQR relaxation rate is shown to be $(T_1 T)^{-1} \sim \chi_s^f(t) \propto t^{-\zeta}$, which also quantitatively agrees with $(T_1 T)^{-1} \sim T^{-0.5}$ in $YbRh_2Si_2$ [8].

6.4.2 Resistivity

We note that the electrical resistivity $\rho(T)$ shows a T-linear dependence in the regime $t \gtrsim 5$ ($y \gtrsim 1$) where Eq. (6.31) is applicable, as shown in Fig. 6.3b. Here, following a formalism of Ref. [38], $\rho(T)$ is calculated as

$$\rho(T) \propto \frac{1}{T}\int_{-\infty}^{\infty} d\omega\omega n(\omega)[n(\omega)+1]\int_{0}^{q_c} dqq^3 \mathrm{Im}\chi_{\mathrm{v}}^{\mathrm{R}}(q,\omega) \tag{6.33}$$

with $n(\omega) = 1/(e^{\beta\omega}-1)$ being the Bose distribution function, and $\chi_{\mathrm{v}}^{\mathrm{R}}(q,\omega) = (\eta + Aq^2 - iC_q\omega)^{-1}$, a retarded valence susceptibility. Here, $y(t)$ in Fig. 6.3a is used for the clean system $C_q = C/q$, and the normalization constant is taken as 1 in the $\rho(t)$ plot. The emergence of $\rho(t) \propto t$ behavior can be understood from the locality of valence fluctuations: In the system with a small coefficient A, where the local character is strong, the dynamical exponent may be regarded as $z = \infty$ when we write C_q in a general form as $C_q = C/q^{z-2}$. By using this expression in $\chi_{\mathrm{v}}^{\mathrm{R}}(q,\omega)$ in the calculation of $\rho(T)$ for $z = \infty$, we obtain $\rho(T) \propto T$ toward $T \to 0$ K. This result indicates that the locality of valence fluctuations causes the T-linear resistivity. The emergence of $\rho(T) \propto T$ by valence fluctuations was shown theoretically on the basis of the valence susceptibility χ_{v} which has an approximated form for $z = \infty$ in Ref. [39], as discussed in §4.2. The present formulation and the renormalization analysis provide a reasonable ground for the χ_{v} introduced phenomenologically in Ref. [39].

As noted below Eq. (6.14), the dynamical valence susceptibility $\chi_{\mathrm{v}}(\boldsymbol{q}, i\omega_l)$ has different form in the clean system and the dirty system via C_q in Eq. (6.61). Hence, for $T \lesssim T_0$, the resistivity behaves as $\rho \sim T^{3/2}$ in the clean system while $\rho \sim T^{5/3}$ in the dirty system (see the $t \lesssim 1$ regime in Fig. 6.3b).

In β-$YbAlB_4$ and α-$YbAl_{0.896}Fe_{0.014}B_4$, the rounding in the resistivity at low temperatures was observed as $\rho \sim T^{1.5}$ [19] and $\rho \sim T^{1.6}$ [23] respectively (see Table 6.1). This implies that the characteristic temperature of critical valence fluctuations T_0 is located around the measured lowest temperature $T \sim 30$ mK. Indeed, by the Mössbauer measurement, very slow valence fluctuation time scale $\tau \sim 2$ ns was observed in β-$YbAlB_4$ [41], which corresponds to $T_0 \sim 24$ mK as discussed in Sect. 6.5. In case that T_0 is well below the measured lowest temperature, the T-linear resistivity is observed in the whole low-T regime as observed in $YbRh_2Si_2$ and the quasicrystal $Yb_{15}Al_{34}Au_{51}$ (see Table 6.1).

6.4.3 Specific Heat

The specific heat is derived from the free energy for critical valence fluctuations [37]. The entropy $S = -\left(\frac{\partial \tilde{F}}{\partial T}\right)_V$ is obtained from the free energy $\tilde{F}(y)$ in Eq. (6.17) with the stationary condition of the SCR theory being satisfied, which results in

$$S = -3N \int_0^{x_c} dx x^2 \left\{ \ln\sqrt{2\pi} - u + \left(u - \frac{1}{2}\right) \ln u - \ln\Gamma(u) \right\}$$
$$+ 3N \int_0^{x_c} dx x^2 u \left\{ \ln u - \frac{1}{2u} - \psi(u) \right\}. \quad (6.34)$$

Here, $\Gamma(u)$ is the Gamma function.

Then, the specific heat is derived from $C_V = T\left(\frac{\partial S}{\partial T}\right)_V$ as

$$C_V = C_a - C_b, \quad (6.35)$$

where C_a and C_b are given by

$$C_a = -3N \int_0^{x_c} dx x^2 u^2 \left\{ \frac{1}{u} + \frac{1}{2u^2} - \psi'(u) \right\}, \quad (6.36)$$

$$C_b = \tilde{C}_b \left(\frac{\partial y}{\partial t}\right)_V, \quad (6.37)$$

respectively [43]. In Eq. (6.36), $\psi'(u)$ is the trigamma function. The explicit form of $\tilde{C}_b$ is given by

$$\tilde{C}_b = -3N \int_0^{x_c} dx x^3 u \left\{ \frac{1}{u} + \frac{1}{2u^2} - \psi'(u) \right\}. \quad (6.38)$$

The explicit form of the temperature-dependent factor $\left(\frac{\partial y}{\partial t}\right)_V$ is obtained by differentiating Eq. (6.31) with respect to t under a constant volume as

$$\left(\frac{\partial y}{\partial t}\right)_V = \frac{3}{2} y_1 \frac{\frac{I_1}{y} + \frac{t I_2}{12 y^2}}{1 + \frac{3}{2} y_1 t \left(\frac{I_1}{y^2} + \frac{t I_2}{12 y^3}\right)}, \quad (6.39)$$

where I_1 is given by $I_1 = \frac{x_c^3}{6y} - \frac{1}{2y}\int_0^{x_c} dx \frac{x^3}{x + \frac{t}{6y}}$ and I_2 is given by $I_2 = \int_0^{x_c} dx \frac{x^3}{\left(x + \frac{t}{6y}\right)^2}$.

By solving the valence SCR equation (6.31) and Eq. (6.39), we obtain the temperature dependence of y and $\left(\frac{\partial y}{\partial t}\right)_V$ respectively. Then, we calculate C_a in Eq. (6.36) and C_b in Eq. (6.37) and finally we obtain the specific heat C_V in Eq. (6.35). The temperature dependence of the specific heat at the valence QCP is plotted in Fig. 6.4 for $y_0 = 0.0$ and $y_1 = 1.0$. As temperature decreases, the specific-heat coefficient C_V/Nt increases. For $t \gtrsim 1$, i.e., $T \gtrsim T_0$, the logarithmic-like increase appears. It is noted that for $t \ll 1$, the specific heat coefficient behaves as $C_V/Nt \sim -\log t$. Hence, asymptotic behavior appears in $C_V(t)$ for $t \gtrsim 1$ region. The least-square fit of C_V/t for $1 \le t \le 10^2$ gives $C_V/t \sim t^{-0.79}$. The dominant contribution of C_V comes from C_a. It is noted that the least square fit gives $C_a/t \sim t^{-0.66} \sim t^{-2/3}$. The

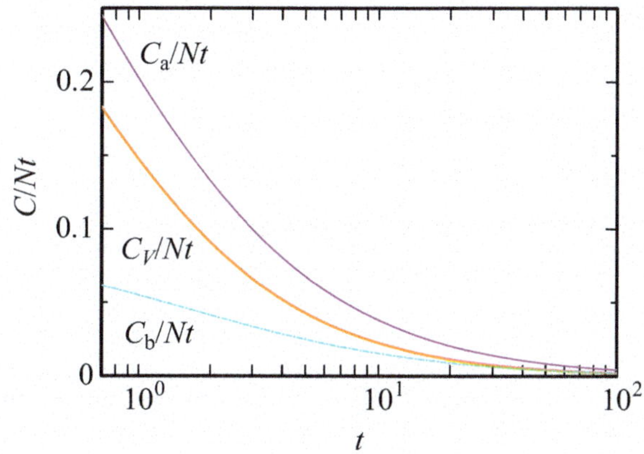

Fig. 6.4 Specific heat coefficient versus scaled temperature $t = T/T_0$. C_V/Nt (bold solid line), C_a/Nt (thin solid line), C_b/Nt (dash-dotted line) at the valence QCP

exponent is the same as the valence criticality 2/3, which was obtained for $y \gg t$ in the valence SCR equation (6.31).

6.4.4 Thermal Expansion Coefficient

The thermal-expansion coefficient α is expressed as

$$\alpha = -\frac{1}{V}\left(\frac{\partial S}{\partial P}\right)_T, \tag{6.40}$$

where V is the molar volume and P is the pressure. The detailed derivation is explained in Appendix D. By differentiating the entropy in Eq. (6.34) with respect to the pressure, we obtain

$$\alpha = \alpha_a + \alpha_b, \tag{6.41}$$

where α_a and α_b are given by

$$\alpha_a = \frac{1}{V}\frac{C_a}{T_0}\left(\frac{\partial T_0}{\partial P}\right)_T, \tag{6.42}$$

$$\alpha_b = \frac{1}{V}\frac{\tilde{C}_b}{t}\left(\frac{\partial y}{\partial P}\right)_T, \tag{6.43}$$

respectively.

By solving the valence SCR equation (6.31) and Eq. (6.39) for $y_0 = 0.0$ and $y_1 = 1.0$, we obtain the temperature dependence of y and $(\partial y/\partial t)_V$ respectively. Then, we calculate C_a in Eq. (6.36) and $\tilde{C}_b$ in Eq. (6.38) and substitute them

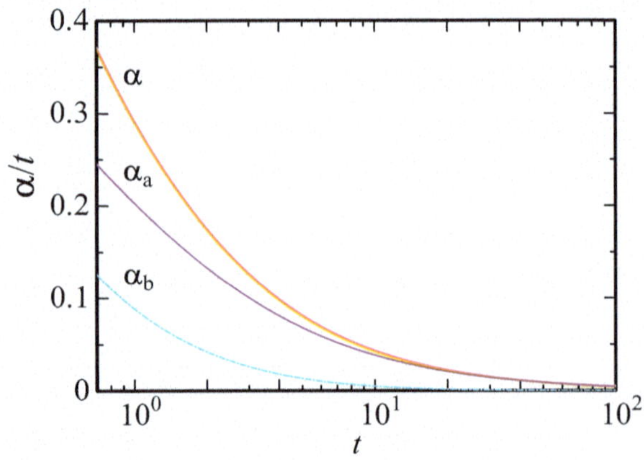

Fig. 6.5 Thermal-expansion coefficient α/t (thick line), $\alpha_{\rm a}/t$ (thin line), and $\alpha_{\rm b}/t$ (dash-dotted line) versus scaled temperature $t = T/T_0$ at the valence QCP

into Eq. (6.42) and Eq. (6.43). As for $\alpha_{\rm b}$, $(\partial y/\partial P)_T$ is obtained by calculating Eq. (5.89) for $d = 3$ and $z = 3$ with setting $(\partial y_0/\partial P)_T = 1$, $(\partial y_1/\partial P)_T = 1$, and $(\partial T_0/\partial P)_T/T_0 = 1$ as representative values. Then, we obtain $\alpha_{\rm a}$ and $\alpha_{\rm b}$, which results in α in Eq. (6.41). The obtained temperature dependence of α, $\alpha_{\rm a}$, and $\alpha_{\rm b}$ are plotted in Fig. 6.5. For $t \gtrsim 1$, i.e., $T \gtrsim T_0$, $\alpha_{\rm a}$ contributes dominantly to α, where the least-square fit of $\alpha_{\rm a}/t$ gives $\alpha_{\rm a}/t \sim t^{-0.66}$. The least-square fit of α/t for $1 \le t \le 10^2$ gives $\alpha/t \sim t^{-0.81}$. For $t \ll 1$, $\alpha_{\rm b}$ contributes dominantly to α so that α/t behaves as $\alpha/t \sim t^{-2/3}$.

6.4.5 Grüneisen Parameter

The Grüneisen parameter is expressed as

$$\Gamma = \frac{\alpha V}{C_V \kappa_T}, \tag{6.44}$$

where κ_T is isothermal compressibility (see Appendix D.2). By substituting α obtained in Sect. 6.4.4 and C_V obtained in Sect. 6.4.3 into Eq. (6.44), the Grüneisen parameter is obtained.

$$\Gamma = \Gamma_{\rm a} + \Gamma_{\rm b} \tag{6.45}$$

where $\Gamma_{\rm i}({\rm i} = {\rm a}, {\rm b})$ is given by

$$\Gamma_{\rm a} = \frac{C_{\rm a}}{C_{\rm a} - C_{\rm b}} \frac{1}{\kappa_T} \frac{1}{T_0} \left(\frac{\partial T_0}{\partial P}\right)_T, \tag{6.46}$$

$$\Gamma_{\rm b} = \frac{\tilde{C}_{\rm b}}{C_{\rm a} - C_{\rm b}} \frac{1}{\kappa_T} \frac{1}{t} \left(\frac{\partial y}{\partial P}\right)_T. \tag{6.47}$$

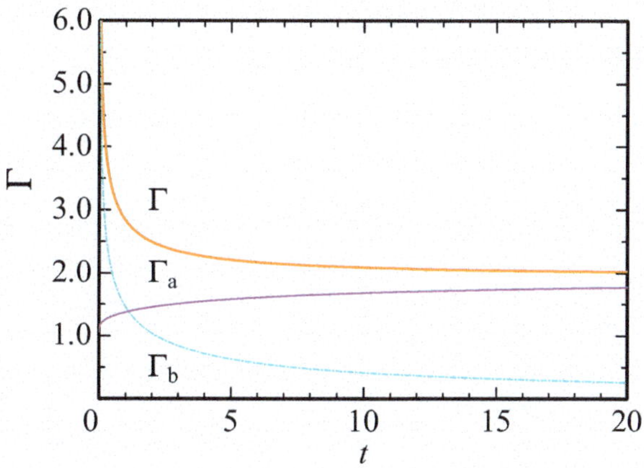

Fig. 6.6 Grüneisen parameter Γ (thick line), Γ_a (thin line), and Γ_b (dash-dotted line) versus scaled temperature $t = T/T_0$ at the valence QCP

At the valence QCP, the charge compressibility is considered to exhibit no singularity such as the divergence. Actually, by the DMRG calculation applied to the extended Anderson lattice model it was confirmed that the charge compressibility

$$\kappa_T = \frac{\partial \bar{n}}{\partial \mu} \tag{6.48}$$

with $\bar{n}$ being the total charge $\bar{n} \equiv \bar{n}_f + \bar{n}_c$ and μ being the Fermi energy does not diverge at the QCP of the first-order valence transition. Hence, here we set $\kappa_T = 1$ as a representative value. The result is plotted in Fig. 6.6. At high temperatures, Γ is dominated by Γ_a. As t decreases, Γ_a shows a slight decrease while Γ_b increases. The least-square fit of Γ_b for $1 \leq t \leq 10^2$ gives $\Gamma_b \sim t^{-0.60}$. As a result, Γ increases as t decreases. The least-square fit of Γ for $1 \leq t \leq 10^2$ gives $\Gamma \sim t^{-0.73}$. For $t \ll 1$, Γ is dominated by Γ_b [see Eq. (6.47)], which behaves as $\Gamma \sim -t^{-2/3}/\ln t$, which diverges for $t \to 0$. It is remarked that Γ_a is reflected in the pressure derivative of the characteristic temperature of the critical valence fluctuations [see Eq. (6.46)].

6.5 T/B Scaling in Temperature and Magnetic Field of Magnetization

As noted in Table 6.1, the heavy fermion metal β-$YbAlB_4$ exhibits non-Fermi-liquid behavior at low temperatures, which do not follow any class of the magnetic criticality in Table 1.1 but follows the quantum valence criticality as discussed in Sect. 6.1. Interestingly, from the magnetization data for $T \lesssim 3$ K and the magnetic field

$B \lesssim 2$ T, in β-YbAlB$_4$ it has been discovered that the magnetic susceptibility χ shows the following T/B scaling behavior over four decades of T/B [20]:

$$\chi^{-1} = (\mu_B B)^{1/2} \varphi \left(\frac{k_B T}{\mu_B B} \right), \tag{6.49}$$

where μ_B and k_B are the Bohr magneton and Boltzmann constant, respectively, and φ is the function $\varphi(x) = \Lambda(\Gamma^2 + x^2)^{1/4}$ with Λ and Γ being constants [20]. Namely, $\chi^{-1}/(\mu_B B)^{1/2}$ is expressed as a single scaling function of the ratio T/B.

This striking behavior of Eq. (6.49) calls for theoretical explanation, and it has so far been proposed that anisotropic hybridization between f and conduction electrons is the key origin of the emergence of T/B scaling [44]. However, this theory requires an assumption that the renormalized f level is pinned at the hybridized band edge, and it also seems unclear whether the unconventional criticality observed in C_e/T and the resistivity can be explained by the anisotropic hybridization.

It has been shown theoretically that a new type of quantum criticality emerges near the quantum critical point (QCP) of the first-order valence transition in Yb- and Ce-based heavy-electron systems [30]. Critical valence fluctuations of Yb or Ce cause the quantum criticality in physical quantities such as χ, C_e/T, resistivity, and the NMR/NQR relaxation rate $(T_1 T)^{-1}$, which give a unified explanation for the measured unconventional criticality in β-YbAlB$_4$ [30, 45]. Hence, it is interesting to examine whether the critical Yb-valence fluctuation can account for the T/B scaling observed in β-YbAlB$_4$.

In Sect. 6.5.1, it is discussed that the T/B scaling can be understood from the viewpoint of the quantum valence criticality. By developing a theoretical framework for the quantum critical phenomena of Yb-valence fluctuations under a magnetic field, it is shown that the T/B scaling emerges near the QCP of the valence transition. It is demonstrated that the emergence of the T/B scaling is a hallmark of the presence of the small characteristic energy scale of the critical Yb-valence fluctuations.

6.5.1 *Mode Coupling Theory of Valence Fluctuations Under Magnetic Field for β-YbAlB$_4$*

The mode coupling theory of valence fluctuations under magnetic field is formulated [46]. A key point is that the parameters in the mode-coupling equation, i.e., the SCR equation are directly derived from the extended Anderson lattice model. This framework has made it possible not only to describe the hierarchical energy scales (bandwidth of conduction electrons $D >$ Kondo temperature $T_K >$ characteristic temperature of critical valence fluctuations T_0) correctly but also to compare the experimental T-B phase diagram of β-YbAlB$_4$ with the theoretical phase diagram quantitatively.

We employ the theoretical framework developed in Sect. 6.2, whose formulation is extended so as to describe the effect of a magnetic field [46]. Hereafter, we take the energy units of $k_{\mathrm{B}} = 1$, $\hbar = 1$, and $\mu_{\mathrm{B}} = 1$ unless otherwise noted. We consider the simplest minimal model

$$H = H_{\mathrm{PAM}} + H_{U_{\mathrm{fc}}} + H_{\mathrm{Zeeman}} \tag{6.50}$$

as the starting Hamiltonian, where

$$\begin{aligned} H_{\mathrm{PAM}} &= \sum_{k\sigma} \varepsilon_k c^{\dagger}_{k\sigma} c_{k\sigma} + \varepsilon_{\mathrm{f}} \sum_{i\sigma} n^{\mathrm{f}}_{i\sigma} + \sum_{k\sigma} \left(V_k f^{\dagger}_{k\sigma} c_{k\sigma} + \mathrm{h.c.} \right) + U \sum_i n^{\mathrm{f}}_{i\uparrow} n^{\mathrm{f}}_{i\downarrow}, \\ H_{U_{\mathrm{fc}}} &= \sum_{i\sigma\sigma'} n^{\mathrm{f}}_{i\sigma} n^{\mathrm{c}}_{i\sigma'}, \\ H_{\mathrm{Zeeman}} &= -h \sum_i S^{\mathrm{f}z}_i \end{aligned} \tag{6.51}$$

with $n^{a}_{i\sigma} \equiv a^{\dagger}_{i\sigma} a_{i\sigma}$ for $a = f$ or c and $S^{\mathrm{f}z}_i \equiv \frac{1}{2}(n^{\mathrm{f}}_{i\uparrow} - n^{\mathrm{f}}_{i\downarrow})$.

Performing the procedure described in Sect. 6.2, we obtain the distribution function $Z = \int \mathcal{D}\varphi \exp(-S[\varphi])$ with

$$\begin{aligned} S[\varphi] = \sum_{\sigma} \Bigg[& \frac{1}{2} \sum_{\bar{q}} \Omega_{2\sigma}(\bar{q}) \varphi_{\sigma}(\bar{q}) \varphi_{\sigma}(-\bar{q}) + \sum_{\bar{q}_1, \bar{q}_2, \bar{q}_3} \Omega_{3\sigma}(\bar{q}_1, \bar{q}_2, \bar{q}_3) \\ & \times \varphi_{\sigma}(\bar{q}_1) \varphi_{\sigma}(\bar{q}_2) \varphi_{\sigma}(\bar{q}_3) \delta\left(\sum_{i=1}^{3} \bar{q}_i \right) + \sum_{\bar{q}_1, \bar{q}_2, \bar{q}_3, \bar{q}_4} \Omega_{4\sigma}(\bar{q}_1, \bar{q}_2, \bar{q}_3, \bar{q}_4) \\ & \times \varphi_{\sigma}(\bar{q}_1) \varphi_{\sigma}(\bar{q}_2) \varphi_{\sigma}(\bar{q}_3) \varphi_{\sigma}(\bar{q}_4) \delta\left(\sum_{i=1}^{4} \bar{q}_i \right) + \cdots \Bigg], \end{aligned} \tag{6.52}$$

where the abbreviation $\bar{q} \equiv (\boldsymbol{q}, i\omega_l)$ with $\omega_l = 2l\pi T$ is used. Since the long wavelength around $\boldsymbol{q} = \boldsymbol{0}$ and the low-frequency regions play dominant roles in the critical phenomena, $\Omega_{i\sigma}$ for $i = 2, 3$, and 4 are expanded for q and ω_l around $(\boldsymbol{0}, 0)$:

$$\Omega_{2\sigma}(q, i\omega_l) \approx \eta_{0\sigma} + A_{\sigma} q^2 + C_{\sigma} \frac{|\omega_l|}{q}, \tag{6.53}$$

where $\eta_{0\sigma}$ is given by

$$\eta_{0\sigma} = U_{\mathrm{fc}} \left[1 - 2U_{\mathrm{fc}} \left\{ \chi^{\mathrm{ffcc}}_{0\sigma}(\boldsymbol{0}, 0) + \chi^{\mathrm{cfcf}}_{0\sigma}(\boldsymbol{0}, 0) \right\} \right]. \tag{6.54}$$

Here, $\chi_{0\sigma}^{\alpha\beta\gamma\delta}(\boldsymbol{q}, i\omega_l)$ is given by

$$\chi_{0\sigma}^{\alpha\beta\gamma\delta}(\boldsymbol{q}, i\omega_l) = -\frac{T}{N_s}\sum_{\boldsymbol{k},n} G_{\boldsymbol{k}+\boldsymbol{q}\sigma}^{\alpha\beta}(i\varepsilon_n + i\omega_l) G_{\boldsymbol{k}\sigma}^{\gamma\delta}(i\varepsilon_n), \tag{6.55}$$

where $G_{\boldsymbol{k}\sigma}^{\rm ff}(i\varepsilon_n)$, $G_{\boldsymbol{k}\sigma}^{\rm cc}(i\varepsilon_n)$, and $G_{\boldsymbol{k}\sigma}^{\rm cf}(i\varepsilon_n)$ are given by

$$G_{\boldsymbol{k}\sigma}^{\rm ff}(i\varepsilon_n) = \frac{1}{i\varepsilon_n - \bar{\varepsilon}_{\rm f\sigma} - \bar{V}_{\boldsymbol{k}}^2/(i\varepsilon_n - \bar{\varepsilon}_{\mathbf{k}\sigma})}, \tag{6.56}$$

$$G_{\boldsymbol{k}\sigma}^{\rm cc}(i\varepsilon_n) = \frac{1}{i\varepsilon_n - \bar{\varepsilon}_{\boldsymbol{k}\sigma} - \bar{V}_{\boldsymbol{k}}^2/(i\varepsilon_n - \bar{\varepsilon}_{\rm f\sigma})}, \tag{6.57}$$

$$G_{\boldsymbol{k}\sigma}^{\rm cf}(i\varepsilon_n) = \frac{\bar{V}_{\boldsymbol{k}}}{(i\varepsilon_n - \bar{\varepsilon}_{\rm f\sigma})(i\varepsilon_n - \bar{\varepsilon}_{\boldsymbol{k}\sigma}) - \bar{V}_{\boldsymbol{k}}^2}, \tag{6.58}$$

respectively, with $\varepsilon_n = (2n+1)\pi T$. Here, $\bar{\varepsilon}_{\boldsymbol{k}\sigma}$, $\bar{\varepsilon}_{\rm f\sigma}$, and $\bar{V}_{\boldsymbol{k}}$ are defined as $\bar{\varepsilon}_{\boldsymbol{k}\sigma} \equiv \varepsilon_{\boldsymbol{k}} + \frac{U_{\rm fc}}{2}$, $\bar{\varepsilon}_{\rm f\sigma} \equiv \varepsilon_{\rm f} + \frac{U_{\rm fc}}{2} + \frac{\lambda}{\sqrt{N_s}} - P(\sigma)\frac{h}{2}$, and $\bar{V}_{\mathbf{k}} \equiv \frac{V_{\boldsymbol{k}} b}{\sqrt{N_s}}$, respectively, with $P(\uparrow) \equiv +1$ and $P(\downarrow) \equiv -1$. Since $\chi_{0\sigma}^{\rm ffcc}(\mathbf{0}, 0) \gg \chi_{0\sigma}^{\rm cfcf}(\mathbf{0}, 0)$, as shown in Sect. 6.2, hereafter we use the approximated form $\eta_{0\sigma} \approx U_{\rm fc}\left[1 - 2U_{\rm fc}\chi_{0\sigma}^{\rm ffcc}(\mathbf{0}, 0)\right]$ for simplicity of calculation. For $\Omega_{3\sigma}$ and $\Omega_{4\sigma}$, expansion up to the zeroth order is performed as $\Omega_{3\sigma}(\bar{q}_1, \bar{q}_2, \bar{q}_3) \approx v_{3\sigma}/\sqrt{\beta N_s}$ and $\Omega_{4\sigma}(\bar{q}_1, \bar{q}_2, \bar{q}_3, \bar{q}_4) \approx v_{4\sigma}/(\beta N_s)$, respectively. The mode-coupling constant $v_{4\sigma}$ is derived as

$$v_{4\sigma} = \frac{U_{\rm fc}^4}{4}\left[\frac{T}{N_s}\sum_n\sum_{\boldsymbol{k}} G_{\boldsymbol{k}\sigma}^{\rm cf}(i\varepsilon_n)^2 G_{\boldsymbol{k}\sigma}^{\rm cc}(i\varepsilon_n) G_{\boldsymbol{k}\sigma}^{\rm ff}(i\varepsilon_n) + \frac{T}{2N_s}\sum_n\sum_{\boldsymbol{k}} G_{\boldsymbol{k}\sigma}^{\rm cc}(i\varepsilon_n)^2 G_{\boldsymbol{k}\sigma}^{\rm ff}(i\varepsilon_n)^2\right], \tag{6.59}$$

where the first and second terms are expressed by a Feynman diagram in Figs. 6.7a and 6.7b, respectively.

Since renormalization-group analysis has shown that higher order terms v_i ($i \geq 3$) are irrelevant for the $d = 3$ spatial dimension as discussed in Sect. 6.3, we construct the action for the Gaussian fixed point. Taking account of the mode-coupling effects up to the fourth order in $S[\varphi]$ in Eq. (6.52), we employ Feynman's inequality for the free energy [47]: $F \leq F_{\rm eff} + T\langle S - S_{\rm eff}\rangle_{\rm eff} \equiv \tilde{F}(\eta)$, where $S_{\rm eff}$ is the effective action for the best Gaussian,

$$S_{\rm eff}[\varphi] = \frac{1}{2}\sum_\sigma\sum_{\boldsymbol{q},l} \chi_{v\sigma}(\boldsymbol{q}, i\omega_l)^{-1}|\varphi_\sigma(\boldsymbol{q}, i\omega_l)|^2 \tag{6.60}$$

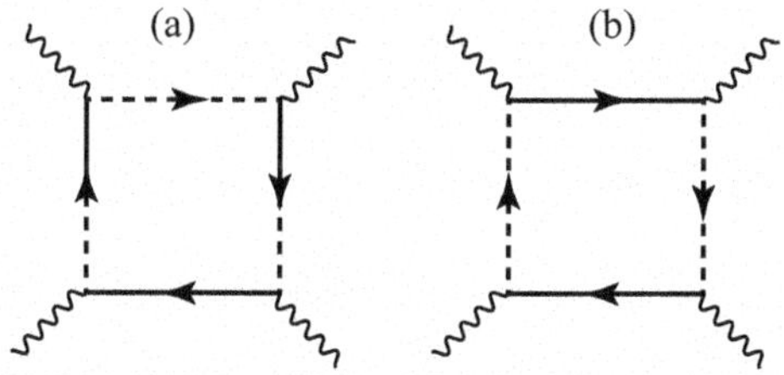

Fig. 6.7 Feynman diagrams for the **a** first term and **b** second term in the mode-coupling constant $v_{4\sigma}$ given by Eq. (6.59). The solid and dashed lines with an arrow represent f and conduction-electron Green functions $G^{\rm ff}_{k\sigma}(i\varepsilon_n)$ and $G^{\rm cc}_{k\sigma}(i\varepsilon_n)$, respectively. The half-dashed and solid line with an arrow represents the off-diagonal Green function $G^{\rm cf}_{k\sigma}(i\varepsilon_n)$. The wiggly line represents critical valence fluctuations [46]. Reprinted from Ref. [46]. Copyright 2014 The Physical Society of Japan

Here, $\chi_{{\rm v}\sigma}(\boldsymbol{q}, i\omega_l)$ is the valence susceptibility defined as

$$\chi_{{\rm v}\sigma}(\boldsymbol{q}, i\omega_l)^{-1} \approx \eta + A_\sigma q^2 + C_\sigma \frac{|\omega_l|}{q}, \tag{6.61}$$

where the notation follows in Sect. 6.2. Under the optimal condition $\frac{d\tilde{F}(\eta)}{d\eta} = 0$, the self-consistent renormalization (SCR) equation under a magnetic field in the $A_\sigma q_{\rm B}^2 \lesssim \eta$ regime is obtained:

$$\begin{aligned} &\sum_\sigma A_\sigma q_{{\rm B}\sigma}^4 \frac{T_{0\sigma}}{T_{{\rm A}\sigma}^2}\left(1 + \frac{v_{4\sigma} q_{{\rm B}\sigma}^3}{\pi^2}\frac{T_{0\sigma}}{T_{{\rm A}\sigma}^2}\right) \\ &\times \left[y_{0\sigma} - \tilde{y}_\sigma + \frac{3}{2} y_{1\sigma} t_\sigma \left\{\frac{x_{\rm c}^3}{6\tilde{y}_\sigma} - \frac{1}{2\tilde{y}_\sigma}\int_0^{x_{\rm c}} \frac{x^3}{x + \frac{t_\sigma}{6\tilde{y}_\sigma}} dx\right\}\right] \\ &\times \left[C_{2\sigma} + \frac{x_{\rm c}^3}{3}\frac{t_\sigma}{\tilde{y}_\sigma^2}\int_0^{x_{\rm c}} \frac{x^4}{\left(x + \frac{t_\sigma}{6\tilde{y}_\sigma}\right)^2} dx\right] = 0, \end{aligned} \tag{6.62}$$

where $\tilde{y}_\sigma = y\frac{A}{A_\sigma}\left(\frac{q_{\rm B}}{q_{{\rm B}\sigma}}\right)^2$, $t_\sigma = \frac{T}{T_{0\sigma}}$,

$$T_{0\sigma} = \frac{A_\sigma q_{{\rm B}\sigma}^3}{2\pi C_\sigma} \tag{6.63}$$

and $T_{{\rm A}\sigma} = \frac{A q_{{\rm B}\sigma}^2}{2}$ with $q_{{\rm B}\sigma}$ being the Brillouin zone for "spin" σ. Note that A, C, and $q_{\rm B}$ are the zero-field values of A_σ, C_σ, and $q_{{\rm B}\sigma}$, respectively. Here, y is defined as $y \equiv \frac{\eta}{A q_{\rm B}^2}$, and the dimensionless integral variable and its cutoff are defined as $x \equiv q/q_{\rm B}$ and $x_{\rm c} \equiv q_{\rm c}/q_{\rm B}$, respectively. The parameters $y_{0\sigma}$ and $y_{1\sigma}$ are given by

$$y_{0\sigma} = \frac{\frac{\eta_{0\sigma}}{A_\sigma q_{B\sigma}^2} + v_{4\sigma} \frac{T_{0\sigma}}{T_{A\sigma}^2} \frac{q_{B\sigma}^3}{\pi^2} C_{1\sigma}}{1 + v_{4\sigma} \frac{T_{0\sigma}}{T_{A\sigma}^2} \frac{q_{B\sigma}^3}{\pi^2} C_{2\sigma}}, \tag{6.64}$$

$$y_{1\sigma} = \frac{v_{4\sigma} \frac{T_{0\sigma}}{T_{A\sigma}^2} \frac{4 q_{B\sigma}^3}{3\pi^2}}{1 + v_{4\sigma} \frac{T_{0\sigma}}{T_{A\sigma}^2} \frac{q_{B\sigma}^3}{\pi^2} C_{2\sigma}}, \tag{6.65}$$

respectively, where $C_{1\sigma}$ and $C_{2\sigma}$ are constants given by

$$C_{1\sigma} = \int_0^{x_c} dx x^3 \ln \left| \frac{(A_\sigma q_{B\sigma}^2 x^3)^2 + (C_\sigma \omega_c / q_{B\sigma})^2}{(A_\sigma q_{B\sigma}^2 x^3)^2} \right| \tag{6.66}$$

and

$$C_{2\sigma} = 2(C_\sigma \omega_c)^2 \int_0^{x_c} dx \frac{x}{(A_\sigma q_{B\sigma}^3 x^3)^2 + (C_\sigma \omega_c)^2}, \tag{6.67}$$

respectively.

Note that in the zero-field case, $h = 0$, Eq. (6.62) reproduces the SCR equation for critical valence fluctuations Eq. (6.31) in Sect. 6.2. It is noted that at the QCP of the valence transition, the magnetic susceptibility diverges, whose singularity is the same as the valence susceptibility $\chi \propto \chi_v(\mathbf{0}, 0) \propto y^{-1}$ since the main contribution to χ and χ_v comes from the common many-body effects caused by U_{fc}, which can be expressed by the common Feynman diagrams near the QCP as discussed in Sect. 6.2 [30].

Here, we demonstrate that the T/B scaling behavior appears when the characteristic temperature of critical valence fluctuations T_0 is smaller than (or at least comparable to) the measured lowest temperature. Hence, we here set the coefficient A_σ in Eq. (6.61) as a small input parameter to discuss the effect of a small T_0 on physical quantities. The procedure of our calculation is summarized as follows.

First, we solve the saddle-point solution for $\exp(-S_0)$ at $T = 0$ for given parameters of ε_f, $V_{\boldsymbol{k}}$, $U = \infty$, and h at the filling

$$n \equiv \frac{1}{2N_s} \sum_{i\sigma} \langle n_{i\sigma}^f + n_{i\sigma}^c \rangle \tag{6.68}$$

by using the slave-boson mean-field theory.

Second, we calculate $\chi_{0\sigma}^{ffcc}(\mathbf{0}, 0)$ and the $[\ldots]$ part in Eq. (6.59) by using the saddle-point solution. Then we obtain $\eta_{0\sigma}$ and $v_{4\sigma}$ for a given U_{fc}.

Third, by using $y_{0\sigma}$ and $y_{1\sigma}$ obtained from Eqs. (6.64) and (6.65), respectively, we solve the valence SCR equation [Eq. (6.62)] and finally obtain $y(t)$.

We note that the crystalline electronic field (CEF) ground state of β-$YbAlB_4$ has been suggested to be the Kramers doublet, which is well separated from the excited CEF levels [19, 48]. Since the analysis of the CEF-level scheme, which well reproduces the anisotropy of the magnetic susceptibility, deduces that a hybridization node exists along the c-axis in β-YbAlB$_4$ [48–50], we employ the anisotropic hybridization in the form of

$$V_{k} = V\left(1 - \hat{k}_z^2\right) \tag{6.69}$$

with $\hat{\boldsymbol{k}} \equiv \boldsymbol{k}/|\boldsymbol{k}|$ to simulate β-YbAlB$_4$ most simply.

For evaluation of the saddle-point solution, we employ the typical parameter set for heavy-electron systems: $D = 1$, $V = 0.65$, and $U = \infty$ at the filling $n = 0.8$. Here, D is the half bandwidth of conduction electrons given by $\varepsilon_{k} = \boldsymbol{k}^2/(2m_0) - D$, which is taken as the energy unit. The mass m_0 is set such that the integration from $-D$ to D of the density of states of conduction electrons per "spin" and site is equal to 1.

Following the argument in Sect. 6.2, we discuss the general property at the QCP of the valence transition by defining it as the point with the solution of Eq. (6.62) y being zero at $T = 0$, which is identified to be $(\varepsilon_{\rm f}, U_{\rm fc}) = (-0.7, 0.700328652)$ for $A = 5 \times 10^{-6}$ at $h = 0$. This $U_{\rm fc}$ is larger than $U_{\rm fc}^{\rm RPA} \equiv 1/\chi_0^{\rm ffcc}(\boldsymbol{0}, 0) = 0.62404$ for $\varepsilon_{\rm f} = -0.7$, which reflects the mode-coupling effect of critical valence fluctuations. Namely, a positive $v_{4\sigma}$ overcomes a negative $\eta_{0\sigma}$ for $U_{\rm fc} > U_{\rm fc}^{\rm RPA}$ [see Eq. (6.64)], giving rise to $0.700328652 > U_{\rm fc}^{\rm RPA}$.

It is noted that here we set a rather large c-f hybridization strength V to simulate β-YbAlB$_4$ with a large fundamental characteristic energy scale ≈ 200 K [19]. Actually, the characteristic energy for heavy electrons, which is defined as the Kondo temperature $T_{\rm K} \equiv \bar{\varepsilon}_{\rm f} - \mu$ within the saddle-point solution for $\exp(-S_0)$, is estimated to be $T_{\rm K} = 0.02437$.

To examine the magnetic-field dependence of $y(t)$ at the QCP, we solve the valence SCR equation [Eq. (6.62)] for $U_{\rm fc} = 0.700328652$. To make a comparison with experiments where a magnetic field from the order of $B = 10^{-4}$ T to $B = 2$ T is applied, we apply a magnetic field ranging from $h = 10^{-8}$ to $h = 10^{-4}$[1]. Here, we note that the energy unit of our theory is the conduction bandwidth $D = 1$, which is of the order of 10^4 T ($\approx 10^4$ K). To compare with experiments measured in the temperature range from the order of $T = 10^{-2}$ K to $T = 3$ K, we solve the valence SCR equation [Eq. (6.62)] for $6 \times 10^{-6} \leq T \leq 3 \times 10^{-4}$. As noted above, A is set as $A = 5 \times 10^{-6}$, which gives $T_0 = 3 \times 10^{-6}$, slightly smaller than the lowest temperature but of the same order. Owing to the smallness of A_σ, hereafter we neglect its field dependence and set $A = A_\sigma$ for $h \neq 0$.

---

[1] When a magnetic field is applied, the Fermi surface of the lower hybridized band with the majority spin expands, while the Fermi surface with a minority spin shrinks in the periodic Anderson model. It is known that at a magnetic field $h = h_0$ comparable to $T_{\rm K}$, the Fermi surface with the majority spin reaches the Brillouin zone, i.e., a Lifshitz transition occurs [51, 52].

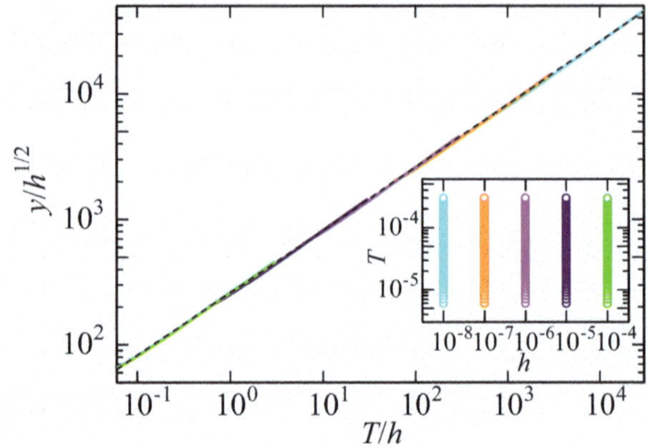

Fig. 6.8 Scaling of the data for $T \leq 3 \times 10^{-4}$ and $h \leq 10^{-4}$. The inset shows the T-h range where the scaling applies. The dashed line represents the fitting function $a(T/h)^{1/2}$. The data were obtained by solving the valence SCR equation [Eq. (6.62)] for $D = 1$, $V = 0.65$, $\varepsilon_{\mathrm{f}} = -0.7$, $U = \infty$, and $U_{\mathrm{fc}} = 0.700328652$ at $n = 0.8$ [46]. Reprinted from Ref. [46]. Copyright 2014 The Physical Society of Japan

The results are shown in Fig. 6.8. Intriguingly, we find that all the data over four decades of the magnetic field fall down to a single scaling function of the ratio T/h:

$$y = h^{1/2}\varphi\left(\frac{T}{h}\right). \tag{6.70}$$

The least-squares fit of the scaling function $\varphi(x) = ax^{1/2}$ to the data for $10^1 \leq T/h \leq 10^4$ shows that the data are well fitted by the dashed line in Fig. 6.8. Namely, $y/h^{1/2} \approx a(T/h)^{1/2}$, i.e., $y \approx T^{1/2}$. This implies that the quantum criticality of Yb-valence fluctuations is dominant, giving rise to the non-Fermi liquid regime [30]. This behavior coincides with the measured scaling function Eq. (6.49) for $x = T/h \gg \Gamma$. It is noted that as x decreases the data tend to deviate from $ax^{1/2}$, i.e., there is a tendency of upward deviation from the dashed line toward $x = T/h \ll 1$ in the smaller T/h region than that shown in Fig. 6.8. This reflects the suppression of the valence susceptibility by applying the magnetic field. Namely, as $x = T/h$ decreases, the crossover from the non-Fermi-liquid regime with the quantum valence criticality to the Fermi liquid regime with suppressed valence fluctuation occurs. As noted above, the uniform magnetic susceptibility χ has the same temperature dependence as the valence susceptibility $\chi \propto \chi_{\mathrm{v}}(\mathbf{0}, 0) \propto y^{-1}$ [30]. This indicates that general tendency of the T/B scaling observed in the magnetization data of β-$YbAlB_4$ can be reproduced by the solutions of the valence SCR equation [Eq. (6.62)] under a magnetic field.

To analyze how the T/h scaling behavior appears in the present theoretical framework, let us rewrite the valence SCR equation [Eq. (6.62)] with the scaled form of $y/t^{1/2}$ and t/h:

$$
\begin{aligned}
&\sum_{\sigma} A_{\sigma} q_{\mathrm{B}\sigma}^{4} \frac{T_{0\sigma}}{T_{\mathrm{A}\sigma}^{2}} \left(1 + \frac{v_{4\sigma} q_{\mathrm{B}\sigma}^{3}}{\pi^{2}} \frac{T_{0\sigma}}{T_{\mathrm{A}\sigma}^{2}}\right) \left[\frac{y_{0\sigma}}{h^{1/2}} - \left(\frac{\tilde{y}_{\sigma}}{h^{1/2}}\right) \right. \\
&\left. + \frac{3}{2} y_{1\sigma} \left(\frac{t_{\sigma}}{h}\right) \left\{ \frac{x_{\mathrm{c}}^{3}}{6\left(\frac{\tilde{y}_{\sigma}}{h^{1/2}}\right)} - \frac{1}{2\left(\frac{\tilde{y}_{\sigma}}{h^{1/2}}\right)} \int_{0}^{x_{\mathrm{c}}} \frac{x^{3}}{x + h^{1/2} \frac{\left(\frac{t_{\sigma}}{h}\right)}{6\left(\frac{\tilde{y}_{\sigma}}{h^{1/2}}\right)}} dx \right\} \right] \\
&\times \left[C_{2\sigma} + \frac{x_{\mathrm{c}}^{3}}{3} \frac{\left(\frac{t_{\sigma}}{h}\right)}{\left(\frac{\tilde{y}_{\sigma}}{h^{1/2}}\right)^{2}} \int_{0}^{x_{\mathrm{c}}} \frac{x^{4}}{\left\{ x + h^{1/2} \frac{\left(\frac{t_{\sigma}}{h}\right)}{6\left(\frac{\tilde{y}_{\sigma}}{h^{1/2}}\right)} \right\}^{2}} dx \right] = 0.
\end{aligned}
\tag{6.71}
$$

We see that most terms can be expressed in the form of $y/h^{1/2}$ and t/h, except for the term with the x-integration in each square bracket [ ]. Namely, extra $h^{1/2}$ factors appear in the denominators of the integrands. This implies that *the T/h scaling does not hold exactly*. From Eq. (6.71), however, it turns out that in the case of $t_{\sigma}/\tilde{y}_{\sigma} \gg 1$, the denominators of the integrands become large, which make the x-integrations negligibly small. We confirmed that this is the case when T_0 is below (or at least comparable to) the measured lowest temperature. In the present calculation, we set $T_0 = 3.0 \times 10^{-6}$ and the lowest temperature for the data in Fig. 6.8 is $T = 6.0 \times 10^{-6}$, i.e., T_0 is a few times smaller than the lowest temperature. Note that T_0 is in the same order as the lowest temperature.

From these results, the T/B scaling observed in β-YbAlB$_4$ suggests that a small characteristic temperature of critical valence fluctuations T_0 exists. Since the measured lowest temperature is on the order of 10^{-2} K in β-YbAlB$_4$, T_0 is considered to be of the same order or smaller.

As shown in Fig. 6.2, because of the strong local-correlation effect by $U \gg D$, an almost dispersionless critical valence-fluctuation mode appears, giving rise to the extremely small q^2-coefficient A in the momentum space. This almost flat mode is reflected in the emergence of the extremely small characteristic temperature T_0. Owing to the extremely small T_0, the temperature at the low-T measurement can be regarded as a "high" temperature in the scaled temperature $t = T/T_0 \gtrsim 1$, where unconventional quantum criticality emerges in physical quantities such as χ, $(T_1 T)^{-1}$, C_{e}/T, and resistivity, which well account for the behavior of β-YbAlB$_4$ [30]. Our results show that observation of the T/B scaling indicates the presence of the small characteristic temperature T_0. In other words, quantum valence criticality gives a unified explanation for the unconventional criticality in physical quantities as well as the T/B scaling in β-YbAlB$_4$.

To verify the existence of such a small T_0 experimentally, the measurement of the dynamical valence susceptibility $\chi_{\mathrm{v}}(\boldsymbol{q}, \omega)$ is desirable as a direct observation. Mössbauer measurement has detected very slow valence fluctuation whose time scale is about $\tau \approx 2$ ns [41]. This corresponds to about 24 mK, which is indeed located around the measured lowest temperature ~ 30 mK. As shown in Fig. 6.3, in the vicinity of T_0, the resistivity shows rounding as $\rho \sim T^{3/2}$ and $\rho \sim T^{5/3}$. Hence,

$T_0 = 24$ mK is consistent with the fact that the resistivity shows the rounding around the lowest temperature $T \sim 30$ mK $\rho \sim T^{1.5}$ observed in β-YbAlB$_4$ and $\rho \sim T^{1.6}$ in α-YbAl$_{0.896}$Fe$_{0.014}$B$_4$ and the T-linear resistivity at higher temperatures are also explained (see Table 6.1).

Although Eq. (6.49) shows that $\chi \approx (\mu_B B)^{-1/2}$ for $x = \frac{k_B T}{\mu_B B} \ll \Gamma$, it should be noted that a very narrow range of experimental data is used to derive this limiting behavior: A large magnetic field of $B = 2$ T and intermediate temperatures of $0.2\,\mathrm{K} \le T \le 0.5$ K (but *not* the lowest temperature) are used [20]. Namely, the scaling form in the $x \ll \Gamma$ regime is an outcome of the transient behavior of the magnetization, where χ is greatly suppressed to be almost constant around $B = 1 - 2$ T [20]. Furthermore, we should be careful about the fact that the whole scaling range of $10^{-1} \le T/B \le 10^3$ is *not* covered by a series of experimental data as a function of T for a fixed B. From these circumstances, it seems to be appropriate to consider that $\chi^{-1}/(\mu_B B)^{1/2} \approx (\frac{k_B T}{\mu_B B})^{1/2}$ for the $x = \frac{k_B T}{\mu_B B} \gg \Gamma$ regime in Eq. (6.49), derived from the experimental data for the wide T and B range is a substantial scaling function.

6.5.2 *Direct Observation of Quantum Valence Criticality in* α*-YbAl*$_{1-x}$*Fe*$_x$*B*$_4$ **($x = 0.014$)**

In α-YbAlB$_4$ which is the sister compound of β-YbAlB$_4$, the Fermi-liquid behavior is observed [23]. However, by replacing Fe with Al 1.4 %, the common quantum criticality to that observed in β-YbAlB$_4$ appers in the physical quantities such as the magnetic susceptibility, the specific heat, and the resistivity (see Table 6.1). Interestingly, the doping dependence of Fe on the Yb valence in α-YbAl$_{1-x}$Fe$_x$B$_4$ is observed by the hard X-ray photoemission spectroscopy measurement. This has revealed that a sharp change in the Yb valence occurs at $x = x_c \equiv 0.014$ as shown in Fig. 6.9. By the X-ray diffraction measurement, the doping dependence of Fe on

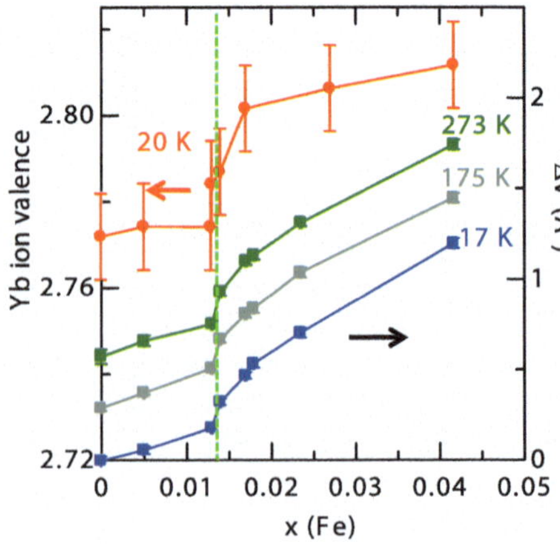

Fig. 6.9 The x dependence of the Yb valence in α-YbAl$_{1-x}$Fe$_x$B$_4$ at $T = 20$ K (left axis). The x dependence of the volume change in the unit cell ($\Delta V < 0$) at $T = 273$ K, 175 K, and 17 K (right axis). At $x = 0.014$ (dashed line), the quantum critical phenomena and the T/B scaling common to those observed in β-YbAlB$_4$ are observed [23]. Reprinted with permission from Ref. [23]. Copyright 2018 AAAS

the lattice constant is observed, which has revealed that the sharp change in the unit-cell volume occurs at $x = x_c$, corresponding to the sharp change in the Yb valence, as shown in Fig. 6.9. Moreover, at $x = x_c$, the magnetic susceptibility χ at the low temperature and magnetic field exhibits the T/B scaling shown in Eq. (6.49) whose behavior is common to β-$YbAlB_4$. These results indicate that the quantum critical point of the valence transition or the valence crossover in the vicinity, corresponding to Fig. 1.4b or Fig. 1.4a respectively with the horizontal axis g being the Fe doping x, is directly detected experimentally. These results also suggest strongly that the origin of the quantum criticality as well as the T/B scaling observed in β-$YbAlB_4$ is the Yb-valence fluctuations.

6.5.3 *On the Yb Valence in $Yb(Rh_{1-x}Co_x)_2Si_2$*

It is noted that the valence of Yb in $Yb(Rh_{1-x}Co_x)_2Si_2$ was observed for a few sets of Co doping $x = 0.00, 0.07, 0.12, 0.38, 0.68$, and 1.00 [53]. In Ref. [53], the sharp change in the Yb valence was not remarkably detected for the discrete sets of x and the authors asserted that the possibility of the valence fluctuations is excluded as the origin of the unconventional quantum criticality observed in $YbRh_2Si_2$. However, it should be pointed out that the doping interval in Ref. [53] is too sparse. As shown in Fig. 6.9, a sharp change in the Yb valence $\Delta n_f \approx 0.005$ is detected in α-$YbAl_{1-x}Fe_xB_4$ between $x = 0.013$ and $x = 0.014$. Namely, the interval of the doping x is $\Delta x = 0.001$. On the other hand, the doping interval around $x = 0.00$ in $Yb(Rh_{1-x}Co_x)_2Si_2$ used in Ref. [53] was $\Delta x = 0.07$, which is too large. As clearly seen in Fig. 6.9, if the doping interval was sparse as $\Delta x = 0.07$ in Ref. [53], the sharp change in the Yb valence would not be detected in α-$YbAl_{1-x}Fe_xB_4$. Therefore, in order to conclude whether the sharp Yb-valence change occurs in $Yb(Rh_{1-x}Co_x)_2Si_2$, the Yb valence for the closer doping rate x with $\Delta x = 0.001$ as performed in α-$YbAl_{1-x}Fe_xB_4$ should be examined, which is left for the future study.

6.6 Case Study of Critical Valence Transition: Quasicrystal Heavy Fermion $Yb_{15}Al_{34}Au_{51}$

In periodic crystals, a new class of unconventional quantum criticality has been observed as listed in Table 6.1, which is explained by the theory of critical valence fluctuations as discussed in Sect. 6.2. Interestingly, the common quantum critical phenomena have been discovered in the Yb-based quasicrystal[2] $Yb_{15}Al_{34}Au_{51}$ at ambient pressure $P = 0$ and zero magnetic field $B = 0$ [26, 27] (see Table 6.1). Surprisingly,

[2] Quasicrystal has no periodicity but has a unique rotational symmetry forbidden in periodic crystals.

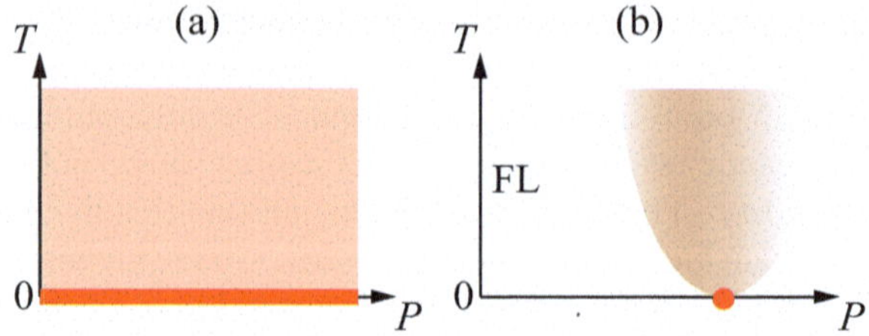

Fig. 6.10 **a** Temperature-pressure phase diagram of the quasicrystal $Yb_{15}Al_{34}Au_{51}$. **b** Temperature-pressure phase diagram of the 1/1 approximant $Yb_{14}Al_{35}Au_{51}$. In **a** and **b**, red regions indicate the quantum-critical regime. In **a**, the red line at $T = 0$ indicates the quantum critical line (see text in Sect. 6.6.1). In **b**, the filled circle at $T = 0$ indicates the QCP and FL denotes the Fermi liquid regime [37]

the quantum criticality persists even under pressure at least up to $P = 1.6$ GPa [26], as shown in Fig. 6.10a. The common quantum criticality has also been observed in the approximant crystal[3] $Yb_{14}Al_{35}Au_{51}$ when pressure is tuned to $P \approx 2$ GPa as shown in Fig. 6.10b [54]. Interestingly, a new type of scaling called "T/B scaling" has been discovered in the quasicrystal [55], where the magnetic susceptibility is expressed as a single scaling function of the ratio of the temperature T to the magnetic field B [54]. This behavior is essentially the same as that expressed in Eq. (6.49) observed in the periodic crystal β-$YbAlB_4$ [20] as discussed in Sect. 6.5. Furthermore, the T/B scaling behavior has also been observed even in the approximant crystal when pressure is tuned to $P \approx 2$ GPa [54]. In this section, it is discussed that these striking phenomena are explained by the theory of critical valence fluctuations in a unified way.

Before going into detail, first the relation between the quasicrystal and the approximant crystal is explained. The quasicrystal $Yb_{15}Al_{34}Au_{51}$ and the approximant crystal consist of the Tsai-type cluster shown in Figs. 6.11a-6.11e [56]. The approximant crystal retains the periodicity as well as the local atomic configuration common to the QC. There exists a series of approximant crystals such as 1/1 approximant crystal, 2/1 approximant crystal, 3/2 approximant crystal, $\cdots$, where the $n \to \infty$ limit in the F_{n-1}/F_{n-2} approximant crystal corresponds to the quasicrystal with F_n being the Fibonacci number [28, 57, 58]. In the rare-earth-based 1/1 approximant crystal composed of the Tsai-type cluster, there exist two icosahedrons in the (expanded) unit cell of the body-center-cubic (bcc) lattice, where the rare-earth atoms are located at the 12 vertices of each icosahedron.

[3] The periodic crystal with the local configuration of atoms common to that in the quasicrystal is referred to as the approximant crystal.

6.6.1 Robust Quantum Criticality Under Pressure

In this subsection, the mechanism of the robustness of the quantum criticality against pressure is discussed from the viewpoint of the critical Yb-valence fluctuations [59]. Let us start with analyzing the lattice structure of the quasicrystal $Yb_{15}Al_{34}Au_{51}$ and approximant crystal. Fig. 6.11 shows concentric shell structures of Tsai-type cluster in the Yb-Al-Au approximant, which is the basic structure of the quasicrystal $Yb_{15}Al_{34}Au_{51}$ [26]. The 1/1 approximant has periodic arrangement of the bcc structure whose unit cell contains the shell structures shown in Figs. 6.11a-e. In the second shell, 12 sites are the Al/Au mixed sites (the sites framed in red in Fig. 6.11b) where the Al or Au atom exists with the rate of 62 %/38 %, respectively [56]. In the fourth shell, six sites are the Al/Au mixed sites (the sites framed in red in Fig. 6.11d) where the Al or Au atom exists with the rate of 59 %/41 %, respectively [56]. These rates are average values of the whole crystal. Hence the location of the Al and Au sites and the existence ratio can be different at the next-to-next concentric shells each other both in the quasicrystal and approximant. Thus the second and fourth shells illustrated in Figs. 6.11b and 6.11d, respectively, are such examples.

As discussed in Sect. 6.2, almost dispersionless critical valence fluctuation mode is the key origin of emergence of the new type of quantum criticality shown in Table 6.1. This implies that locality of valence fluctuation is essentially important. Namely, charge transfer between the Yb site and surrounding atoms is considered to play a key role, which is basically local. Hence, we concentrate on the Tsai-type cluster shown in Fig. 6.11.

We consider the extended Anderson lattice model on the Tsai-type cluster which consists of the 4f electrons at Yb and conduction electrons. As conduction electrons, the main contribution is considered to come from the 3p electrons at Al and the 5d electrons at Yb. Here, as a first step of analysis, we consider the following two orbital model

$$\mathcal{H} = H + H_{U_{\rm fc}}, \tag{6.72}$$

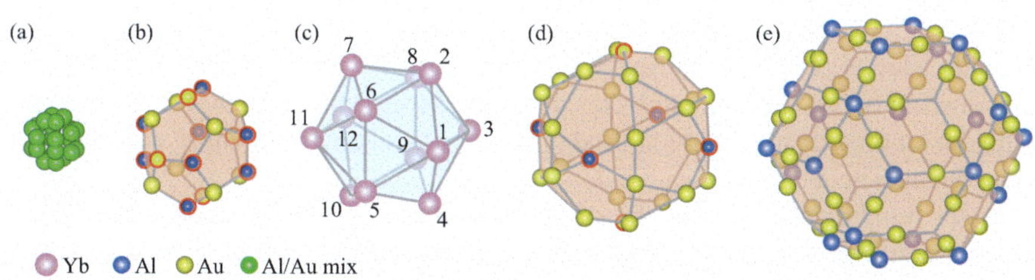

Fig. 6.11 Concentric shell structures of Tsai-type cluster in the Yb-Al-Au quasicrystal: **a** first shell, **b** second shell, **c** third shell, **d** fourth shell, and **e** fifth shell. The number in **c** indicates the i-th Yb site [59]

where H is the Anderson lattice model

$$H = -\sum_{\langle \xi,\nu \rangle \sigma} \left(t_{\xi,\nu} c^{\dagger}_{\xi\sigma} c_{\nu\sigma} + \text{h.c.} \right) + \varepsilon_{\mathrm{f}} \sum_{i=1\sigma}^{12} n^{\mathrm{f}}_{i\sigma} + \sum_{\langle i,\xi \rangle \sigma} \left(V_{i,\xi} f^{\dagger}_{i\sigma} c_{\xi\sigma} + \text{h.c.} \right)$$
$$+ U \sum_{i=1}^{12} n^{\mathrm{f}}_{i\uparrow} n^{\mathrm{f}}_{i\downarrow}. \quad (6.73)$$

The first term represents the transfer of conduction electrons and in the second term, ε_{f} is the 4f energy level. The third term represents the hybridization between the 4f and conduction electrons and the fourth term describes the onsite Coulomb repulsion between the 4f electrons.

In Eq. (6.72), $H_{U_{\mathrm{fc}}}$ denotes the Coulomb repulsion U_{fc} between the 4f and conduction electrons

$$H_{U_{\mathrm{fc}}} = U_{\mathrm{fc}} \sum_{\langle i,\xi \rangle} \sum_{\sigma,\sigma'} n^{\mathrm{f}}_{i\sigma} n^{\mathrm{c}}_{\xi\sigma'}. \quad (6.74)$$

Since Yb^{3+} ion has $4f^{13}$ configuration and Yb^{2+} ion has $4f^{14}$ configuration which is the closed shell, we consider $\mathcal{H}$ in the hole picture. In Eq. (6.72), the conduction orbital is regarded as a Wannier orbital composed of Al 3p and Yb 5d orbitals.

By using the slave-boson mean-field theory for $U = \infty$, we determined the ground-state phase diagram of the Tsai-type cluster for a typical parameter set of heavy electrons, i.e., $t_{\xi,\nu} = 1$, $V_{i,\xi} = 0.3$, and $U = \infty$ at the total electron number $(N_{\uparrow}, N_{\downarrow}) = (24, 24)$ [59]. Here we show the result for the configuration of atoms illustrated in Figs. 6.11a-6.11e as a representative case. The result of the contour plot of the valence susceptibility for each f site indicated as the i-th site in Fig. 6.11c,

$$\chi_{\mathrm{v}i} = -\frac{\partial \langle n^{\mathrm{f}}_i \rangle}{\partial \varepsilon_{\mathrm{f}}} \quad (6.75)$$

is shown in Figs. 6.12a-6.12f, representing the critical valence fluctuations in the phase diagram of the f-level ε_{f} and interorbital Coulomb repulsion U_{fc}. We see that the valence quantum critical points (QCPs) appear as spots, which are located inside the white islands. This is because the strength of the f-c hybridization at each f site differs owing to the Al/Au mixed sites. Namely, as shown in Fig. 6.11, the presence of the Al/Au mixed sites framed in red makes the local environment of each Yb site inequivalent. The Yb site surrounded by the larger number of the close Al sites earns the larger number of paths for the 4f-3p hybridization. This promotes the charge transfers between the Yb 4f hole and the surrounded Al 3p holes, which makes the location of the QCP at the larger U_{fc} position in the ε_{f}-U_{fc} phase diagram. For instance, let us focus on the f-c hybridization between the second and third shells in the Tsai-type cluster (see Figs. 6.11b and 6.11c). The 4f orbital at the i = 1st Yb site hybridizes the conduction orbitals at the three Al sites on the second shell (see

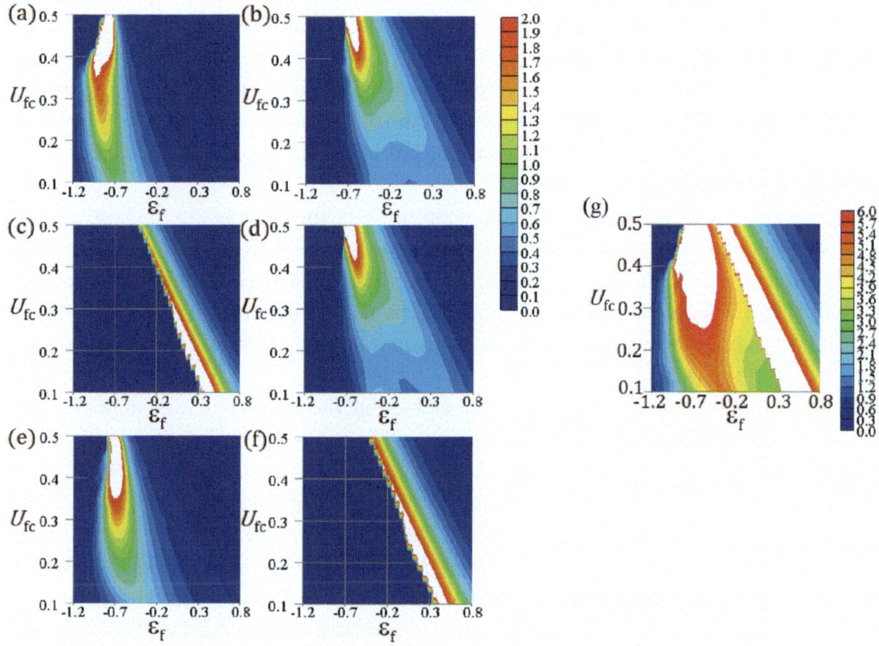

Fig. 6.12 Contour plot of valence susceptibility χ_{vi} for **a** $i = 1$, **b** $i = 2$, **c** $i = 3$, **d** $i = 4$, **e** $i = 5$, and **f** $i = 6$ calculated for the Tsai-type cluster [59]. In each white region, the critical valence fluctuations diverge at the QCP of valence transition. **g** Total valence susceptibility $\chi_v = \sum_{i=1}^{12} \chi_{vi}$ [59]. Reprinted from Ref. [59]. Copyright 2013 The Physical Society of Japan

Fig. 6.11b). On the other hand, the 4f orbital at the $i = 6$th Yb site only hybridizes with the conduction orbital at one Al site on the second shell. Other 4f orbitals at the ith Yb site for $i = 2, 3, 4,$ and 5 have the hybridization paths on the second shell in between. These are reflected in the locations of the white island in Figs. 6.12a-6.12f. In Fig. 6.12a, the white region is located at the upper (larger U_{fc})-left position while in Fig. 6.12f, the white region is located at lower (smaller U_{fc})-right position.

The contour plot of the total valence susceptibility

$$\chi_v = \sum_{i=1}^{12} \chi_{vi} \tag{6.76}$$

is shown in Fig. 6.12g. An important result is that the QCPs appear as spots in the phase diagram, whose critical regions are overlapped to be unified, giving rise to a wide quantum critical region.

Namely, there appear many spots of the QCPs but not a single quantum critical point (QCP) well known in periodic crystals. In the 1/1 approximant crystal, there exist 24 Yb sites in the unit cell. As the approximation degree increases as in the 2/1 approximant crystal, 3/2 approximant crystal, $\cdots$, the size of the unit cell increases and the number of the Yb sites increases. The quasicrystal corresponds to the infinite limit of the unit-cell size in the F_{n-1}/F_{n-2} approximant crystal for $n \to \infty$. Hence, many QCPs are expected to be condensed in the ground-state phase diagram of the quasicrystal giving rise to the wide quantum critical regime. This implies that the quantum critical regime in the quasicrystal is much larger than that in the 1/1 approximant crystal.

When pressure is applied to Yb-based compounds, the f level ε_{f} in the hole picture decreases and U_{fc} increases generally. Hence, when applying pressure follows the line located in the enhanced critical-valence-fluctuation region, a robust quantum criticality appears under pressure. The emergence of a wide critical region also gives a natural explanation for why quantum criticality appears without tuning the control parameters in the quasicrysal $Yb_{15}Al_{34}Au_{51}$ [59].

In ref [59], it was pointed out that it is possible that the same criticality appears even in the 1/1 approximant crystal $Yb_{14}Al_{35}Au_{51}$ which shows Fermi-liquid behavior at ambient pressure if pressure is tuned. Actually, the common quantum criticality has been observed at $P = P_{\mathrm{c}} = 1.96$ GPa in the 1/1 approximant crystal [54], which has confirmed the theoretical prediction. When pressure is applied to the 1/1 approximant crystal $Yb_{14}Al_{35}Au_{51}$ and is tuned to $P = P_{\mathrm{c}}$, the magnetic susceptibility has been discovered to exhibit the quantum critical behavior $\chi \sim T^{-0.5}$, which is the same as that in the quasicrystal $Yb_{15}Al_{34}Au_{51}$. It has also been observed that the quantum critical region in the T-P phase diagram of the 1/1 approximant crystal is limited around $P = P_{\mathrm{c}}$ [54], which is in sharp contrast to the wide quantum critical region in the quasicrystal.

6.6.2 *Origin of Quantum Criticality and T/B Scaling*

In Sect. 6.6.1, it is discussed that the theory of critical Yb-valence fluctuations explains the fact that the qnasicrystal and the 1/1 approximant crystal exhibit the same quantum criticality where the former shows the vast extent of quantum critical region under pressure.

To clarify the origin of quantum critical phenomena commonly observed in the quasicrystal and pressurized approximant crystal, we construct the Anderson lattice model for the 1/1 approximant crystal with 4f electrons at Yb sites and 3p electrons at Al sites [60].

$$H = H_{\mathrm{f}} + H_{\mathrm{c}} + H_{\mathrm{hyb}}. \tag{6.77}$$

Namely, we consider the system with the periodic arrangement of the bcc structure of the Yb-Al-Au cluster (see Fig. 6.11), which was discussed in Sect. 6.6.1. As the first step of analysis, we consider the case where Al/Au mixed sites framed in red in Figs. 6.11b and 6.11d are occupied by Al.

In Eq. (6.77), the 4f-hole part is given by

$$H_{\mathrm{f}} = \sum_{j=1}^{N_{\mathrm{L}}} \left[\varepsilon_{\mathrm{f}} \sum_{i=1\sigma}^{24} n^{\mathrm{f}}_{ji\sigma} + U \sum_{i=1}^{24} n^{\mathrm{f}}_{ji\uparrow} n^{\mathrm{f}}_{ji\downarrow} \right], \tag{6.78}$$

where $f^{\dagger}_{ji\sigma}$ ($f_{ji\sigma}$) is the creation (anihilation) operator for f hole at the i-th site in the j-th unit cell with spin σ, and $n^{\mathrm{f}}_{ji\sigma} \equiv f^{\dagger}_{ji\sigma} f_{ji\sigma}$. Here N_{L} is the number of unit cells and i specifies the Yb site on the Yb_{12} cluster (see Fig. 6.11c) at the body center ($i = 1 \sim 12$) and at the 8 corners of the bcc unit cell ($i = 13 \sim 24$). The first term represents the f level and the second term represents the on-site Coulomb repulsion of the 4f holes at the Yb sites.

In Eq. (6.77), the conduction-hole part is given by

$$H_{\mathrm{c}} = - \sum_{\langle j\xi, j'\nu \rangle \sigma} \left(t_{j\xi, j'\nu} c^{\dagger}_{j\xi\sigma} c_{j'\nu\sigma} + \mathrm{h.c.} \right), \tag{6.79}$$

where $c^{\dagger}_{j\xi\sigma}$ ($c_{j\xi\sigma}$) is the creation (annihilation) operator for conduction hole at the ξ-th site in the j-th unit cell with spin σ. Here $\langle j\xi, j'\nu \rangle$ denotes the pairs between the $j\xi$-th Al site and the $j'\nu$-th Al site. The transfer integrals $t_{j\xi, j'\nu}$ are set for the nearest-neighbor (N.N.) Al sites on the second shell as t_2 and between the second and fourth shells as t_{2-4}, and are set up to the next N.N. Al sites between the fourth and fifth shells as t_{4-5}. In reality, there may also exist conduction orbital at the Au site. To take into account this effect, we consider the effective transfer via the Au site as a parameter, which is expressed as t'_5 between the N.N. Al sites on the fifth shell since the existence ratio of the Al's is 100% [56].

In Eq. (6.77), the hybridization between 4f-hole and conduction-hole orbitals is given by

$$H_{\mathrm{hyb}} = \sum_{\langle ji, j'\xi \rangle \sigma} \left(V_{ji, j'\xi} f^{\dagger}_{ji\sigma} c_{j'\xi\sigma} + \mathrm{h.c.} \right), \tag{6.80}$$

where $\langle ji, j'\xi \rangle$ denotes the pairs between the ji-th site and the $j'\xi$-th site and the hybridization matrix element is given by $V_{ji, j'\xi}$. Here i specifies the Yb site ($i = 1 \sim 24$) in the j-th unit cell and ξ specifies the N.N. Al sites on the fourth and fifth shells and up to the next N.N. Al sites on the second shell in the j'-th unit cell for each i-th site. Corresponding $V_{ji, j'\xi}$ is expressed as V_{3-4}, V_{3-5}, and V_{2-3}, respectively.

To set transfer integrals, we employ the relation $t_{j\xi, j'\nu} \propto 1/r^{\ell+\ell'+1}$, where r is the distance between the $j\xi$-th site and $j'\nu$-th site with wave functions with azimuthal quantum numbers ℓ and ℓ', respectively [61–63]. By inputting $\ell = \ell' = 1$ and r obtained from Ref. [56], we have $t_{2-4} = 0.357 t_2$ and t_{4-5} is set to be either of $0.173 t_2$ or $0.052 t_2$ corresponding to Al-Al distances (for instance, the former is set between the $\xi = 13$th and 19th Al sites and the latter is set between the $\xi = 13$th and 20th Al sites in Figs. 6.11d and 6.11e, respectively). As for the f-c hybridization, the similar relation $V_{ji,j'\xi} \propto 1/r^{\ell+\ell'+1}$ holds. Hence by inputting $\ell = 3$ and $\ell' = 1$ and the Yb-Al distance r, V_{3-4} and V_{3-5} are set as $V_{3-4} = 1.070V_0$ and $V_{3-5} = 0.714V_0$, respectively. V_{2-3} is set to be either of V_0 or $0.767V_0$ corresponding to Yb-Al distances [for instance, the former is set between the i = 1st and ξ = 1st sites and the latter is set between the i = 1st and ξ = 3rd sites in Figs. 6.11c and 6.11b, respectively]. Hereafter, the energy unit is taken as t_2, i.e., $t_2 = 1.0$, and V_0 and t'_5 are set as parameters.

The heavy electron behavior observed in the 1/1 approximant crystal $Yb_{14}Al_{35}Au_{51}$ [26] is considered to be originated from strong onsite Coulomb repulsion U between the 4f holes at Yb. To clarify the band structure for the heavy quasiparticles in the approximant crystal, we apply the slave-boson mean-field theory [32, 64] to Eq. (6.77) for $U = \infty$ [60]. We obtained the ground state at half filling $\bar{n} = 1$. Here, the filling is defined by the hole number per site, which is given by $\bar{n} = \bar{n}_f + \bar{n}_c$ where

$$\bar{n}_f \equiv \frac{1}{N_L}\sum_{j=1}^{N_L}\frac{1}{24}\sum_{i=1}^{24}\sum_{\sigma}\langle n^f_{ji\sigma}\rangle, \tag{6.81}$$

$$\bar{n}_c \equiv \frac{1}{N_L}\sum_{j=1}^{N_L}\frac{1}{48}\sum_{\xi=1}^{48}\sum_{\sigma}\langle n^c_{j\xi\sigma}\rangle \tag{6.82}$$

with $n^c_{j\xi\sigma} \equiv c^\dagger_{j\xi\sigma}c_{j\xi\sigma}$. The result of the band structure and the density of states (DOS) is shown in Fig. 6.13 for $V_0 = 0.13$, $\varepsilon_f = -0.4$ and $t'_5 = 0.2$ calculated in $N_L = 8 \times 8 \times 8$ as a typical case for the approximant crystal.

The renormalized f level is raised up to $\varepsilon \approx 0.2$, where the Fermi level ε_F is located, giving rise to the heavy quasiparticle band in Fig. 6.13a. This is reflected in the sharp peak of the total DOS $\rho_{tot}(\varepsilon)$ around ε_F, so-called Kondo peak, in Fig. 6.13b. Here, $\rho_{tot}(\varepsilon)$ is given by $\rho_{tot}(\varepsilon) = \rho_f(\varepsilon) + \rho_c(\varepsilon)$ with

$$\rho_f(\varepsilon) \equiv \sum_{i=1}^{24} A^{ff}_{ii}(\varepsilon), \tag{6.83}$$

$$\rho_c(\varepsilon) \equiv \sum_{\xi=1}^{48} A^{cc}_{\xi\xi}(\varepsilon), \tag{6.84}$$

where the spectral function $A^{aa}_{bb}(\varepsilon)$ is defined as

$$A^{aa}_{bb}(\varepsilon) \equiv -\frac{1}{\pi N_L}\sum_{\boldsymbol{k}} \mathrm{Im}G^{aa\ R}_{bb,\sigma}(\boldsymbol{k}, \varepsilon) \tag{6.85}$$

with the retarded Green function for quasiparticles $\hat{G}^R_\sigma(\boldsymbol{k}, \varepsilon) \equiv (\varepsilon + i\delta - \tilde{H}_{\boldsymbol{k}\sigma})^{-1}$ and $\delta = 0.01$, where $\tilde{H}_{\boldsymbol{k}\sigma}$ is the slave-boson Hamiltonian in the momentum space [60]. Since ε_f is located at rather deep position from the Fermi level, the present state is

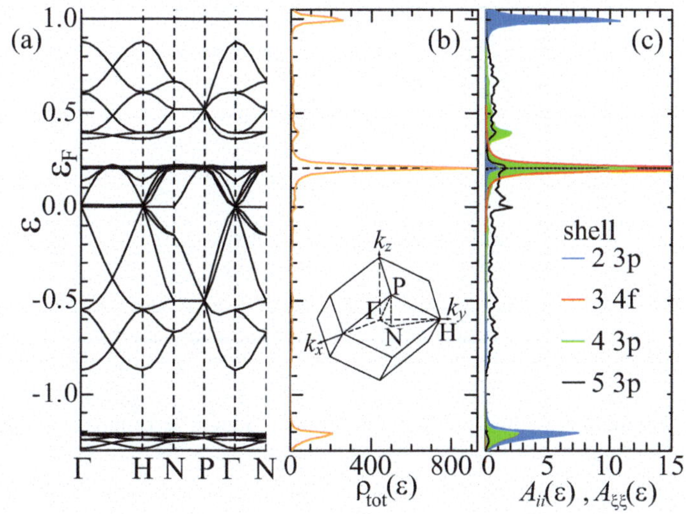

Fig. 6.13 **a** Energy band along high-symmetry lines for $t_2 = 1.0$, $t_5' = 0.2$, $V_0 = 0.13$, $\varepsilon_f = -0.4$, and $U = \infty$ at $\bar{n} = 1$ calculated in $N_L = 8^3$. **b** Total density of states. Inset shows bcc Brillouin zone. **c** Spectral functions $A_{ii}(\varepsilon)$ for $i = 1$ (red) and $A_{\xi\xi}(\varepsilon)$ for $\xi = 1$ (blue), $\xi = 13$ (green), and $\xi = 19$ (black). Horizontal dashed lines indicate Fermi level ε_F [60]. Reprinted from Ref. [60]. Copyright 2016 The Physical Society of Japan

considered to correspond to pressurized approximant crystal. Actually, at the Fermi level, the f-component is dominant with $\rho_f(\varepsilon_F)$ being sharing 79.5% of $\rho_{tot}(\varepsilon_F)$. The conduction bands of 3p holes on the fourth, second, and fifth shells hybridize with 4f holes with each DOS at ε_F sharing 15.7%, 2.8%, and 2.0% of $\rho_{tot}(\varepsilon_F)$, respectively, forming three hybridized bands at ε_F below pseudo-gap-like DOS around $\varepsilon \sim 0.3$ in Fig. 6.13b. Almost flat dispersions around $\varepsilon \approx -1.2$ and 1.0 in Fig. 6.13a are reflected in marked DOS's in Fig. 6.13b, respectively, which are mainly due to splitting of 3p bands on the second shell (see Fig. 6.13c).

The dynamical susceptibility for the charge transfer between 4f and conduction holes

$$\chi^{\rm ffcc}_{ii\xi\xi,\sigma}(\boldsymbol{q}, i\omega_m) = -\frac{T}{N_L}\sum_{n}\sum_{\boldsymbol{k}} G^{\rm ff}_{ii,\sigma}(\boldsymbol{k}+\boldsymbol{q}, i\varepsilon_n + i\omega_m) G^{\rm cc}_{\xi\xi,\sigma}(\boldsymbol{k}, i\varepsilon_n) \quad (6.86)$$

is calculated by using $\hat{G}_\sigma(\boldsymbol{k}, i\varepsilon_n) \equiv (i\varepsilon_n - \tilde{H}_{\boldsymbol{k}\sigma} + \mu)^{-1}$ with $\varepsilon_n = (2n+1)\pi T$ and the chemical potential μ, and $\omega_m = 2m\pi T$. The $\boldsymbol{q}$ dependence of $\chi^{\rm ffcc}_{ii\xi\xi,\sigma}(\boldsymbol{q}, 0)$ calculated at $T = 0.0001$ with the number of Matsubara frequency $n_M = 2^{15}$ being kept is shown in Fig. 6.14. A remarkable result is that almost flat-momentum dependence appears in the charge-transfer mode between the i = 1st Yb site (see Fig. 6.11c) and the ξ = 1st, 2nd, 3rd Al sites (see Fig. 6.11b), ξ = 13th Al site (see Fig. 6.11d), and ξ = 19th Al site (see Fig. 6.11e), which are connected via $H_{\rm hyb}$ in Eq. (6.80). Emer-

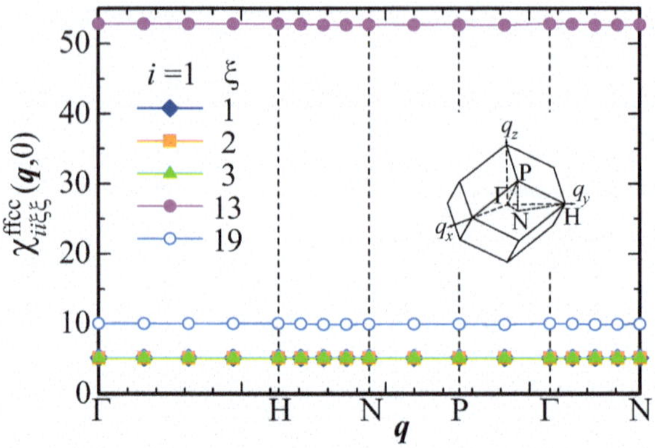

Fig. 6.14 $\chi^{\rm ffcc}_{ii\xi\xi,\sigma}(\mathbf{q}, 0)$ along high-symmetry lines for $i = 1$ and $\xi = 1$ (filled diamond), 2 (filled square), 3 (filled triangle), 13 (filled circle), and 19 (open circle) at $T = 0.0001$ calculated in $N_{\rm L} = 8^3$ and $n_{\rm M} = 2^{15}$ for $t_2 = 1.0$, $t'_5 = 0.2$, $V_0 = 0.13$, $\varepsilon_{\rm f} = -0.4$, and $U = \infty$ at $\bar{n} = 1$. Note that data for $\xi = 1$ and 2 are degenerated. Inset shows bcc Brillouin zone [60]. Reprinted from Ref. [60]. Copyright 2016 The Physical Society of Japan

gence of almost flat mode is considered to be ascribed to strong local correlation effect by $U = \infty$ [30, 45, 65].

The reason why the charge-transfer mode between the third and fourth shells is extraordinary enhanced [see the $(i, \xi) = (1, 13)$ data in Fig. 6.14] is due to the strongest f-c hybridization, $|V_{3-4}| > |V_{2-3}|, |V_{3-5}|$, arising from the shortest Yb-Al distance and existence of the DOS around $\varepsilon_{\rm F}$ (see $A^{\rm cc}_{\xi\xi}(\varepsilon_{\rm F})$ for $\xi = 13$ in Fig. 6.13c). Maximum of $\chi^{\rm ffcc}_{ii\xi\xi}(\boldsymbol{q}, 0)$ for $i = 1$ and $\xi = 13$ is located at the Γ point, $\boldsymbol{q} = \mathbf{0}$.

To clarify how the locality of the charge-transfer fluctuation affects the quantum criticality, and also to get insight into the mechanism of emergence of the T/B scaling, let us focus on the charge-transfer mode between the N.N. Yb on the third shell and Al on the fourth shell since it is overwhelmingly dominant in Fig. 6.14 [see the $(i, \xi) = (1, 13)$ data]. Then, we apply the recently-developed mode-coupling theory for critical valence fluctuations under magnetic field [46] to the present system, which starts from the Hamiltonian

$$\mathcal{H} = H + H_{U_{\rm fc}} + H_{\rm Zeeman}, \tag{6.87}$$

where the charge-transfer fluctuation, i.e., Yb-valence fluctuation, is caused by the inter-orbital Coulomb repulsion [30, 32, 65]

$$H_{U_{\rm fc}} = U_{\rm fc} \sum_{\langle ji, j'\xi\rangle} \sum_{\sigma\sigma'} n^{\rm f}_{ji\sigma} n^{\rm c}_{j'\xi\sigma'}. \tag{6.88}$$

Here, $\langle ji, j'\xi\rangle$ denotes the N.N. pair between Yb on the third shell and Al on the fourth shell. In reality, it is expected that $H_{U_{\rm fc}}$ mainly comes from the Coulomb repulsion between 4f and 5d orbitals at Yb because of onsite interaction. Since 5d wave function is widespread to certain extent, which is considered to hybridize with conduction orbitals at surrounding atoms. To express this effect most simply we set $H_{U_{\rm fc}}$ as Eq. (6.88) instead of introducing Yb 5d orbital explicitly. It is noted that the model calculation including the 4f-5d Coulomb repulsion is also performed, which is discussed in Sect. 6.6.5.

The Zeeman term is given by

$$H_{\rm Zeeman} = -h \sum_{j=1}^{N_{\rm L}} \left[\sum_{i=1}^{24} S_{ji}^{\rm fz} + \sum_{\xi=1}^{48} S_{j\xi}^{\rm cz} \right], \tag{6.89}$$

where h is magnetic field, and $S_{ji}^{\rm fz} \equiv \frac{1}{2}(n_{ji\uparrow}^{\rm f} - n_{ji\downarrow}^{\rm f})$ and $S_{j\xi}^{\rm cz} \equiv \frac{1}{2}(n_{j\xi\uparrow}^{\rm c} - n_{j\xi\downarrow}^{\rm c})$.

Taking into account the mode-coupling effects of the charge-transfer fluctuation up to the fourth order of $U_{\rm fc}$ in the action $S[\varphi]$ for $\mathcal{H}$, we construct the best Gaussian. Following the same procedure described in Sect. 6.5, we have derived the mode-coupling equation of the valence fluctuations, i.e., the valence SCR equation under the magnetic field [60].

First, we have identified the valence QCP at zero magnetic field. Then, by using the parameters, we have solved the valence SCR equation under the magnetic filed and have obtained the solution y, where y^{-1} is proportional to the valence susceptibility $\chi_{\rm v} \propto y^{-1}$ [60]. The result is shown in Fig. 6.15 for $0.0001 \le T \le 0.0014$ and $10^{-8} \le h \le 10^{-5}$ for the parameter of the QCP. We find that the data seem to fall down to a single scaling function of T/h over four decades

$$y = h^{1/2}\phi\left(\frac{T}{h}\right). \tag{6.90}$$

The least-square fit for the large T/h regime of $10^2 \le T/h \le 1.4 \times 10^5$ gives

$$\frac{y}{h} = c\left(\frac{T}{h}\right)^{\zeta} \tag{6.91}$$

with $\zeta = 0.503$, which is shown as a dashed line. This indicates that scaling function for $x \gg 1$ has the form of $\phi(x) = cx^{1/2}$. In this $T/h \gg 1$ region, we have $y/h^{1/2} \approx c(T/h)^{1/2}$, i.e., $y \approx T^{1/2}$. This is caused by critical valence fluctuation with strong locality, giving rise to the non-Fermi-liquid regime. This result indicates that the magnetic susceptibility behaves as $\chi \approx T^{-1/2}$ for $h \to 0$, which has been observed in quasicrystal $Yb_{15}Al_{34}Au_{51}$ [26]. Our result implies that even in the approximant crystal, the same behavior is expected to appear when pressure is applied and is tuned to $P \approx 2$ GPa.

As T/h decreases, the data tend to show deviation from the dashed line as seen in Fig. 6.15. This reflects suppression of valence susceptibility by magnetic field and indicates crossover to the Fermi liquid regime toward $T/h \ll 1$.

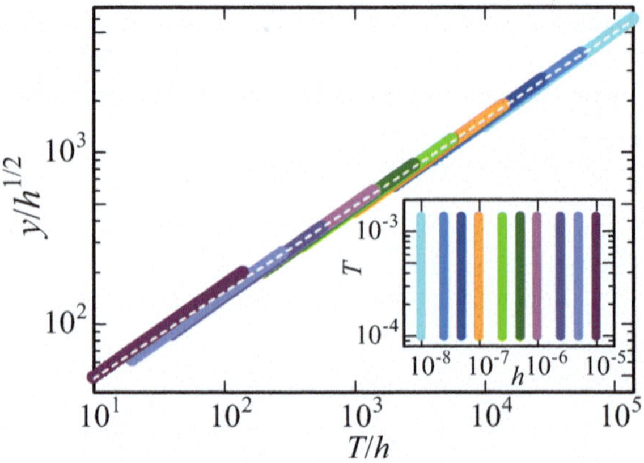

Fig. 6.15 Scaling of the data for $0.0001 \le T \le 0.0014$ and $10^{-8} \le h \le 10^{-5}$. Inset shows the T-h range where the scaling applies. The dashed line represents the fitting function $c(T/h)^{\zeta}$ with $\zeta = 0.503$. The data were obtained for $t_2 = 1.0$, $t_5' = 0.2$, $V_0 = 0.13$, $\varepsilon_{\rm f} = -0.4$, $U_{\rm fc} = 0.0192$, and $U = \infty$ at $\bar{n} = 1$ in $N_{\rm L} = 8^3$ and $n_{\rm M} = 2^{15}$ [60]. Reprinted from Ref. [60]. Copyright 2016 The Physical Society of Japan

Emergence of the T/h scaling behavior is ascribed to presence of small characteristic temperature of critical valence fluctuation T_0. The characteristic temperature of critical valence fluctuations

$$T_0 = \frac{Aq_{\rm B}^3}{2\pi C} \tag{6.92}$$

is estimated to be $T_0 = 1.4 \times 10^{-4}$, which is around the lowest temperature (see the inset of Fig. 6.15). Since T_0 is proportional to the coefficient A in the valence-fluctuation mode

$$\chi_{ii\xi\xi}^{\rm ffcc}(\boldsymbol{q}, 0) \approx \chi_{ii\xi\xi}^{\rm ffcc}(\boldsymbol{0}, 0) - Aq^2, \tag{6.93}$$

almost flat dispersion around the Γ point shown in Fig. 6.14 gives small A, giving rise to the small T_0. The smallness of T_0 is assisted by the small size of the Brillouin zone $q_{\rm B}$ reflecting the large size of the unit cell of the 1/1 approximant crystal with the lattice constant $a = 14.5$ Å. In case that T_0 is below (or at least comparable to) the lowest temperature, $t/y \gg 1$ holds where all the terms with y and t in the valence SCR equation [60], as discussed in Sect. 6.5. We have confirmed that this is the case for all the data used in Fig. 6.15. Hence, the realization of small T_0 is the main reason for emergence of the T/h scaling as well as $\chi \sim T^{-1/2}$ for the zero-field limit.

Since the characteristic temperature of critical valence fluctuations T_0 is extremely low, which is located around the measured temperature, the experimentally accessible low temperature T satisfies $t = T/T_0 \gtrsim 1$. As shown in Sect. 6.2, in this temperature range, the T-linear resistivity appears $\rho \sim T$. As explained in Sect. 6.6.3, the specific heat is shown to behave $C/T \sim -\log T$ at low temperatures. Thus, the quantum criticality in the quasicrytal $Yb_{15}Al_{34}Au_{51}$ in Table 6.1 is explained. Since the local atomic environment around the Yb site in the Tsai-type cluster is common between the approximant crystal and quasicrystal and the infinite limit of the unit-cell size corresponds to the quasicrystal as noted in Sect. 6.6, the quantum criticality observed in the quasicrystal and 1/1 approximant crystal at $P \approx 2$ GPa as well as the T/B scaling is explained by the theory of critical valence fluctuations in a unified way.

6.6.3 *Non-divergent Grüneisen Parameter in Quantum Critical Quasicrystal*

As discussed in Sect. 6.4.3, the theoretical framework for calculating the specific heat C_V, the thermal-expansion coefficient α, and the Grüneisen parameter Γ by enhanced critical valence fluctuations near the valence QCP in the thermodynamically consistent way has been developed [37]. This framework is applied to the quasicrystal $Yb_{15}Al_{34}Au_{51}$ and the results are discussed in this subsection. An important point is that by using the *same* parameters for the valence QCP in the mode-coupling theory of critical valence fluctuations, C_V, α, and Γ are calculated in the unified framework together with the magnetic as well as valence susceptibility. First, the background for the Grüneisen parameter is explained below.

6.6.3.1 Grüneisen Parameter at the QCP in Heavy-Fermion Metals

As for the general property of the QCP, the Grüneisen parameter [66]

$$\Gamma = \frac{\alpha V}{C_V \kappa_T} \tag{6.94}$$

where C_V is the specific heat under a constant volume V, α is the thermal-expansion coefficient, and κ_T is the isothermal compressibility, has been used widely as a clear indicator of whether the material is located at the QCP since it was asserted that the Grüneisen parameter diverges at *any* QCPs by the renormalization-group theory [67, 68]. Divergence of the Grüneisen parameter at the QCP was observed in several materials [69] such as $CeNi_2Ge_2$ [70], $YbRh_2(Si_{0.95}Ge_{0.05})_2$ [70], $CeCu_{5.8}Ag_{0.2}$ [71], and $CeCu_{5.9}Au_{0.1}$ [72]. However, recent preliminary measurement in the QC $Yb_{15}Al_{34}Au_{51}$ has revealed that $|\Gamma|$ does not show such a divergence as lowering the temperature but rather decreases and finally converges into a finite value $\Gamma \approx -55$ at the lowest temperature $T = 70$ mK [74]. This poses a serious challenge to theories ever proposed. Furthermore, the Grüneisen parameter in the 1/1 approximant crystal $Yb_{14}Al_{35}Au_{51}$ with the periodic lattice structure showing the Fermi-liquid behavior has been observed as $\Gamma \approx -130$ at $T = 70$ mK, whose absolute value is much larger than that in the quasicrystal [74].

6.6.3.2 Specific Heat, Thermal-Expansion and Grüneisen Parameter in Quasicrystal

To discuss the low-T properties of $C_V(T)$, $\alpha(T)$, and $\Gamma(T)$ in the quasicrystal, we solve the valence SCR equation [Eq. (6.31)] at the QCP. As the input parameters, we employ $y_0 = 0$ and $y_1 = 0.195$ which were evaluated in Eqs. (6.64) and (6.65) respectively on the basis of the extended Anderson lattice model $H + H_{U_{\rm fc}}$ in Eq. (6.87) on the approximant crystal for the pressure-tuned valence QCP [60], as discussed in Sect. 6.6.2.

By solving Eq. (6.31) as a function of t, the specific-heat coefficient C_V/Nt is obtained in Eq. (6.35) with $\left(\frac{\partial y}{\partial t}\right)_V$ being calculated, which is shown in Fig. 6.16. Here, we plot the $t = T/T_0 \gtrsim 1$ regime, where $\chi \propto y^{-1} \sim T^{-0.5}$ and the T/B scaling were shown to appear [60]. Note that T is in the low-T region even if $t \gtrsim 1$ because T_0 is now extremely small reflecting the smallness of A in Eq. (6.23), which was explicitly shown in the Anderson lattice model in the approximant crystal [60]. This is due to the localness of the 4f-3p charge-transfer fluctuation (see Fig. 6.14). As t decreases, $C_{\rm a}$ becomes dominant in C_V, which shows the logarithmic behavior $C_{\rm a}/T \sim -\ln T$ in Fig. 6.16. Hence, C_V/t at the lower t region starts to show logarithmic increase toward $t \ll 1$ as observed in the quasicrystal [26]. On the other hand, the data in the region $1 < t \lesssim 10$ can be fit with a power-law form, giving $C_V/t \sim t^{-0.66}$ as shown in the inset of Fig. 6.16, which is also favorably compared with the experimental report in Ref. [27].

Next, we calculate the thermal-expansion coefficient. As shown in Sect. 6.4.4, the thermal-expansion coefficient α is expressed by Eq. (6.41) near the valence QCP, in general. However, the quasicrytal $Yb_{15}Al_{34}Au_{51}$ exhibits the quantum critical "line" but not "point" as shown in Fig. 6.10a. Hence, let us carefully analyze the pressure dependence of the magnetic susceptibility.

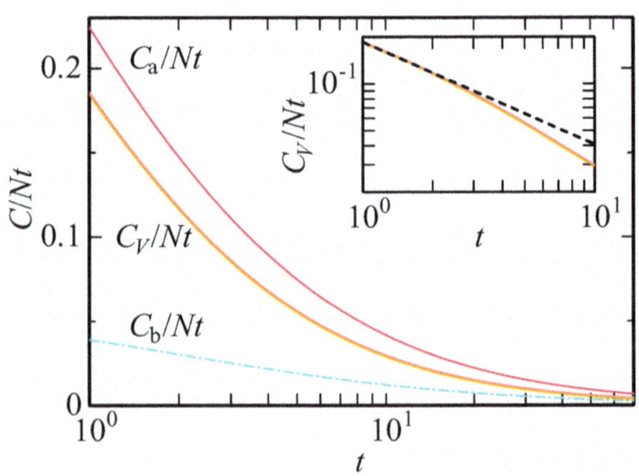

Fig. 6.16 Specific heat coefficient versus scaled temperature $t = T/T_0$. C_V/Nt (bold solid line), $C_{\rm a}/Nt$ (thin solid line), $C_{\rm b}/Nt$ (dash dotted line) at the valence QCP. Dashed line is a fit by const. $\times\, t^{-0.66}$ [37]. Reprinted from Ref. [37]. Copyright 2020 Elsevier

The experimental data in the χ^{-1} versus $T^{0.5}$ plot at low temperatures in the quasicrystal show the straight lines from the origin with almost the same slopes from $P = 0$ to $P = 1.6$ GPa [26, 28]. This indicates that

$$\left(\frac{\partial \chi^{-1}}{\partial P}\right)_T = 0 \tag{6.95}$$

holds for low T giving rise to

$$\left(\frac{\partial y}{\partial P}\right)_T = 0 \tag{6.96}$$

since the relation $y \propto \chi_v^{-1} \propto \chi^{-1}$ holds near the valence QCP [30]. This is also natural consequence of the theory of critical valence fluctuations for the quasicrystal that has shown the condensation of the valence QCPs in the ground-state phase diagram [59]. If we focus on the vicinity of each one of the condensed QCPs, the concentration is shown to be reflected in $\left(\frac{\partial y}{\partial P}\right)_{T=0} = 0$ in α_b. Namely, the ground state is singular as $1/y \to \infty$ for $T \to 0$ while the pressure derivative of y becomes zero in the bulk limit of the quasicrystal. Then, α is expressed by only α_a in the quasicrystal $Yb_{15}Al_{34}Au_{51}$.

Here we input $\frac{1}{T_0}\left(\frac{\partial T_0}{\partial P}\right)_T = -0.44$ GPa^{-1} into Eq. (6.42) so as to reproduce the measured lowest temperature value of $\Gamma = -55$ in the quasicrystal (see Fig. 6.18 below). The result is shown in Fig. 6.17. As t decreases, $-\alpha/t$ divergingly increases. Here, to plot α in the unit of K^{-1}, we restored the Boltzmann constant k_B in Eq. (6.42) and used the lattice constant $a = 14.5$ Å of the bcc lattice of the Tsai-type cluster in the approximant crystal. The low-t part of $-\alpha/t$ is fitted with $-\alpha/t \sim t^{-0.60}$ as shown in the inset in Fig. 6.17, which is slightly different from $C_V/t \sim t^{-0.66}$ for the low-t part. This is due to the contribution from C_b in C_V, but not α_b in α in the quasicrystal.

Then, we calculate the Grüneisen parameter. The Grüneisen parameter in the quasicrystal is obtained by substituting $\alpha = \alpha_a$ and $C_V = C_a - C_b$ into Eq. (6.94) as

$$\Gamma = \frac{C_a}{C_a - C_b}\frac{1}{\kappa_T}\frac{1}{T_0}\left(\frac{\partial T_0}{\partial P}\right)_T. \tag{6.97}$$

Since any indication of the phase separation has not been observed in the quasicrystal, here we use the isothermal compressibility observed at room temperature $\kappa_T = 9.6 \times 10^{-3}$ GPa^{-1} in the quasicrystal [73]. The resultant $\Gamma(t)$ is shown in Fig. 6.18. As t decreases, $|\Gamma(t)|$ decreases as observed in the quasicrystal. This decrease reflects the decrease in the ratio of the specific heat $C_a/(C_a - C_b)$ in Eq. (6.97) since $\frac{1}{\kappa_T}\frac{1}{T_0}\left(\frac{\partial T_0}{\partial P}\right)_T$ has no temperature dependence. This result gives a qualitative explanation for why $|\Gamma(T)|$ decreases on cooling in the quasicrystal.

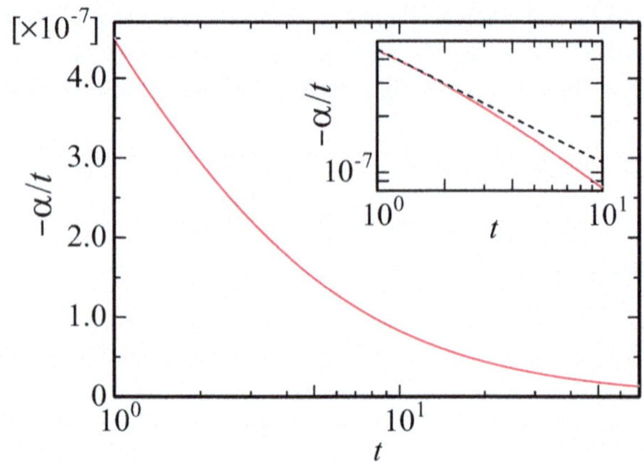

Fig. 6.17 Thermal-expansion coefficient α/t versus scaled temperature $t = T/T_0$ at the valence QCP for the quasicrystal. Inset shows log-log plot of α/t versus t. Dashed line is a fit by const. $\times\, t^{-0.60}$ [37]. Reprinted from Ref. [37]. Copyright 2020 Elsevier

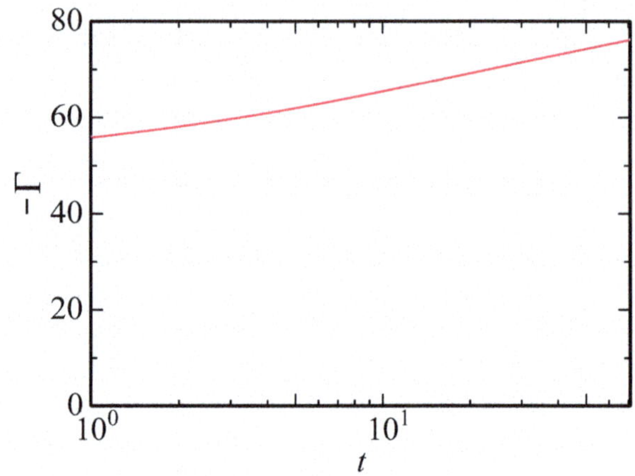

Fig. 6.18 Grüneisen parameter Γ versus scaled temperature t at the valence QCP for the quasicrystal [37]. Reprinted from Ref. [37]. Copyright 2020 Elsevier

Non-divergent Grüneisen parameter is ascribed to the condensation of valence QCPs in the quasicrystal. Emergence of many spots of the QCPs and their condensation in the ground-state phase diagram of the quasicrystal [59] make the critical term in Γ giving the divergence vanish.

6.6.3.3 Comparison with Approximant Crystal

Next we discuss the meaning of the difference between the Grüneisen parameters in the quasicrystal and approximant crystal at ambient pressure. To analyze the Grüneisen parameter in the quasicrystal in Eq. (6.97), we derive the microscopic expression of the characteristic temperature of critical valence fluctuations on the basis of the Anderson lattice model in Eq. (6.77) for 4f and 3p electrons as

$$T_0 = \frac{16}{3\pi^2 \bar{n}^{1/3} \bar{n}_c^{1/3}} \frac{\tilde{A}}{\bar{r}^2} T_K, \tag{6.98}$$

where the Kondo temperature T_K in lattice systems is defined as $T_K \equiv \bar{z} V_{fc}^2 N_{cF}$ with $\bar{z}$ being renormalization factor of quasiparticles and N_{cF} being density of states of conduction electrons at the Fermi level. Here, we have assumed that the 3p electron band is the free-electron band and the Fermi surface is spherical. In Eq. (6.98), $\tilde{A}$ is the dimensionless coefficient in the most dominant charge-transfer (4f-3p) fluctuation defined as

$$\chi_{ii\xi\xi}^{\text{ffcc}}(\boldsymbol{q}, 0) \approx \chi_{ii\xi\xi}^{\text{ffcc}}(\boldsymbol{0}, 0) \left[1 - \tilde{A} \left(\frac{q}{k_F} \right)^2 \right] \tag{6.99}$$

with k_F being the Fermi wavenumber. Here, the irreducible susceptibility $\chi_{ii\xi\xi}^{\text{ffcc}}(\boldsymbol{q}, i\omega_m)$ is defined by Eq. (6.86). In Eq. (6.98), $\bar{n}$ and $\bar{n}_c$ are fillings of total and conduction electrons in the hole picture, respectively, and $\bar{r}$ is a dimensionless constant defined as $\varepsilon_{k_F} - \bar{\varepsilon}_f = \bar{r} D$, where $\bar{\varepsilon}_f$ is the renormalized f level by onsite 4f-4f Coulomb repulsion, $2D$ is the width of the conduction band ε_k. Since the pressure dependence arises from $\tilde{A}$, $\bar{r}$, and T_K in Eq. (6.98), we obtain

$$\frac{1}{T_0} \left(\frac{\partial T_0}{\partial P} \right) = \frac{1}{\tilde{A}} \left(\frac{\partial \tilde{A}}{\partial P} \right) - \frac{2}{\bar{r}} \left(\frac{\partial \bar{r}}{\partial P} \right) + \frac{1}{T_K} \left(\frac{\partial T_K}{\partial P} \right). \tag{6.100}$$

By substituting Eq. (6.100) to Eq. (6.97), we obtain

$$\Gamma = -\frac{C_a}{C_a - C_b} \left[\frac{V}{\tilde{A}} \left(\frac{\partial \tilde{A}}{\partial V} \right)_T - \frac{2V}{\bar{r}} \left(\frac{\partial \bar{r}}{\partial V} \right)_T + \frac{V}{T_K} \left(\frac{\partial T_K}{\partial V} \right)_T \right]. \tag{6.101}$$

On the other hand, the approximant crystal at ambient pressure is located in the heavy-electron Fermi-liquid regime characterized by a small T_K [26, 27]. The Grüneisen parameter is expressed as [75–77]

$$\Gamma_{FL} = -\frac{V}{T_K} \left(\frac{\partial T_K}{\partial V} \right)_S. \tag{6.102}$$

For strong 4f-4f Coulomb repulsion, the Kondo temperature is expressed as

$$T_K = D \exp\left(-\frac{1}{J N_{cF}} \right), \tag{6.103}$$

where J is the 4f-3p exchange coupling in lattice systems [78, 79]. Then, Eq. (6.102) is expressed as

$$\Gamma_{\mathrm{FL}} \approx -\frac{1}{JN_{\mathrm{cF}}}\left\{\frac{V}{J}\left(\frac{\partial J}{\partial V}\right)_S + c\right\}, \tag{6.104}$$

where c is a constant of $O(1)$ (e.g., $c = 2/3$ for free conduction electrons). Since $1/(JN_{\mathrm{cF}}) \approx O(10)$ holds in heavy-electron systems, $|\Gamma_{\mathrm{FL}}|$ is enhanced by this factor. Hence, the large Grüneisen parameter observed in the approximant crystal at $P = 0$, $\Gamma \approx -130$, is understood from Eq. (6.104).

Since Eq. (6.102) is included in the r.h.s. of Eq. (6.101), the Grüneisen parameter in the quasicrystal is also enhanced. By comparing Eq. (6.97) and Eq. (6.102), it turns out that the difference between Γs in the quasicrystal ($\Gamma \approx -55$) and approximant crystal for the measured lowest temperature at $P = 0$ reflects the volume derivative of characteristic temperatures T_0 and T_{K}, respectively. The microscopic origin of the difference is attributed to the volume derivative of the q^2 coefficient of the 4f-3p charge-transfer mode $\tilde{A}$ and the relative position of the renormalized 4f level to the conduction band $\bar{r}$ from the viewpoint of the approximant crystal, which leads to the quasicrystal for the infinite limit of the unit-cell size, as seen in the 1st and 2nd terms in the r.h.s. of Eq. (6.101), respectively.

6.6.3.4 Theoretical Predictions to Experiments

By applying pressure to the approximant crystal, the same quantum criticality as the quasicrystal $\chi \sim T^{-0.5}$ was observed at $P = P_{\mathrm{c}} = 1.96$ GPa [54]. Since the $\chi^{-1}(T)$ versus $T^{0.5}$ plot changes its slope and intercept as P varies and the intercept becomes zero only at $P = P_{\mathrm{c}}$, the approximant crystal is sensitive to pressure in sharp contrast to the quasicrystal. This implies that $\left(\frac{\partial \chi^{-1}}{\partial P}\right)_T \neq 0$ and hence $\left(\frac{\partial y}{\partial P}\right)_T \neq 0$ in the approximant crystal. Then, α_{b} in Eq. (6.43) contributes to α in addition to α_{a}, which also contibutes to Γ in Eq. (6.94) [see Eq. (6.47)], making $|\Gamma(t)|$ increase as t decreases. Therefore, $|\alpha(T)|$ and $|\Gamma(T)|$ in the approximant crystal at $P = P_{\mathrm{c}}$ are expected to be larger than those observed in the quasicrystal. On the other hand, the specific heat C_V is contributed from C_{b} as in Eq. (6.35) in both the quasicrystal and approximant crystal. Hence, the specific heat in the approximant crystal at P=P_{c} is expected to show the similar temperature dependence to that in the quasicrystal as shown in Fig. 6.16.

6.6.4 Elastic Softening in Quantum Critical Yb-Al-Au Approximant Crystal and Quasicrystal

In this section, the elastic property of the quantum critical approximant crystal $Yb_{14}Al_{35}Au_{51}$ and quasicrystal is discussed [80]. To clarify the temperature dependence of the elastic constants, it is necessary to consider the wavefuctions of the 4f and 5d electrons in the ground state of the crystalline electric field (CEF), which correctly reflects the symmetry of the crystal. Here we construct the tight-binding model which consist of the 4f and 5d orbitals at Yb and 3p orbitals at Al. In the quasicrystal $Yb_{15}Al_{34}Au_{51}$, from the entropy analysis by integrating the specific heat by temperature, the first excited CEF energy is at least larger than $\Delta = 20$ K [28]. We focus on the elastic property of the approximant crystal and quasicrystal at low temperatures for $T \ll \Delta$.

6.6.4.1 Construction of the Model

The effective model for the low-energy electronic states in the AC, which consists of Yb 4f and 5d orbitals and Al 3p orbitals in the hole picture is constructed as

$$H = H_{\rm p} + H_{\rm f} + H_{\rm d} + H_{U_{\rm fd}} + H_{\rm fp} + H_{\rm dp}. \tag{6.105}$$

In Eq. (6.105), the 3p part is given by

$$H_{\rm p} = \sum_{\langle j\xi, j'\xi'\rangle} \sum_{\sigma=\uparrow,\downarrow} \sum_{m,m'=z,\pm} t^{\rm pp}_{j\xi m, j'\xi' m'} p^{\dagger}_{j\xi m\sigma} p_{j'\xi' m'\sigma}, \tag{6.106}$$

where the $p_{j\xi m\sigma}$ $(p^{\dagger}_{j\xi m\sigma})$ operators annihilate (create) 3p holes at the ξth Al site in the jth unit cell with $m = p_z, p_{\pm}$, where $p_{\pm}$ is defined as $p_{\pm} \equiv (p_x \pm i p_y)/\sqrt{2}$, and spin $\sigma = \uparrow, \downarrow$. The energy levels of the $m = z, \pm$ states are set to be the same for simplicity of analysis, which is taken as the origin of the energy. Here, $\langle j\xi, j'\xi'\rangle$ denotes the pair of the Al sites, where we set the transfer integral $t^{\rm pp}_{j\xi m, j'\xi' m'}$ for up to the fifth N.N. Al sites. Since the existence ratio of the atoms in the first shell is small (Al/Au: 7.8%/8.9%) (see Fig. 6.11a) [56], here we consider the second–fifth shells. As a first step of analysis, the Al/Au mixed sites shown in the sites framed in red in Figs. 6.11b and 6.11d are assumed to be all occupied by Al. There are 48 Al sites in the bcc unit cell of the AC.

The 4f part is given by

$$H_{\rm f} = \sum_{j=1}^{N_{\rm L}} \sum_{i=1}^{24} \left[\varepsilon_{\rm f} \sum_{\eta=\pm} n^{\rm f}_{ji\eta} + U n^{\rm f}_{ji+} n^{\rm f}_{ji-} \right] \tag{6.107}$$

with $N_{\rm L}$ being the number of the unit cell and $n^{\rm f}_{ji\eta} \equiv f^{\dagger}_{ji\eta} f_{ji\eta}$, where the $f_{ji\eta}$ $(f^{\dagger}_{ji\eta})$ operators annihilate (create) 4f holes with the Kramers state η and the CEF ground energy $\varepsilon_{\rm f}$ at the ith Yb site in the jth unit cell.

The 5d part is given by

$$H_{\rm d} = \varepsilon_{\rm d} \sum_{j=1}^{N_{\rm L}} \sum_{i=1}^{24} \sum_{\eta=\pm} n^{\rm d}_{ji\eta} + \sum_{\langle ji, j'i'\rangle} \sum_{\eta,\eta'=\pm} t^{\rm dd}_{ji\eta, j'i'\eta'} d^{\dagger}_{ji\eta} d_{j'i'\eta'} \tag{6.108}$$

with $n^{\rm d}_{ji\eta} \equiv d^{\dagger}_{ji\eta} d_{ji\eta}$, where $d_{ji\eta}$ $(d^{\dagger}_{ji\eta})$ operators annihilate (create) 5d holes with the Kramers state η and the CEF ground energy $\varepsilon_{\rm d}$ at the ith Yb site in the jth unit cell.

Here, $\langle ji, j'i' \rangle$ denotes the pair of the N.N. Yb sites (see Fig. 6.11c) and $t^{\rm dd}_{ji\eta, j'i'\eta'}$ is the transfer integral between the 5d states.

The 4f-5d interaction is given by

$$H_{U_{\rm fd}} = U_{\rm fd} \sum_{j=1}^{N_{\rm L}} \sum_{i=1}^{24} \sum_{\eta=\pm} \sum_{\eta'=\pm} n^{\rm f}_{ji\eta} n^{\rm d}_{ji\eta'}, \tag{6.109}$$

where $U_{\rm fd}$ is the onsite Coulomb repulsion between the 4f and 5d holes at the Yb site.

The 4f-3p hybridization is given by

$$H_{\rm fp} = \sum_{\langle ji, j'\xi \rangle} \sum_{\eta=\pm} \sum_{\sigma=\uparrow,\downarrow} \left(V^{\rm fp}_{ji\eta, j'\xi\sigma} f^{\dagger}_{ji\eta} p_{j'\xi\sigma} + h.c. \right), \tag{6.110}$$

where $\langle ji, j'\xi \rangle$ denotes the pairs between the ith Yb site in the jth unit cell and the ξth Al site in the j'th unit cell. We consider up to the 4th N.N. Yb-Al sites (for example, the pairs between the $i = 1$st Yb site and the $\xi = 1, 2, 3, 13$ and 19th Al sites as shown in Fig. 6.11f).

The 5d-3p hybridization is

$$H_{\rm dp} = \sum_{\langle ji, j'\xi \rangle} \sum_{\eta=\pm} \sum_{\sigma=\uparrow,\downarrow} \left(V^{\rm dp}_{ji\eta, j'\xi\sigma} d^{\dagger}_{ji\eta} p_{j'\xi\sigma} + h.c. \right), \tag{6.111}$$

where $\langle ji, j'\xi \rangle$ denotes the same Yb-Al pairs as noted in $H_{\rm fp}$.

In the calculations of the transfer integrals $t^{\rm pp}_{j\xi m, j'\xi' m'}$, $t^{\rm dd}_{ji\eta, j'i'\eta'}$ and hybridizations $V^{\rm fp}_{ji\eta, j'\xi\sigma}$, $V^{\rm dp}_{ji\eta, j'\xi\sigma}$, it is necessary to input the Slater-Koster parameters [81, 82]. Here, we use the general relations $(pp\pi) = (pp\sigma)/2$, $(pd\pi) = -(pd\sigma)/\sqrt{3}$, $(pf\pi) = -(pf\sigma)/\sqrt{3}$, $(dd\pi) = -2(dd\sigma)/3$, and $(dd\delta) = (dd\sigma)/6$, which were shown to hold in the linear muffin-tin orbital (LMTO) method [61]. In this study, we set $(pp\sigma) = -1$ as the energy unit and set $(dd\sigma) = 0.4$, $(pf\sigma) = -0.3$, and $(pd\sigma) = 0.6$ in the hole picture as typical values. The distance dependences of transfer integrals and hybridizations between the l and l' orbitals where l is the orbital angular momentum are set so as to follow $\sim 1/r^{l+l'+1}$ with r being the distance of the atoms, following the LMTO argument [61].

To calculate $t^{\rm dd}_{ji\eta, j'i'\eta'}$, $V^{\rm fp}_{ji\eta, j'\xi\sigma}$, and $V^{\rm dp}_{ji\eta, j'\xi\sigma}$, the 4f and 5d wave functions at each Yb site, $\Psi^{\rm 4f}(\hat{\boldsymbol{r}}_i)$ and $\Psi^{\rm 5d}(\hat{\boldsymbol{r}}_i)$, are necessary, which can be obtained, as follows: In the beginning, suppose that the local xyz coordinate (see Figs. 6.19f and 6.19g) coincides with the $x'y'z'$ coordinate in Fig. 6.19a. In the $x'y'z'$ coordinate, each axis is parallel to each side of the bcc unit cell. First, rotate the $x'y'z'$ coordinate around the z' axis with the angle α: $x'y'z' \to x''y''z'$. Second, rotate the coordinate around the y'' (x'') axis with the angle β, giving rise to the local xyz coordinate at each Yb site. These procedures are expressed in the so-called zyz (zxz) rotation. The transformation of the $|J, J_z\rangle$ state in the zyz rotation is performed as

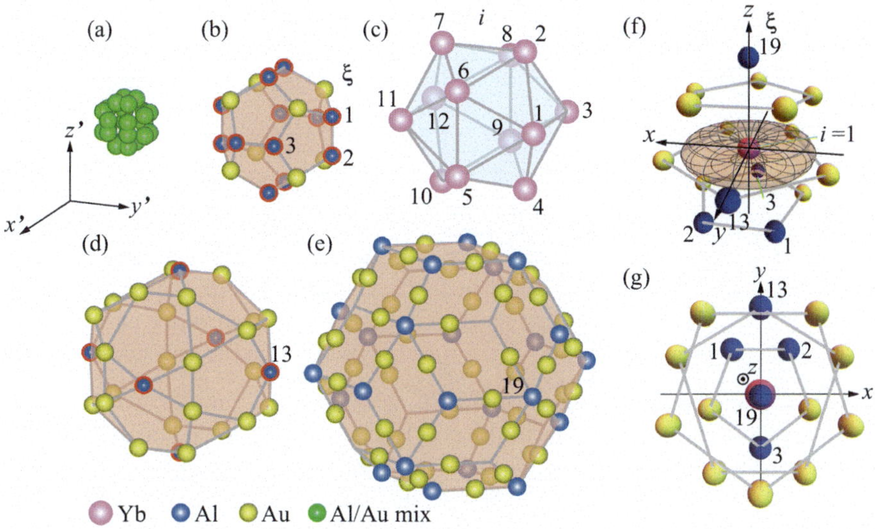

Fig. 6.19 Concentric shell structures of Tsai-type cluster in the Yb-Al-Au approximant crystal: **a** first shell, **b** second shell **c** third shell, **d** fourth shell, and **e** fifth shell. The number in **b**, **d**, **e**, **f**, and **g** indicates the ξth Al site and number in **c** and **f** specifies the ith Yb site. **f** Local configuration around the Yb atom is shown in a side view. The pseudo fivefold axis, passing from the center of the icosahedron in **c** through the Yb atom, is illustrated as the z axis. The square of absolute value of spherical part of the 4f wave function $\Psi^{4\mathrm{f}}(\hat{\boldsymbol{r}})$ at the 1st Yb site is shown. **g** Top view around the 1st Yb site along the pseudo fivefold axis [80]

$$\hat{R}(\alpha,\beta,0)|J J_z\rangle = \sum_{J_z'} D^J_{J_z' J_z}(\alpha,\beta,0)|J J_z'\rangle, \tag{6.112}$$

where $\hat{R}(\alpha,\beta,\gamma) = e^{-i\alpha J_z}e^{-i\beta J_y}e^{-i\gamma J_z}$ is the rotation operator for the Euler angles α, β, and γ, and $D^J_{J_z'J_z}(\alpha,\beta,\gamma)$ is the Wigner's D matrix [83]:

$$D^J_{J_z'J_z}(\alpha,\beta,\gamma) = e^{-i\alpha J_z'} d^J_{J_z'J_z}(\beta) e^{-i\gamma J_z}. \tag{6.113}$$

Here, $d^J_{J_z'J_z}(\beta)$ is the Wigner's small d matrix [83], whose explicit form can be found in literatures [84]. For the zxz rotation, $\hat{R}$ is given by $\hat{R}(\alpha,\beta,\gamma) = e^{-i\alpha J_z}e^{-i\beta J_x}e^{-i\gamma J_z}$ and $d^J_{J_z'J_z}(\beta)$ in Eq. (6.113) should be replaced to $\tilde{d}^J_{J_z'J_z}(\beta)$, whose explicit form is given by

$$\tilde{d}^J_{J_z'J_z}(\beta) = \sum_k \frac{\sqrt{(J+J_z)!(J-J_z)!(J+J_z')!(J-J_z')!}}{(J+J_z-k)!k!(k-J_z+J_z')!(J-k-J_z')!}(-i)^{2k+J_z'-J_z} \times \left(\cos\frac{\beta}{2}\right)^{2J-2kJ_z-J_z'} \left(\sin\frac{\beta}{2}\right)^{2k+J_z'-J_z} \tag{6.114}$$

Table 6.2 Euler angles for each Yb site (see Fig. 6.11c). * indicates the Yb site performed by the zxz rotation and otherwise by the zyz rotation

i	α_i	β_i	i	α_i	β_i
1	ϕ	$\pi/2$	7*	0	θ
2*	π	θ	8*	$\pi/2$	$-\phi$
3	$\pi-\phi$	$\pi/2$	9*	$-\pi/2$	$\pi-\phi$
4*	0	$-\pi+\theta$	10*	π	$-\pi+\theta$
5*	$\pi/2$	$\pi-\phi$	11	$-\phi$	$\pi/2$
6*	$-\pi/2$	$-\phi$	12	$-\pi+\phi$	$\pi/2$

Here the summation for k is taken over $\max(0, J_z - J_z') \le k \le \min(J - J_z', J + J_z)$.

By applying the zyz rotation to the $i = 1, 3, 10$, and 12th Yb site and the zxz rotation to the $i = 2, 4, 5, 6, 7, 8, 9$, and 11th Yb site with α_i, β_i, and $\gamma_i = 0$ listed in Table 6.2 with $\theta = 31.97°$ and $\phi = 58.03°$, the local wave function of $\Psi^{4\mathrm{f}}(\hat{\boldsymbol{r}}_i)$ and $\Psi^{5\mathrm{d}}(\hat{\boldsymbol{r}}_i)$ at each Yb site is obtained. In the rotated xyz coordinate, the z axis passing from the cluster center of the icosahedron (see Fig. 6.11c) through the Yb atom is the quantization axis of the CEF, which is regarded as the pseudo fivefold axis (see Fig. 6.11g).

6.6.4.2 QCP of Valence Transition

To analyze electronic states in the approximant crystal $Yb_{14}Al_{35}Au_{51}$, we apply the slave-boson mean-field (MF) theory to Eq. (6.50) [85]. To describe the state for $U = \infty$ responsible for heavy electrons, we consider $f^{\dagger}_{ji\eta}b_{i\alpha}$ instead of $f^{\dagger}_{ji\eta}$ in Eq. (6.50) by introducing the slave-boson operator b_{ji} to describe the f^0-hole state and requiring the constraint $\sum_{ji}\lambda_{ji}(\sum_{\eta=\pm} n^{\mathrm{f}}_{ji\eta} + b^{\dagger}_{ji}b_{ji} - 1)$ with λ_{ji} being the Lagrange multiplier. For $H_{U_{\mathrm{fd}}}$ in Eq. (6.109), we employ the MF decoupling as $U_{\mathrm{fd}}n^{\mathrm{f}}_{ji\eta}n^{\mathrm{d}}_{ji\eta'} \approx U_{\mathrm{fd}}\bar{n}^{\mathrm{f}}_i n^{\mathrm{d}}_{ji\eta'} + \bar{R}_i n^{\mathrm{f}}_{ji\eta} - \bar{R}_i\bar{n}^{\mathrm{f}}_i$, with $\bar{R}_i \equiv U_{\mathrm{fd}}\bar{n}^{\mathrm{d}}_i$ and $\bar{n}^{\mathrm{f(d)}}_i \equiv \sum_{j\eta}\langle n^{\mathrm{f(d)}}_{ji\eta}\rangle/N_{\mathrm{L}}$. For simplicity, we neglected the Fock term which has a finite value in the present locally-inversion symmetry broken system at Yb since the previous study including the term did not change the main result [86]. Here, we have assumed that MFs are uniform since we now focus on the paramagnetic-metal phase. Hence, we have $\bar{b}_i = \langle b_{ji}\rangle$ and $\bar{\lambda}_i = \lambda_{ji}$. Then, by optimizing the Hamiltonian as $\partial\langle H\rangle/\partial\bar{b}_i = 0$, $\partial\langle H\rangle/\partial\bar{\lambda}_i = 0$, and $\partial\langle H\rangle/\partial\bar{R}_i = 0$, we obtain the set of the MF equations:

$$\frac{1}{N_{\mathrm{L}}}\sum_{k\eta}\langle f^{\dagger}_{ki\eta}f_{ki\eta}\rangle + \bar{b}_i^2 = 1, \tag{6.115}$$

$$\frac{1}{2N_{\mathrm{L}}}\sum_{k}\left[\sum_{\eta\xi m\sigma} V^{\mathrm{fp}}_{k,\xi m\sigma,i\eta}\langle f^{\dagger}_{ki\eta}p_{k\xi m\sigma}\rangle + h.c.\right] + \bar{\lambda}_i\bar{b}_i = 0, \tag{6.116}$$

$$\frac{1}{N_{\mathrm{L}}}\sum_{k\eta}\langle f^{\dagger}_{ki\eta}f_{ki\eta}\rangle = \bar{n}^{\mathrm{f}}_i. \tag{6.117}$$

The MF equations together with the equation for the half filling $\bar{n} = 3$ are solved self-consistently, where the total filling $\bar{n}$ is defined by

$$\bar{n} \equiv \frac{1}{24}\sum_{i=1}^{24}\left(\bar{n}_i^{\rm f} + \bar{n}_i^{\rm d}\right) + \frac{1}{48}\sum_{\xi=1}^{48}\bar{n}_\xi^{\rm p} \tag{6.118}$$

with

$$\bar{n}_\xi^{\rm p} \equiv \frac{1}{3N_{\rm L}}\sum_{km\sigma}\langle p_{k\xi m\sigma}^\dagger p_{k\xi m\sigma}\rangle. \tag{6.119}$$

Below, the results for $\varepsilon_{\rm d} = -0.4$ in the $N_{\rm L} = 8^3$ system will be shown as a typical case. Figure 1a shows the $\varepsilon_{\rm f}$ dependence of $\bar{n}^{\rm f}$ $(= \bar{n}_i^{\rm f}$ for $i = 1 - 24)$ for the ground state. As $U_{\rm fd}$ increases, the $\bar{n}^{\rm f}$ change becomes sharp and the slope $-\partial\bar{n}^{\rm f}/\partial\varepsilon_{\rm f}$ diverges at $\varepsilon_{\rm f} = -0.5992$ for $U_{\rm fd} = 0.364$. For $U > 0.364$, a jump in $\bar{n}_{\rm f}$ appears, which indicates the first-order valence transition. From these results, the QCP of the valence transition is identified to be $(\varepsilon_{\rm f}^{\rm QCP}, U_{\rm fd}^{\rm QCP}) \approx (-0.5992, 0.364)$. At the QCP, $\bar{n}^{\rm f} = 0.69$ is realized, which is favorably compared with the intermediate valence $Yb^{+2.67}$ observed in the QC $Yb_{15}Al_{34}Au_{51}$ [27]. The mass renormalization factor is obtained as $z_i \equiv \bar{b}_i^2 \approx 0.30$ at the QCP, which reflects the quasiparticle state with the intermediate valence within the MF theory [87].

Here we estimate the value of $(pp\sigma)$ as $(pp\sigma) \approx 4.01$ eV by inputting the N.N. Al-Al distance $d = 2.48$ Å into the Harrison's formula $(pp\sigma) = \eta_{pp\sigma}\frac{\hbar^2}{md^2}$ with $\eta_{pp\sigma} = 3.24$ [63, 86]. This gives $U_{\rm fd}^{\rm QCP} \approx 1.46$ eV in Eq. (6.109). Direct examination of the $U_{\rm fd}$ value by experiments such as the partial-fluorescence-yield method in the X ray [88] in the QC and also in the approximant crystal is an interesting future subject.

It is noted that onsite 4f-5d hybridization occurs at Yb via the 4f-3p and 3p-5d hybridizations in Eq. (6.50). This is nothing but the odd-parity CEF due to the local violation of the inversion symmetry at the Yb site by the fivefold configuration of the surrounding Au and Al sites (see Fig. 6.19g).

6.6.4.3 Elastic Property Near the QCP

Next, let us discuss the elastic property near the QCP. By using the random phase approximation (RPA) with respect to $U_{\rm fd}$ as the corrections to the MF state, electric quadrupole fluctuations near the QCP can be analyzed. The RPA susceptibility is given by [89]

$$\chi^{ii'}_{\ell_1\eta_1\ell_2\eta_2\ell_3\eta_3\ell_4\eta_4}(\boldsymbol{q},\omega) = \bar{\chi}^{ii'}_{\ell_1\eta_1\ell_2\eta_2\ell_3\eta_3\ell_4\eta_4}(\boldsymbol{q},\omega)$$
$$-\sum_{i''}\sum_{\tau\tau'}\sum_{m\neq m'} \bar{\chi}^{ii''}_{\ell_1\eta_1\ell_2\eta_2 m\tau m\tau}(\boldsymbol{q},\omega)U_{\mathrm{fd}}\chi^{i''i}_{m'\tau'm'\tau'\ell_3\eta_3\ell_4\eta_4}(\boldsymbol{q},\omega)$$
$$+\sum_{i''}\sum_{\tau\tau'}\sum_{m\neq m'} \bar{\chi}^{ii''}_{\ell_1\eta_1\ell_2\eta_2 m\tau m'\tau'}(\boldsymbol{q},\omega)U_{\mathrm{fd}}\chi^{i''i'}_{m\tau m'\tau'\ell_3\eta_3\ell_4\eta_4}(\boldsymbol{q},\omega). \quad (6.120)$$

Here, the susceptibility is defined by

$$\chi^{ii'}_{\ell_1\eta_1\ell_2\eta_2\ell_3\eta_3\ell_4\eta_4}(\boldsymbol{q},\omega) \equiv \frac{i}{N_{\mathrm{L}}}\sum_{\boldsymbol{k}\boldsymbol{k}'}\int_0^{\infty} dt e^{i\omega t}$$
$$\times\langle[c^{\dagger}_{\boldsymbol{k}i\ell_1\eta_1}(t)c_{\boldsymbol{k}+\boldsymbol{q}i\ell_2\eta_2}(t), c^{\dagger}_{\boldsymbol{k}'+\boldsymbol{q}i'\ell_4\eta_4}c_{\boldsymbol{k}'i'\ell_3\eta_3}]\rangle, \quad (6.121)$$

where $\ell = 1(2)$ specifies the f (d) orbital and $\bar{\chi}^{ii''}_{\ell_1\eta_1\ell_2\eta_2\ell_3\eta_3\ell_4\eta_4}(\boldsymbol{q},\omega)$ denotes the susceptibility in the MF state. Here the susceptibility is expressed in the retarded form, which is obtained by analytic continuation from the Matsubara form [89].

The valence transition is caused by the onsite Coulomb repulsion U_{fd} which affects heavy quasiparticles realized by large $U(\gg U_{\mathrm{fd}})$ [90]. To describe this, we first obtained the QCP of the valence transition within the MF theory for $U \to \infty$ and then to take account of critical fluctuations caused by U_{fd} beyond the MF theory, Eq. (6.120) is employed. This approach is based on the physical picture discussed in the mode-mode-coupling theory of the critical valence fluctuations [30], which describes the quantum valence criticality observed in experiments [19, 20, 23, 26, 29] correctly. The detail of the formulation was discussed in Ref [91, 92]. The RPA susceptibility in Eq. (6.120) is expressed by the 384 × 384 matrix $\hat{\chi}$, $\hat{\bar{\chi}}$ and $\hat{U}$ in the symmetrized form as

$$\hat{\chi} = \hat{\bar{\chi}}^{1/2}\left(\hat{1} - \hat{\bar{\chi}}^{1/2}\hat{U}\hat{\bar{\chi}}^{1/2}\right)^{-1}\hat{\bar{\chi}}^{1/2}, \quad (6.122)$$

where $\hat{\bar{\chi}}^{1/2}$ is the matrix satisfying $\hat{\bar{\chi}} = \hat{\bar{\chi}}^{1/2}\hat{\bar{\chi}}^{1/2}$ and $\hat{1}$ is the identity matrix [92]. The interaction matrix $\hat{U}$ has the elements of U_{fd} for $\ell_1 = \ell_3 \neq \ell_2 = \ell_4$, $\eta_1 = \eta_3$, and $\eta_2 = \eta_4$ and $-U_{\mathrm{fd}}$ for $\ell_1 = \ell_2 \neq \ell_3 = \ell_4$, $\eta_1 = \eta_2$, and $\eta_3 = \eta_4$.

The critical point in this RPA formalism is determined by

$$\det\left(\hat{1} - \hat{\bar{\chi}}^{1/2}\hat{U}\hat{\bar{\chi}}^{1/2}\right) = 0. \quad (6.123)$$

By using the MF states obtained in the calculation in Fig. 6.20a, we calculate $\hat{\bar{\chi}}$ on the basis of Eq. (6.121). Then, by solving Eq. (6.123), the critical point in this RPA formalism is obtained as $(\varepsilon_{\mathrm{f}}^{\mathrm{cRPA}}, U_{\mathrm{fd}}^{\mathrm{cRPA}}) = (-0.5992, 01116)$. We confirmed that the maximum appears at $\boldsymbol{q} = \boldsymbol{0}$ in $\hat{\bar{\chi}}(\boldsymbol{q}, \omega = 0)$ with weak $\boldsymbol{q}$ dependence for finite $\boldsymbol{q}$, as shown in Ref. [60]. This is the reflection of the uniform critical valence fluctuations with the local nature as discussed in Ref. [30].

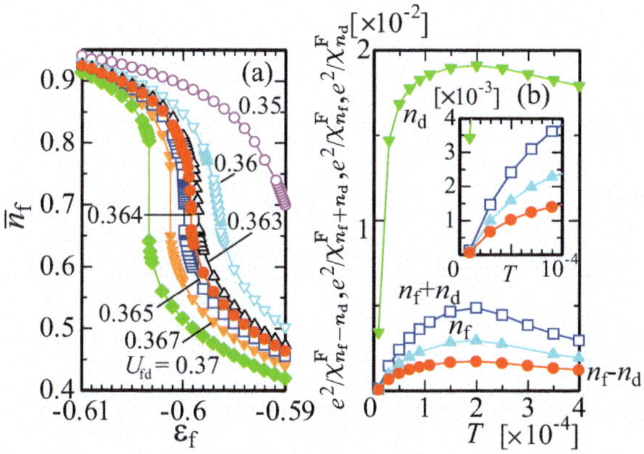

Fig. 6.20 **a** $\bar{n}_{\mathrm{f}}$ versus ε_{f} for $U_{\mathrm{fd}} = 0.350$ (open circle), 0.360 (open inverted triangle), 0.363 (open triangle), 0.364 (filled circle), 0.365 (open square), 0.367 (filled inverted triangle), and 0.370 (filled diamond). **b** Temperature dependence of $e^2/\chi^{\mathrm{F}}_{n_{\mathrm{f}}-n_{\mathrm{d}}}$ (filled circle), $e^2/\chi^{\mathrm{F}}_{n_{\mathrm{f}}+n_{\mathrm{d}}}$ (open square), $e^2/\chi^{\mathrm{F}}_{n_{\mathrm{f}}}$ (filled triangle), and $e^2/\chi^{\mathrm{F}}_{n_{\mathrm{d}}}$ (filled inverted triangle) for ($\varepsilon_{\mathrm{f}}^{\mathrm{cRPA}}$, $U_{\mathrm{fd}}^{\mathrm{cRPA}}$). Inset shows enlargement for $0 \le T \le 10^{-4}$ [80]. Reprinted from Ref. [80]. Copyright 2021 Elsevier

6.6.4.4 Relative and Total Charge Fluctuations

At the valence QCP, the relative-charge fluctuations between the 4f and 5d orbitals, i.e., the critical valence fluctuations diverge, as shown in Fig. 6.20b. Here, we plot the temperature dependence of the uniform relative- and total-charge susceptibility $\chi^{\mathrm{F}}_{n_{\mathrm{f}}\pm n_{\mathrm{d}}}$ given by $\chi^{\mathrm{F}}_{n_{\mathrm{f}}\pm n_{\mathrm{d}}} = \lim_{\boldsymbol{q}\to\boldsymbol{0}} \chi^{\mathrm{F}}_{n_{\mathrm{f}}\pm n_{\mathrm{d}}}(\boldsymbol{q},\omega=0)$ with $\chi^{\mathrm{F}}(\boldsymbol{q},\omega) = \sum_{ii'} \chi^{ii'}_{n_{\mathrm{f}}\pm n_{\mathrm{d}}}(\boldsymbol{q},\omega)$ for ($\varepsilon_{\mathrm{f}}^{\mathrm{cRPA}}$, $U_{\mathrm{fd}}^{\mathrm{cRPA}}$). Here, $\chi^{ii'}_{n_{\mathrm{f}}\pm n_{\mathrm{d}}}(\boldsymbol{q},\omega)$ is defined by

$$\chi^{ii'}_{n_{\mathrm{f}}\pm n_{\mathrm{d}}}(\boldsymbol{q},\omega) = \frac{ie^2}{N_{\mathrm{L}}}\int_0^\infty dt e^{i\omega t}\langle[\delta n^{\mathrm{f}}_{\boldsymbol{q}i}(t) \pm \delta n^{\mathrm{d}}_{\boldsymbol{q}i}(t), \delta n^{\mathrm{f}}_{-\boldsymbol{q}i'}(0) \pm \delta n^{\mathrm{d}}_{-\boldsymbol{q}i'}(0)]\rangle, \tag{6.124}$$

with $+(-)$ denoting the total (relative) charge susceptibility and e being an elementary charge ($e > 0$). Here $\delta\hat{O}$ is defined as $\delta\hat{O} \equiv \hat{O} - \langle\hat{O}\rangle$ and $n^{\mathrm{f(d)}}_{\boldsymbol{q}i}$ is given by $n^{\mathrm{f(d)}}_{\boldsymbol{q}i} = \sum_j e^{-i\boldsymbol{q}\cdot\boldsymbol{r}_j} n^{\mathrm{f(d)}}_{ji}$. The divergence of $\chi^{\mathrm{F}}_{n_{\mathrm{f}}-n_{\mathrm{d}}}$ for $T \to 0$ is confirmed by the eigenvalue analysis of the kernel $\hat{\bar{\chi}}^{1/2}\hat{U}\hat{\bar{\chi}}^{1/2}$, as shown in Fig. 6.21a. For $T \to 0$, the largest eigenvalue is identified to be that of the relative-charge fluctuation, which approaches 1. This satisfies Eq. (6.123) at $T = 0$, giving rise to the valence QCP within the RPA.

Although the total charge fluctuation is also enhanced on cooling in Fig. 6.20b, it turns out that $\chi^{\mathrm{F}}_{n_{\mathrm{f}}+n_{\mathrm{d}}}$ does not diverge at $T = 0$ from the analysis of the eigenvalue of the kernel, which does not approach 1 as shown in Fig. 6.21a. The difference between $\chi^{\mathrm{F}}_{n_{\mathrm{f}}-n_{\mathrm{d}}}$ and $\chi^{\mathrm{F}}_{n_{\mathrm{f}}+n_{\mathrm{d}}}$ comes from the negative value of the interorbital term in the right hand side (r.h.s.) of Eq. (6.124), as a consequence of U_{fd}, which forces the f and d occupancies to compete (see, for detail, Ref. [92]).

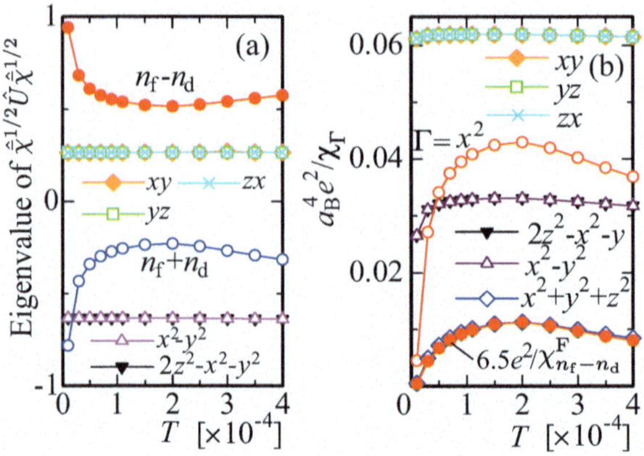

Fig. 6.21 (color online) **a** Temperature dependence of eigenvalues of $\hat{\hat{\chi}}^{1/2}\hat{U}\hat{\hat{\chi}}^{1/2}$ for electric monopoles [relative charge (filled circle) and total charge (open circle)] and electric quadrupoles [$2z^2-x^2-y^2$ (filled inverted triangle), x^2-y^2 (open triangle), xy (filled diamond), yz (open square), and zx (cross)] at (ε_f^{cRPA}, U_{fd}^{cRPA}). **b** Temperature dependence of electric quadrupole susceptibility $a_B^2e^2/\chi_\Gamma$ for $\Gamma = x^2+y^2+z^2$ (open diamond), $2z^2-x^2-y^2$ (filled inverted triangle), and x^2-y^2 (open triangle), xy (filled diamond), yz (open square), zx (cross) and x^2 (open circle) for (ε_f^{RPA}, U_{fd}^{RPA}). The T dependence of $6.5e^2/\chi^F_{n_f-n_d}$ is also shown (filled circle) [80]. Reprinted from Ref. [80]. Copyright 2021 Elsevier

6.6.4.5 Electric Quadrupole Fluctuations

Next, let us discuss the electric quadrupole fluctuation near the QCP of the valence transition. The electric quadrupole susceptibility is expressed as

$$\chi_{O_\Gamma}^{ii'}(\boldsymbol{q},\omega) = \frac{i}{N_L}\int_0^\infty dt e^{i\omega t}\langle[\delta\hat{O}_{\boldsymbol{q}i}^\Gamma(t), \delta\hat{O}_{\boldsymbol{q}i'}^\Gamma(0)^\dagger]\rangle, \tag{6.125}$$

where $\hat{O}_{\boldsymbol{q}i}^\Gamma$ is given by $\hat{O}_{\boldsymbol{q}i}^\Gamma = \sum_j e^{-i\boldsymbol{q}\cdot\boldsymbol{r}_j}\hat{O}_{ji}^\Gamma$ with the irreducible representation Γ. In the approximant crystal $Yb_{14}Al_{35}Au_{51}$, the crystal structure is cubic (space group No. 204, $Im\bar{3}$, T_h^5) [56]. Then, Γ is given by $\Gamma = x^2+y^2+z^2, 2z^2-x^2-y^2, x^2-y^2$, xy, yz, and zx. Here $\hat{O}_{ji}^\Gamma$ is the quadrupole operator

$$\hat{O}_{ji}^\Gamma = \sum_{\ell\ell'}\sum_{\eta\eta'} O_{i\ell\eta,i\ell'\eta'}^\Gamma c_{ji\ell\eta}^\dagger c_{ji\ell'\eta'}, \tag{6.126}$$

where $O_{i\ell\eta,i\ell'\eta'}^\Gamma$ is the form factor

$$O_{i\ell\eta,i\ell'\eta'}^\Gamma = e\langle r^2\rangle_\ell\alpha_\ell\langle i\ell\eta|\hat{O}_\Gamma|i\ell'\eta'\rangle \tag{6.127}$$

with $\langle r^2\rangle_\ell$ being the expectation value of radial part of 4f and 5d wavefunctions. Here, we set $\langle r^2\rangle_1 = 0.1826$ Å^2 following the result of Yb^{+3} [93]. As for the 5d orbital at Yb, we set $\langle r^2\rangle_2 = 0.3246$ Å^2 so as to reproduce the eigenvector of the kernel $\hat{\hat{\chi}}^{1/2}\hat{U}\hat{\hat{\chi}}^{1/2}$ for the electric-quadrupole modes of $\Gamma = 2z^2 - x^2 - y^2$, $x^2 - y^2$, xy, yz, and zx, whose eigenvalues are shown in Fig. 6.21a. Note that the relative value of $\langle r^2\rangle_2/\langle r^2\rangle_1$ is important here and choice of the value of $\langle r^2\rangle_1$ merely shifts the whole scale of χ_Γ as far as the ratio $\langle r^2\rangle_2/\langle r^2\rangle_1$ is maintained, which does not affect the discussions below. In Eq. (6.127), the Stevens factor is given by $\alpha_1 = 2/63$ [94, 95] and $\alpha_2 = -2/15$. Then, Eq. (6.125) leads to [96]

$$\chi^{ii'}_{O_\Gamma}(\boldsymbol{q},\omega) = \sum_{\ell_1\ell_2\ell_3\ell_4}\sum_{\eta_1\eta_2\eta_3\eta_4} O^{\Gamma}_{i\ell_1\eta_1,i\ell_2\eta_2}\chi^{ii'}_{\ell_1\eta_1\ell_2\eta_2\ell_3\eta_3\ell_4\eta_4}(\boldsymbol{q},\omega)O^{\Gamma}_{i'\ell_4\eta_4,i'\ell_3\eta_3}. \quad (6.128)$$

In Eq. (6.127), the operators $\hat{O}_\Gamma$ are expressed by the total angular momentum in the laboratory coordinate as $\hat{O}_{x^2+y^2+z^2} = J_{x'}^2 + J_{y'}^2 + J_{z'}^2$, $\hat{O}_{2z^2-x^2-y^2} = (2J_{z'}^2 - J_{x'}^2 - J_{y'}^2)/2$, $\hat{O}_{x^2-y^2} = \sqrt{3}(J_{x'}^2 - J_{y'}^2)/2$, $\hat{O}_{xy} = \sqrt{3}(J_{x'}J_{y'} + J_{y'}J_{x'})/2$, $\hat{O}_{yz} = \sqrt{3}(J_{y'}J_{z'} + J_{z'}J_{y'})/2$, and $\hat{O}_{zx} = \sqrt{3}(J_{z'}J_{x'} + J_{x'}J_{z'})/2$ [97].

The relation between the total angular momentum in the $x'y'z'$ coordinate and the one in the local xyz coordinate is obtained by using $\hat{R}(\alpha_i,\beta_i,0)$ in Eq. (6.112) for the $i = 1, 3, 11$, and 12th Yb site as $J_{x'} = J_x\cos\alpha_i\cos\beta_i - J_y\sin\alpha_i + J_z\cos\alpha_i\sin\beta_i$, $J_{y'} = J_x\sin\alpha_i\cos\beta_i + J_y\cos\alpha_i + J_z\sin\alpha_i\sin\beta_i$, and $J_{z'} = -J_x\sin\beta_i + J_z\cos\beta_i$. By using these equations and $\Psi^{4f}(\hat{\boldsymbol{r}}_i)$ and $\Psi^{5d}(\hat{\boldsymbol{r}}_i)$, the form factors are calculated as $\langle i\ell\pm|\hat{O}_{x^2+y^2+z^2}|i\ell\pm\rangle = a_\ell(a_\ell+1)$, $\langle i\ell\pm|\hat{O}_{2z^2-x^2-y^2}|i\ell\pm\rangle = a_\ell(2-a_\ell)(\sin^2\beta_i - 2\cos^2\beta_i)/2$, $\langle i\ell\pm|\hat{O}_{x^2-y^2}|i\ell\pm\rangle = \sqrt{3}a_\ell(2-a_\ell)(\sin^2\alpha_i - \cos^2\alpha_i)\sin^2\beta_i/2$, $\langle i\ell\pm|\hat{O}_{xy}|i\ell\pm\rangle = \sqrt{3}a_\ell(-1+2a_\ell)\cos\alpha_i\sin\alpha_i\sin^2\beta_i/2$, $\langle i\ell\pm|\hat{O}_{yz}|i\ell\pm\rangle = \sqrt{3}a_\ell(-1+2a_\ell)\sin\alpha_i\sin\beta_i\cos\beta_i/2$, and $\langle i\ell\pm|\hat{O}_{zx}|i\ell\pm\rangle = \sqrt{3}a_\ell(-1+2a_\ell)\cos\alpha_i\sin\beta_i\cos\beta_i/2$, where a_ℓ is given by $a_1 = 7/2$ for the 4f state and $a_2 = 3/2$ for the 5d state, i.e., the total angular momentum.

For the $i = 2, 4, 5, 6, 7, 8, 9$, and 10th Yb site, we have $J_{x'} = J_x\cos\alpha_i - J_y\sin\alpha_i\cos\beta_i + J_z\sin\alpha_i\beta_i$, $J_{y'} = J_x\sin\alpha_i + J_y\cos\alpha_i\cos\beta_i - J_z\cos\alpha_i\sin\beta_i$, and $J_{z'} = J_y\sin\beta_i + J_z\cos\beta_i$. The form factors for $\Gamma = x^2 - y^2$ and xy are given by multiplying -1 to the r.h.s. of each one shown above. For $\Gamma = yz$ and zx, we have $\langle i\ell\pm|\hat{O}_{yz}|i\ell\pm\rangle = \sqrt{3}a_\ell(1-2a_\ell)\cos\alpha_i\cos\beta_i\sin\beta_i/2$ and $\langle i\ell\pm|\hat{O}_{zx}|i\ell\pm\rangle = \sqrt{3}a_\ell(-1+2a_\ell)\sin\alpha_i\sin\beta_i\cos\beta_i/2$. The other ones are the same as above.

6.6.4.6 Elastic Constants

The elastic constant is given by

$$C_\Gamma = C_\Gamma^{(0)} - g_\Gamma^2\chi_\Gamma, \quad (6.129)$$

where $C_\Gamma^{(0)}$ is the elastic constant of the background and g_Γ is the quadrupole-strain coupling constant. Here, χ_Γ is defined by $\chi_\Gamma = \lim_{\boldsymbol{q}\to\boldsymbol{0}}\chi_\Gamma^{\mathrm{F}}(\boldsymbol{q},\omega=0)$ with

$\chi_{\Gamma}^{F}(\boldsymbol{q},\omega)=\sum_{ii'}\chi_{O_\Gamma}^{ii'}(\boldsymbol{q},\omega)$. Here, $C_B=(C_{11}+2C_{12})/3$ is elastic constant for symmetery strain $\varepsilon_{xx}+\varepsilon_{yy}+\varepsilon_{zz}$, $C_v=(C_{11}-C_{12})/2$ for $(2\varepsilon_{zz}-\varepsilon_{xx}-\varepsilon_{yy})/\sqrt{3}$ and $\varepsilon_{xx}-\varepsilon_{yy}$, and C_{44} for ε_{xy}, ε_{yz}, and ε_{zx} [98].

Figure 6.21b shows the temperature dependence of $a_B^4e^2/\chi_\Gamma$ at the QCP of the valence transition identified by the RPA, i.e., for $(\varepsilon_f^{RPA}, U_{fd}^{RPA})$. Here, a_B is the Bohr radius. Below $T=1.0\times10^{-4}$, which is estimated to be 4.7 K if we assume $(pp\sigma)\approx$ 4.01 eV as above, the enhancement of χ_Γ becomes prominent as seen in $\Gamma=x^2+y^2+z^2$.

Since the positive sign and negative sign appear in the 4f and 5d terms respectively in Eq. (6.126) for $\Gamma=x^2+y^2+z^2$, $\chi_{x^2+y^2+z^2}$ is classified into the relative-charge fluctuation. This is confirmed by the fact that the T dependence of $1/\chi_{x^2+y^2+z^2}$ is well scaled by $1/\chi_{n_f-n_d}^{F}$, i.e., $a_B^4e^2/\chi_{x^2+y^2+z^2}\approx 6.5e^2/\chi_{n_f-n_d}^{F}$, as shown in Fig. 6.21b.

On the other hand, since the eigenvalues for the electric quadrupole modes for $\Gamma=2z^2-x^2-y^2, x^2-y^2, xy, yz$, and zx do not appaoach 1 for $T\to0$ in Fig. 6.21a it turns out that the quadrupole fluctuations χ_Γ do not diverge at $T=0$. However, the enhancement in χ_Γ is induced around the lowest temperatures by enhanced charge fluctuations mainly due to enhanced $\chi_{n_f}^{F}$ arising from the valence QCP. Here, $\chi_{n_f(n_d)}^{F}$ is obtained by setting only the 4f (5d) term in $\chi_{n_f\pm n_d}^{F}$ defined above [see Eq. (6.124)]. As shown in Fig. 6.20b, $\chi_{n_f}^{F}$ is indeed much larger than $\chi_{n_d}^{F}$. Thus it is understandable from the fact that $\chi_{n_f}^{ii'}(\boldsymbol{q},\omega)$ contributes dominantly to the r.h.s. of Eq. (6.128).

Although all the modes show the tendency of increase in χ_Γ for $T\to0$, the enhancement is less weak for $\Gamma=x^2-y^2$ and xy, as shown in Fig. 6.21b. This is due to each value of the corresponding form factors (see Appendix E) where the off-diagonal component with respect to the Yb site in χ_Γ, i.e. $\lim_{\boldsymbol{q}\to\boldsymbol{0}}\chi_{O_\Gamma}^{ii'}(\boldsymbol{q},\omega=0)$ for $i\neq i'$, are cancelled partly for $\Gamma=x^2-y^2$ by their plus and minus contributions and is diminished mostly for $\Gamma=xy$ by vanishment of the form factors except for $i=1, 3, 11$, and 12.

This indicates that softening in the bulk modulus C_B occurs most remarkably and also in C_v at low temperatures for $T\ll\Delta$. Interestingly, in the high-T side of Fig. 6.21b, the enhancement of χ_Γ also occurs as remarkably seen for $\Gamma=x^2+y^2+z^2$, which is also due to the form factors (see Appendix E) where the off-diagonal component $\lim_{\boldsymbol{q}\to\boldsymbol{0}}\chi_{O_\Gamma}^{ii'}(\boldsymbol{q},\omega=0)$ for $i\neq i'$ gives rise to the reduction of χ_Γ around $T=2.0\times10^{-4}$.

It is noted that in the polycrystal the vanishment of the form factors for $\Gamma=xy$ may become incomplete since the quantization axis of the CEF at each Yb site deviates from the one shown in Fig. 6.19f in the different domains, giving rise to finite values of $\langle i\ell\pm|\hat{O}_\Gamma|i\ell\pm\rangle$. This effect is expected to cause the tendency of softening even in the transverse mode $\Gamma=xy$, as shown in the longitudinal mode $\Gamma=x^2$ in Fig. 6.21b. This also applies to the QC. We note that this conclusion does not depend on the detail of the orientation of the $\Psi^{4f}(\hat{\boldsymbol{r}}_i)$ in the CEF ground state (see Fig. 6.19f) since the deviation of the xyz coordinate at each Yb site in the domains in the polycrystal is essential for causing the softening in both modes.

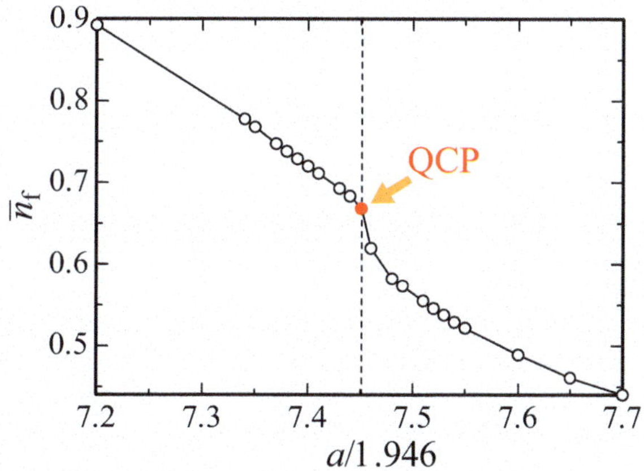

Fig. 6.22 The 4f-hole number $\bar{n}_{\rm f}$ versus the lattice constant $a/1.946$ Å. $\bar{n}_{\rm f}$ at the QCP indicated by an arrow (filled circle) is plotted at $a = 14.5$ Å (dashed line) [86]. Reprinted from Ref. [86]. Copyright 2018 IOP Publishing

6.6.5 Lattice Constant Dependence of Yb Valence

In this section, the lattice constant dependence of the Yb valence in the quasicrystal $Yb_{15}Al_{34}Au_{51}$ and the approximant crystal is discussed [86]. On the basis of the extended Anderson lattice model in the approximant crystal $Yb_{14}Al_{35}Au_{51}$, which consists of the Yb 4f and 5d orbitals and Al 3p orbitals, the QCP of the valence transition is identified in the ground-state phase diagram [86]. Here, the onsite Coulomb repulsion $U_{\rm ff}$ is set to be infinity. In Fig. 6.22, $\bar{n}_{\rm f} = 0.67$ at the valence QCP is plotted as the filled circle, where the dashed line is the guide for the eyes at the lattice constant of the approximant crystal $a = 14.5$ Å. It is noted that the horizontal axis is expressed as the lattice constant of the six spatial dimension a_{6D} and hence the three-dimensional lattice constant a is divided by the scale factor 1.946, i.e., $a_{6D} = a/1.946$.[4] If the relation between the Yb valence ν and $\bar{n}_{\rm f}$ is employed, $\nu = \bar{n}_{\rm f} + 2$, the above value is compatible with the experimentally observed Yb valence $\nu \sim 2.66$ in the quasicrystal $Yb_{15}Al_{34}Au_{51}$ [27].

Then, by changing the lattice constant, the 4f-hole number per Yb site is calculated without inputting any parameters. Namely, the lattice-constant dependence of the transfers, the hybridizations, and the f level are automatically taken into account and the results are plotted by open circles in Fig. 6.22, which are obtained without inputting any adjustable parameters.

As for the transfer integrals of the 3p orbital at Al $t^{\rm pp}_{j\xi m, j'\xi' m'}$, the 5d orbital at Yb $t^{\rm dd}_{ji\eta, ji'\eta'}$, the hybridization between the 4f orbital at Yb and 3p orbital at Al $V^{\rm fp}_{ji\eta, j'\xi\sigma}$, the hybridization between the 5d orbital at Yb and the 3p orbital at Al $V^{\rm dp}_{ji\eta, j'\xi\sigma}$, we employ the distance dependences shown by the linear Muffin-Tin orbital (LMTO) method [61, 62]. Namely, the distance dependence is taken into account as

[4] The quasicrystal is ragared as the projection to the three spatial dimension of the periodic lattice in the six spatial dimension. The $a_{\rm 6D}$ is the lattice parameter in the six dimension.

$$t^{\mathrm{pp}}_{j\xi m,j'\xi'm'},\ t^{\mathrm{dd}}_{ji\eta,j'i'\eta'},\ V^{\mathrm{fp}}_{ji\eta,j'\xi\sigma},\ V^{\mathrm{dp}}_{ji\eta,j'\xi\sigma} \sim \frac{1}{r^{l+l'+1}}, \tag{6.130}$$

where l, l' is the orbital angular momentums for each orbital of 4f ($l = 3$), 5d ($l = 2$), and 3p ($l = 1$).

As for the onsite 4f-4f interaction U_{ff} and the onsite 4f-5d interaction U_{fd}, the effect of shrinkage and expansion of the lattice is considered to be irrelevant. Hence, these values are set to be unchanged from the values at the valence QCP when the lattice constant is varied.

As for the f level, we speculate that the effect of the f-c hybridization is important for the actual crystalline-electric-field (CEF) levels in the following reason. When we apply the pressure to the Yb-Al-Au quasicrystal and approximant crystal, the Yb valence increases [73]. This implies that the effect of the decrease in the f-hole level ε_{f} overcomes the effect of the increase in the f-c hybridization $|V^{\mathrm{fp}}_{ji\eta,j'\xi\sigma'}|$ under pressure. In the point charge model, the CEF ground-state energy depends on the CEF parameter B^m_ℓ, which is the coefficient of the Stevens operator as discussed below [95]. Since B^m_ℓ has the distance dependence as $B^m_\ell \propto R_k^{-\ell-1}$, B^m_ℓ with smaller ℓ usually has a larger value. Actually, the measurement of the CEF level in the approximant crystal $\mathrm{TbCd_6}$ has revealed that B^0_2 is dominant [99, 100]. If this contributes dominantly to the CEF in the quasicrystal $\mathrm{Yb_{15}Al_{34}Au_{51}}$ and approximant crystal, the f level ε_{f} is expected to behave as $\varepsilon_{\mathrm{f}} \propto 1/r^3$.[5] However, this r dependence in ε_{f} cannot overcome the effect of the f-c hybridization under pressure, since the hybridization has the distance dependence as $V^{\mathrm{fp}}_{ji\eta,j'\xi\sigma'} \propto 1/r^5$ as noted in Eq. (6.130).

On the other hand, the CEF level can arise from the f-c hybridization. The CEF level can be evaluated by the second-order perturbation theory with respect to the f-c hybridization [101] and this leads to the distance dependence as

$$\varepsilon_{\mathrm{f}} \propto \frac{|V^{\mathrm{f}}_{ji\eta,j'\xi\sigma'}|^2}{\Delta} \propto \frac{1}{r^{10}}. \tag{6.131}$$

Here, Δ is the excitation energy to the intermediate state with the $\mathrm{f^0}$ or $\mathrm{f^2}$ configuration in the hole picture. The actual CEF level is considered to be contributed from both effects by the point charge model and f-c hybridization. Hence, the present analysis is proceeded as the latter effect is dominant for simplicity.

Taking into account the lattice constant dependence of the model parameters in the extended Anderson lattice model, the ground state is obtained by the slave-boson mean field theory as noted in Sect. 6.6.4.2. The results of $\bar{n}_{\mathrm{f}}$ are plotted by open circles in Fig. 6.22, which show a sudden decrease from critical Yb-valence fluctuations at the valence QCP. As the lattice constant a increases from the value at the QCP, $\bar{n}_{\mathrm{f}}$ first decreases steeply and then turns into a gradual decrease. On the other hand, as a decreases from the value at the QCP, $\bar{n}_{\mathrm{f}}$ increases gradually. The asymmetry

---

[5] Recently, the general formulation of the CEF in the rare-earth base quasicrystal has been developed on the basis of the point charge model and the CEF in the quasicrystal $\mathrm{Yb_{15}Al_{34}Au_{51}}$ has been analyzed in ref. [102].

of the a dependence of $\bar{n}_{\mathrm{f}}$ around the QCP is ascribed to the fact that increasing and decreasing lattice constants are reflected in the changes in the model parameters differently. This will be explained below in the ground-state phase diagram (see Fig. 6.23a) where the first-order valence transition (thick solid line) terminates at the valence QCP (filled circle) and the sharp valence-crossover line (thick dashed line) extends. The 4f-hole number per Yb $\bar{n}_{\mathrm{f}}$ is also plotted as the thin dashed line.

When a increases, the f-hole level ε_{f} increases (see Fig. 6.23b) and the f-c hybridization $|V^{\mathrm{fp}}_{ji\eta,j'\xi\sigma'}|$ decreases. The decrease in the f-c hybridization makes the location of the QCP shift to the ε_{f}-increasing and U_{fd}-decreasing direction, i.e., the first-order valence-transition line extends as in Fig. 6.23c. Hence, as a increases, the increase in ε_{f} (Fig. 6.23b) and the shift of the QCP (Fig. 6.23c) occur simultaneously. Since the contour lines for $\bar{n}_{\mathrm{f}}$ accompanied by the QCP also shift, this implies that the system with expanding a proceeds to the right-upper direction from the QCP in Fig. 6.23d. Since this makes the system go across the relatively dense contour area than the case where the system proceeds to the right-upper direction from the QCP as in Fig. 6.23a, $\bar{n}_{\mathrm{f}}$ changes sharply for a certain change in a (e.g., see a steep decrease in $\bar{n}_{\mathrm{f}}$ from the QCP to $a/1.946 \approx 7.45$ in Fig. 6.22). Hence, the steep decrease in $\bar{n}_{\mathrm{f}}$ immediately appears for a slight increase in a from the vertical dashed line in Fig. 6.22.

On the other hand, when a decreases, ε_{f} decreases (Fig. 6.23e) while $|V^{\mathrm{fp}}_{ji\eta,j'\xi\sigma'}|$ increases. The increase in $|V^{\mathrm{fp}}_{ji\eta,j'\xi\sigma'}|$ makes the location of the QCP shift to the ε_{f}-decreasing and U_{fd}-increasing direction, i.e., the first-order valence-transition line shortens as in Fig. 6.23f. Hence, as a decreases, the decrease in ε_{f} (Fig. 6.23e) and the shift of the QCP (Fig. 6.23f) occur simultaneously. Since the contour lines for $\bar{n}_{\mathrm{f}}$ accompanied by the QCP also shift, this implies that the system with shrinking a

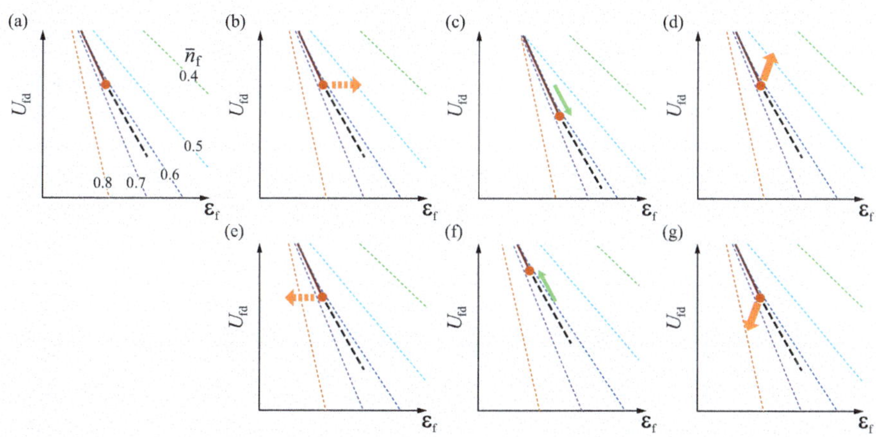

Fig. 6.23 **a** The ground-state phase diagram in the ε_{f}-U_{fd} plane. The first-order valence transition (thick solid line) terminates at the valence QCP (filled circle) and the sharp valence-crossover line (thick dashed line) extends. The 4f-hole number per Yb $\bar{n}_{\mathrm{f}}$ is also plotted as a thin dashed line [86]. **b**, **c**, and **d**: the process of the shrinkage of the lattice. **e**, **f**, and **g**: the process of the expansion of the lattice

proceeds to the left-lower direction from the QCP as in Fig. 6.23g. Since this makes the system proceed toward the relatively-sparse contour area (i.e., the interval of the contour lines becomes spread in the left-lower direction in Fig. 6.23a), $\bar{n}_f$ does not change sharply for a certain change in a (e.g., see a gradual increase in $\bar{n}_f$ from the QCP in Fig. 6.22). Namely, the emergence of the steep decrease in $\bar{n}_f$ just next to the QCP reflects the enhanced critical valence fluctuations χ_v as seen as the divergent slope $-\partial n_f/\partial \varepsilon_f (= \chi_v)$ in the vicinity of the QCP (see Fig. 6.20a).

The results shown in Fig. 6.22 are calculated in the approximant crystal $Yb_{14}Al_{35}Au_{51}$. The local structure around the Yb site is common in both the approximant crystal and the quasicrystal and the valence transition is caused by the onsite 4f-5d Coulomb repulsion at Yb site, which is essentially local. This suggests that the valence QCP is realized in the quasicrystal $Yb_{15}Al_{34}Au_{51}$. If the present theory is correct, it should be able to predict the experiment in the approximant crystal $Yb_{14}Al_{35}Au_{51}$. At ambient pressure, the approximant crystal shows the Fermi-liquid behavior. By applying pressure, it shows the same quantum criticality as that in the quasicrystal at 2 GPa. So, the sharp change in the Yb valence shown in Fig. 6.22 is expected to appear at 2 GPa in the approximant crystal. This prediction [86] has actually been observed by Imura [29], which will be noted in Sect. 6.6.6.

6.6.6 *Direct Observation of Quantum Valence Criticality in Quasicrystal and Pressurized Approximant*

The systematic variation of homologous elements of Au and Al in the quasicrystal $Yb_{15}Al_{34}Au_{51}$, i.e. $Yb_{15}(Al_{1-x}Ga_x)_{34}(Au_{1-y}Cu_y)_{51}$, have been synthesized [29]. The mean valence of Yb of each composition has been observed by the measurement of the partial-fluorescence-yield x-ray absorption spectroscopy (PFY-XAS) at room temperature. The results are plotted as the lattice parameter a_{6D} dependence of the Yb mean valence in Fig. 6.24a. Here, a_{6D} is the lattice parameter in spatial six dimensions. Interestingly, the quasicrystal $Yb_{15}Al_{34}Au_{51}$ (red diamond), which exhibits the quantum criticality, is located at the point where the Yb valence starts to change sharply. As for the 1/1 approximant crystal, $Yb_{14}Al_{35}Au_{51}$ at the pressure of 2 GPa, which exhibits the same quantum criticality, is located at the point where the Yb-valence starts to change sharply. As discussed in Sect. 6.6.5, the valence QCP is located at the point where the Yb valence changes sharply as shown in Fig. 6.22. As the lattice constant changes, the position of the f level ε_f shifts. Hence, the effective parameter corresponding to the horizontal axis in Figs. 6.22 and 6.24 is the f level ε_f, which is the conjugate field of the 4f-hole number per Yb $\bar{n}_f$ [see Eq. (6.107)]. Then, the steep slope in the Yb valence i.e. $\bar{n}_f$ versus ε_f (lattice parameter) plot indicates that the valence susceptibility

$$\chi_v = -\frac{\partial \bar{n}_f}{\partial \varepsilon_f} \tag{6.132}$$

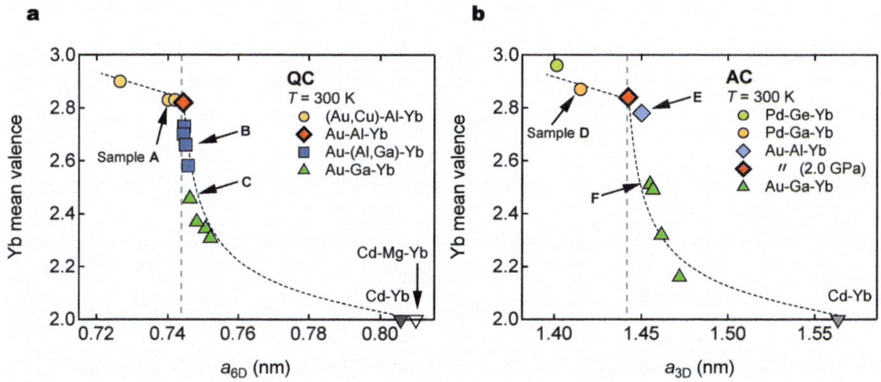

Fig. 6.24 Lattice parameter dependence of Yb mean valence. Dependences of Yb mean valence (ν) on 6D lattice parameter (a_{6D}) of the quasicrystals **a** and 3D lattice parameter (a_{3D}) of the approximant crystals **b**. Vertical dashed lines indicate critical lattice parameters $a^c_{6D} = 0.7443$ nm of Au-Al-Yb quasicrystal and $a^c_{3D} = 1.4424$ nm of Au-Al-Yb AC at $P \approx 2.0$ GPa, respectively. Dotted curves are guides for the eyes. Sample A-F indicate $(Au_{0.5}Cu_{0.1})_{49}Al_{34}Yb_{17}$ quasicrystal, $Au_{49}(Al_{0.7}Ga_{0.3})_{34}Yb_{17}$ quasicrystal, $Au_{49}Ga_{35}Yb_{16}$ QC, $Pd_{30}Ga_{55}Yb_{15}$ approximant crystal, $Au_{49}Al_{36}Yb_{15}$ approximant crystal and $Au_{45}Ga_{40}Yb_{15}$ approximant crystal, respectively [29]. Reprinted with permission from Ref. [29]. Copyright 2020 Springer Nature

diverges. Namely, the critical valence fluctuation diverges. These theoretical and experimental results indicate that the origin of the quantum criticality commonly observed in the quasicrystal and pressurized approximant crystal is the Yb-valence fluctuations originating from critical Yb-valence fluctuations at the valence QCP.

References

1. T. Moriya, *Spin Fluctuations in Itinerant Electron Magnetism* (Springer, Berlin, 1985)
2. J.A. Hertz, Phys. Rev. B **14**, 1165 (1976)
3. A.J. Millis, Phys. Rev. B **48**, 7183 (1993)
4. O. Trovarelli, C. Geibel, S. Mederle, C. Langhammer, F.M. Grosche, P. Gegenwart, M. Lang, G. Sparn, F. Steglich, Phys. Rev. Lett. **85**, 626 (2000)
5. P. Gegenwart, T. Westerkamp, C. Krellner, Y. Tokiwa S. Paschen, C. Geibel, F. Steglich, E. Abrahams, and Q. Si, Science **315**, 969 (2007)
6. P. Gegenwart, J. Custers, Y. Tokiwa, C. Geibel, F. Steglich, Phys. Rev. Lett. **94**, 076402 (2005)
7. J. Custers, P. Gegenwart, H. Wilhelm, K. Neumaier, Y. Tokiwa, O. Trovarelli, C. Geibel, F. Steglich, C. Pépin, P. Coleman, Nature **424**, 524 (2003)
8. K. Ishida, K. Okamoto, Y. Kawasaki, Y. Kitaoka, O. Trovarelli, C. Geibel, F. Steglich, Phys. Rev. Lett. **89**, 107202 (2002)
9. S. Kambe, H. Sakai, Y. Tokunaga, G. Lapertot, T.D. Matsuda, G. Knebel, J. Flouquet, R.E. Walstedt, Nat. Phys. **10**, 840 (2014)
10. Q. Si, S. Rabello, K. Ingersent, J.L. Smith, Nature **413**, 804 (2001)
11. P. Coleman, C. Pépin, Q. Si, R. Ramazashvili, J. Phys.: Condens. Matter **13**, R723 (2001)
12. T. Misawa, Y. Yamaji, M. Imada, J. Phys. Soc. Jpn. **77**, 093712 (2008)
13. V.R. Shaginyan, M.. Ya.. Amusia, A.Z. Msezane, K.G. Popov, Phys. Rep. **492**, 31 (2010)

14. A. Hackl, M. Vojta, Phys. Rev. Lett. **106**, 137002 (2011)
15. E. Abrahams, J. Schmalian, P. Wölfle, Phys. Rev. B **90**, 045105 (2014)
16. S. Paschen, T. Luhmann, S. Wirth, P. Gegenwart, O. Trovarelli, C. Geibel, F. Steglich, P. Coleman, Q. Si, Nature **432**, 881 (2004)
17. M.R. Norman, Phys. Rev B **71**, 220405R (2005)
18. S. Danzenbächer, D.V. Vyalikh, K. Kummer, C. Krellner, M. Holder, M. Höppner, Yu. Kucherenko, C. Geibel, M. Shi, L. Patthey, S.L. Molodtsov, C. Laubschat, Phys. Rev. Lett. **107**, 267601 (2011)
19. S. Nakatsuji, K. Kuga, Y. Machida, T. Tayama, T. Sakakibara, Y. Karaki, H. Ishimoto, S. Yonezawa, Y. Maeno, E. Pearson, G.G. Lonzarich, L. Balicas, H. Lee, Z. Fisk, Nat. Phys. **4**, 603 (2008)
20. Y. Matsumoto, S. Nakatsuji, K. Kuga, Y. Karaki, N. Horie, Y. Shimura, T. Sakakibara, A.H. Nevidomskyy, P. Coleman, Science **331**, 316 (2011)
21. M. Okawa, M. Matsunami, K. Ishizaka, R. Eguchi, M. Taguchi, A. Chainani, Y. Takata, M. Yabashi, K. Tamasaku, Y. Nishino, T. Ishikawa, K. Kuga, N. Horie, S. Nakatsuji, S. Shin, Phys. Rev. Lett. **104**, 247201 (2010)
22. T. Tomita, K. Kuga, Y. Uwatoko, P. Coleman, S. Nakatsuji, Science **349**, 506 (2015)
23. K. Kuga, Y. Matsumoto, M. Okawa, S. Suzuki, T. Tomita, K. Sone, Y. Shimura, T. Sakakibara, D. Nishio-Hamane, Y. Karaki, Y. Takata, M. Matsunami, R. Eguchi, M. Taguchi, A. Chainani, S. Shin, K. Tamasaku, Y. Nishino, M. Yabashi, T. Ishikawa, and S. Nakatsuji, Sci. Adv. **4**, eaao3547 (2018)
24. D.E. MacLaughlin, K. Kuga, L. Shu, O.O. Bernal, P.C. Ho, S. Nakatsuji, K. Huang, Z.F. Ding, C. Tan, J. Zhang, Phys. Rev. B **93**, 214421 (2016)
25. K. Miyake, S. Watanabe, Phys. Rev. B **98**, 075125 (2018)
26. K. Deguchi, S. Matsukawa, N.K. Sato, T. Hattori, K. Ishida, H. Takakura, T. Ishimasa, Nat. Mater. **11**, 1013 (2012)
27. T. Watanuki, S. Kashimoto, D. Kawana, T. Yamazaki, A. Machida, Y. Tanaka, T.J. Sato, Phys. Rev. B **86**, 094201 (2012)
28. N.K. Sato, T. Ishimasa, K. Deguchi, K. Imura, J. Phys. Soc. Jpn. **91**, 072001 (2022)
29. K. Imura, H. Yamaoka, S. Yokota, K. Sakamoto, Y. Yamamoto, T. Kawai, K. Namba, S. Hirokawa, K. Deguchi, N. Hiraoka, H. Ishii, J. Mizuki, T. Ishimasa, N.K. Sato, Sci. Rep. **10**, 17116 (2020)
30. S. Watanabe, K. Miyake, Phys. Rev. Lett. **105**, 186403 (2010)
31. K. Miyake, J. Phys.: Condens. Matter **19**, 125201 (2007)
32. Y. Onishi, K. Miyake, J. Phys. Soc. Jpn. **69**, 3955 (2000)
33. There exists the region where the cubic term in Eq. (6.5) vanishes in the $\varepsilon_{\rm f}$-$U_{\rm fc}$ plane for each V, making the higher order terms v_j $(j \geq 3)$ irrelevant for $d + z > 4$
34. S. Watanabe, A. Tsuruta, K. Miyake, J. Flouquet, Phys. Rev. Lett. **100**, 236401 (2008)
35. T. Sugibayashi, A. Tsuruta, K. Miyake, Physica C **470**, S550 (2010)
36. K. Miyake, O. Narikiyo, Y. Onishi, Physica B **259–261**, 676 (1999)
37. S. Watanabe, K. Miyake, Solid State Commun. **306**, 113774 (2020)
38. K. Ueda, T. Moriya, J. Phys. Soc. Jpn. **39**, 605 (1975)
39. A.T. Holmes, D. Jaccard, K. Miyake, Phys. Rev. B **69**, 024508 (2004)
40. S. Watanabe and K. Miyake, unpublished (arXiv:0906.3986)
41. H. Kobayashi, Y. Sakaguchi, M. Oura, S. Ikeda, K. Kuga, S. Suzuki, S. Nakatsuji, R. Masuda, Y. Kobayashi, M. Seto, Y. Yoda, K. Tamasaku, Y. Komijani, P. Chandra, P. Coleman, Science **379**, 908 (2023)
42. C.M. Varma, Phys. Rev. Lett. **75**, 898 (1995)
43. Y. Takahashi, J. Phys.: Condens. Matter **11**, 6439 (1999)
44. A. Ramires, P. Coleman, A.H. Nevidomskyy, A.M. Tsvelik, Phys. Rev. Lett. **109**, 176404 (2012)
45. S. Watanabe, K. Miyake, J. Phys.: Condens. Matter **24**, 294208 (2012)
46. S. Watanabe, K. Miyake, J. Phys. Soc. Jpn. **83**, 103708 (2014)

47. R. P. Feynman, *Statistical Mechanics* (Addison-Wesley, Reading, Massachusetts, 1990) Sect. 3.4
48. A.H. Nevidomskyy, P. Coleman, Phys. Rev. Lett. **102**, 077202 (2009)
49. H. Ikeda, K. Miyake, J. Phys. Soc. Jpn. **65**, 1769 (1996)
50. Y. Aoki, T. D. Matsuda, H. Sugawara, H. Sato, H. Ohkuni, R. Settai, Y. Ōnuki, E. Yamamoto, Y. Haga, A. V. Andreev, V. Sechovsky, L. Havela, H. Ikeda and K. Miyake, J. Magn. Magn. Mater. **177-181**, 271 (1998)
51. S. Watanabe, J. Phys. Soc. Jpn. **69**, 2947 (2000)
52. K. Miyake, H. Ikeda, J. Phys. Soc. Jpn. **75**, 033704 (2006)
53. K. Kummer, C. Geibel, C. Krellner, G. Zwicknagl, C. Laubschat, N.B. Brookes, D.V. Vyalikh, Nat. Commun. **9**, 2011 (2018)
54. S. Matsukawa, K. Deguchi, K. Imura, T. Ishimasa, N.K. Sato, J. Phys. Soc. Jpn. **85**, 063706 (2016)
55. In Ref. 54, it was reported that the T/B scaling observed in the pressurized approximant crystal is applicable to the Yb-Au-Al quasicrystal
56. T. Ishimasa, Y. Tanaka, S. Kashimoto, Phil. Mag. **91**, 4218 (2011)
57. *The Physics of Quasicrystals* ed. by P. J. Steinhardt and S. Ostlund. (World Scientific, Singapole, 1987)
58. A.I. Goldman, R.F. Kelton, Rev. Mod. Phys. **65**, 213 (1993)
59. S. Watanabe, K. Miyake, J. Phys. Soc. Jpn. **82**, 083704 (2013)
60. S. Watanabe, K. Miyake, J. Phys. Soc. Jpn. **85**, 063703 (2016)
61. O.K. Andersen, O. Jepsen, Physica B **91**, 317 (1977)
62. O.K. Andersen, W. Klose, H. Nohl, Phys. Rev. B **17**, 1209 (1978)
63. W.A. Harrison, *Electronic Structure and the Properties of Solids: The Physics of the Chemical Bond* (W. H. Freeman and Co., San Francisco, CA, 1980)
64. N. Read, D.M. Newns, J. Phys. C: Solid State Phys. **16**, 3273 (1983)
65. K. Miyake, J. Phys.: Condens. Matter **19**, 125201 (2007)
66. E. Grüneisen, Ann. Phys. **344**, 257 (1912)
67. L. Zhu, M. Garst, A. Rosch, Q. Si, Phys. Rev. Lett. **91**, 066404 (2003)
68. M. Garst, A. Rosch, Phys. Rev. B **72**, 205129 (2005)
69. For review, see P. Gegenwart, Rep. Prog. Phys. **79**, 114502 (2016)
70. R. Küchler, N. Oeschler, P. Gegenwart, T. Cichorek, K. Neumaier, O. Tegus, C. Geibel, J.A. Mydosh, F. Steglich, L. Zhu, Q. Si, Phys. Rev. Lett. **91**, 066405 (2003)
71. R. Küchler, P. Gegenwart, K. Heuser, E.-W. Scheidt, G.R. Stewart, F. Steglich, Phys. Rev. Lett. **93**, 096402 (2004)
72. K. Grube, S. Zaum, O. Stockert, Q. Si, and H. v. Löhneysen, Nat. Phys. **13**, 742 (2017)
73. T. Watanuki, private communication
74. P. Gegenwart, Philos. Mag. **97**, 3415 (2017); A. Wörl, S. Matsukawa, N. K. Sato, and P. Gegenwart, in preparation
75. R. Takke, M. Niksch, W. Assmus, B. Lüthi, R. Pott, R. Schefzyk, and D. K. Wohlleben, Z. Phys. B **44**, 33 (1981)
76. P. Thalmeier, P. Fulde, Europhys. Lett. **1**, 367 (1986)
77. A.V. Goltsev, M.M. Abd-Elmeguid, J. Phys.: Condens. Matter **17**, 5813 (2005)
78. Y. Ōno, T. Matsuura, and Y. Kuroda, Physica C **159**, 878 (1989)
79. N. Shibata, T. Nishino, K. Ueda, C. Ishii, Phys. Rev. B **53**, R8828 (1996)
80. S. Watanabe, Solid State Cummun. **336**, 114425 (2021)
81. J.C. Slater, G.F. Koster, Phys. Rev. **94**, 1498 (1954)
82. K. Takegahara, Y. Aoki, A. Yanase, J. Phys.: Solid. St. Phys. **13**, 583 (1980)
83. E.P. Wigner, *Group Theory and its Application to the Quantum Mechanics of Atomic Spectra* (Academic Press, New York, 1959)
84. For example, see A. Messiah, *Quantum Mechanics* Volume 2 (Amsterdam: North-Holland, 1965)
85. N. Read, D.M. Newns, J. Phys. C **16**, 3273 (1983)
86. S. Watanabe, K. Miyake, J. Phys.: Condens. Matter **30**, 184001 (2018)

87. Note that at the valence QCP, the critical valence fluctuations beyond the MF theory contribute to the mass enhancement, giving rise to the enhanced specific-heat coefficient. See, for detail, ref. 37
88. H. Tonai, N. Sasabe, T. Uozumi, N. Kawamura, M. Mizumaki, J. Phys. Soc. Jpn. **86**, 093704 (2017)
89. T. Takimoto, T. Hotta, K. Ueda, Phys. Rev. B **69**, 104504 (2004)
90. S. Watanabe, M. Imada, K. Miyake, J. Phys. Soc. Jpn. **75**, 043710 (2006)
91. S. Watanabe, K. Miyake, J. Phys. Soc. Jpn. **88**, 033701 (2019)
92. S. Watanabe, J. Phys. Soc. Jpn. **89**, 073702 (2020)
93. A.J. Freeman, R.E. Watson, J. Magn. Magn. Mater. **12**, 11 (1979)
94. K.W.H. Stevens, Proc. Phys. Soc. A **65**, 209 (1952)
95. M.T. Hutchings, Solid State Physics **16**, 227 (1964)
96. H. Kontani, T. Saito, S. Onari, Phys. Rev. B **84**, 024528 (2011)
97. S. Hayami, M. Yatsushiro, Y. Yanagi, H. Kusunose, Phys. Rev. B **98**, 165110 (2018)
98. B. Lüthi, *Physical Acoustics in the Solid State* (Springer, Berlin, 2004)
99. S. Jazbec, S. Kashimoto, P. Koželj, S. Vrtnik, M. Jagodič, Z. Jagličić, and J. Dolinšek, Phys. Rev. B **93**, 054208 (2016)
100. P. Das, P.-F. Lory, R. Flint, T. Kong, T. Hiroto, S.L. Bud'ko, P.C. Canfield, M. de Boissieu, A. Kreyssig, A.I. Goldman, Phys. Rev. B **95**, 054408 (2017)
101. H. Takahashi, T. Kasuya, J. Phys. C: Solid State Phys. **18**, 2709 (1985)
102. S. Watanabe, M. Kawamoto, J. Phys. Soc. Jpn. **90**, 063701 (2021)

Chapter 7
Interplay Between Magnetic QCP and Valence QCP

The quantum critical phenomena for the magnetic transition are discussed in Chap. 5. The quantum critical phenomena for the valence transition are discussed in Chap. 6. The interplay of the magnetic order and valence fluctuations offers a key concept for understanding the heavy fermion systems beyond the ordinary concept based on Doniach's phase diagram. In this chapter, the interplay of the magnetic QCP and the valence QCP is discussed. In Sect. 7.1, $CeRhIn_5$ is focused on as a prototypical material to demonstrate how such an interplay plays a critical role in understanding the puzzling behaviors. In Sect. 7.2, it is discussed that the phase diagrams in the heavy fermion systems can be understood comprehensively on the basis of the extended Anderson lattice model.

7.1 Drastic Change of Fermi Surface in $CeRhIn_5$ Under Pressure

7.1.1 Experiments in $CeRhIn_5$

$CeRhIn_5$ undergoes an antiferromagnetic (AF) transition at $T_N = 3.8$ K with the ordered vector $\boldsymbol{Q} = (1/2, 1/2, 0.297)$ at ambient pressure [1]. When pressure is applied under a magnetic field larger than the upper critical field, the AF order is suppressed at $P = P_c \approx 2.35$ GPa [2–8]. Interestingly, a drastic change in Fermi surfaces was discovered at $P = P_c$ by de Haas-van Alphen (dHvA) measurement as shown in Fig. 7.1a [6]: For $0 \le P < P_c$, the main dHvA frequencies are in good agreement with those of $LaRhIn_5$ where 4f electrons are absent. On the other hand, for $P > P_c$, the dHvA frequencies were identified to be approximately the same as those of $CeCoIn_5$, where 4f electrons are itinerant, contributing to the formation of the Fermi surface. The electrons on a typical Fermi surface, the cylindrical β_2 branch, show the mass enhancement from $m^* = 6m_0$ at $P = 0$ to $60m_0$ at $P \approx 2.2$ GPa with

S. Watanabe and K. Miyake, *Quantum Critical Phenomena of Valence Transition*, Springer Tracts in Modern Physics 289,
https://doi.org/10.1007/978-981-99-3518-5_7

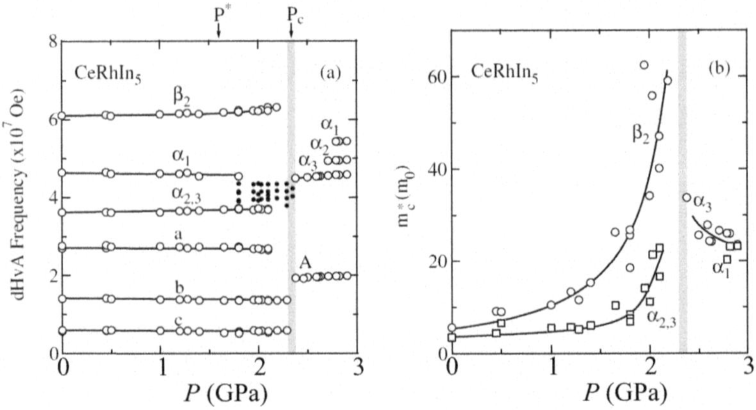

Fig. 7.1 Pressure dependences of **a** the dHvA frequency and **b** cyclotron mass in $CeRhIn_5$ [6]. Reprinted with permission from Ref. [6]. Copyright 2005 The Physical Society of Japan

m_0 being a free-electron mass as shown in Fig. 7.1b [6]. For $P > P_c$, the dHvA signal of the β_2 branch is lost probably because of a too-large effective mass close to $100m_0$, where the heavy-electron state is realized in the paramagnetic metal phase [6].

So far, it has been considered that this drastic change of the Fermi surface might be explained in terms of the "localized" to "itinerant" transition of f electrons [9]. However, this concept encounters a serious difficulty in elucidating the experimental fact that the effective mass of electrons is enhanced even at $P = 0$ with the Sommerfeld constant $\gamma \approx 56$ mJmol^{-1}K^{-2} [2], which is about 10 times larger than that of $LaRhIn_5$ [10, 11].

Furthermore, resistivity measurements revealed that a striking anomaly emerges near $P = P_c$ [4]: The low-T resistivity $\rho(T = 2.25$ K) has a sharp peak at $P = P_c$ as shown in Fig. 7.2a, suggesting that the residual resistivity ρ_0 markedly increases near $P = P_c$ [4, 8, 9]. The low-temperature resistivity also exhibits an anomalous behavior: The T-linear resistivity $\rho(T) \propto T$ emerges most prominently near $P = P_c$ with a wide T range of up to ~10 K [4, 8, 9]. This behavior is different from that of normal metals described by the Fermi liquid $\rho \propto T^2$ as well as the conventional quantum criticality scenario near the AF quantum critical point (QCP) in three dimensions (3D) $\rho \propto T^{1.5}$ [12–14]. The origin and mechanism of these transport anomalies as well as the drastic change of the Fermi surface at $P = P_c$ accompanied by huge mass enhancement need to be explained.

An important measurement has been performed under pressure and magnetic field [8]. Knebel et al. measured the T^2 coefficient A of the resistivity in the $T \to 0$ limit at the magnetic field $H = 15$ T and found that A increases as pressure increases toward $P = P_c$. They have shown that under pressure, $\sqrt{A}$ increases similarly to the effective mass of electrons m^* of the β_2 branch obtained by dHvA measurement [6], satisfying $\sqrt{A}/m^* =$ const. scaling [see Fig. 7.2b]. This indicates that the mass enhancement is *not* caused by the quantum criticality of the AF spin fluctuations, since in the case of the 3D (2D) AF QCP, $\sqrt{A}/m^*$ is expected to show $T^{-1/4}$

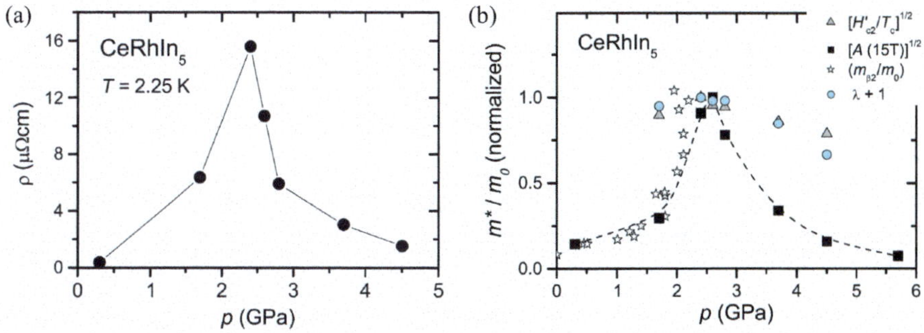

Fig. 7.2 **a** The pressure dependence of resistivity at $T = 2.25$ K in CeRhIn$_5$ at zero magnetic field [8]. **b** Estimation of the pressure dependence of the effective mass m^* determined from (i) the initial slope of the upper critical field, $m^* \propto \sqrt{H_{c2}}/T_c$ with T_c being the superconducting transition temperature (triangles), (ii) the A coefficient measured at 15 T, $m^* \propto \sqrt{A}$ (squares), and (iii) the pressure variation of the cyclotron mass of the β_2 branch from the dHvA experiment from Ref. [6] (stars). All values are normalized to the value at P_c and the dashed line is the guide for the eyes [8]. Reprinted with permission from Ref. [8]. Copyright 2008 The Physical Society of Japan

($T^{-1/2}$) divergence [12–14] as discussed in Sect. 5.3. Namely, the $\sqrt{A}/m^*$ =const. scaling strongly suggests that the mass enhancement near $P = P_c$ purely comes from the effect of the band with electron correlations.

7.1.2 Theory Treating Equal Footing of Magnetic and Valence Transitions

To resolve the outstanding puzzle in CeRhIn$_5$, theoretical analysis which treats equal footing of magnetic and valence transition was performed [15]. The theory on the basis of the extended Anderson lattice model and the comparison with the experiments are discussed in the following subsections.

7.1.2.1 Model and Method

To describe the essential part of the physics of CeRhIn$_5$, we consider the extended Anderson lattice model which consists of f and conduction electrons

$$H = H_c + H_f + H_{hyb} + H_{U_{fc}}. \tag{7.1}$$

The conduction-electron part is

$$H_c = \sum_{k\sigma} \varepsilon_k c^\dagger_{k\sigma} c_{k\sigma} \tag{7.2}$$

with the conduction band ε_k. The f-electron part is

$$H_{\mathrm{f}} = \varepsilon_{\mathrm{f}} \sum_{i\sigma} n^{\mathrm{f}}_{i\sigma} + U \sum_{i=1}^{N} n^{\mathrm{f}}_{i\uparrow} n^{\mathrm{f}}_{i\downarrow} \tag{7.3}$$

with the f level ε_{f} and the onsite Coulomb repulsion for f electrons U_{fc}. The hybridization between f and conduction electrons is

$$H_{\mathrm{hyb}} = V \sum_{i\sigma} \left(f^{\dagger}_{i\sigma} c_{i\sigma} + c^{\dagger}_{i\sigma} f_{i\sigma} \right). \tag{7.4}$$

The Coulomb repulsion between f and conduction electrons is

$$H_{U_{\mathrm{fc}}} = U_{\mathrm{fc}} \sum_{i=1}^{N} n^{\mathrm{f}}_i n^{\mathrm{c}}_i. \tag{7.5}$$

To treat both the effects of the magnetism and valence fluctuations on equal footing, we apply the slave-boson mean-field theory [16] to Eq. (7.1). We use the hybridization form $V Z_{i\sigma} f^{\dagger}_{i\sigma} c_{i\sigma}$ instead of $V f^{\dagger}_{i\sigma} c_{i\sigma}$ in Eq. (7.1) by introducing the bose creation (annihilation) operators $e^{\dagger}_i (e_i)$ and $d^{\dagger}_i (d_i)$ for the empty and doubly occupied states, respectively, and $p^{\dagger}_{i\uparrow} (p_{i\uparrow})$ and $p^{\dagger}_{i\downarrow} (p_{i\downarrow})$ for singly occupied states by requiring the constraint for the completeness condition

$$\sum_i \lambda'_i (e^{\dagger}_i e_i + p^{\dagger}_{i\uparrow} p_{i\uparrow} + p^{\dagger}_{i\downarrow} p_{i\downarrow} + d^{\dagger}_i d_i - 1) \tag{7.6}$$

and

$$\sum_{i\sigma} \lambda_{i\sigma} (f^{\dagger}_{i\sigma} f_{i\sigma} - p^{\dagger}_{i\sigma} p_{i\sigma} - d^{\dagger}_i d_i) \tag{7.7}$$

with λ'_i and $\lambda_{i\sigma}$ being the Lagrange multipliers. Here, the renormalization factor for hybridization is defined as

$$Z_{i\sigma} \equiv (1 - d^{\dagger}_i d_i - p^{\dagger}_{i\sigma} p_{i\sigma})^{-1/2} (e^{\dagger}_i p_{i\sigma} + p^{\dagger}_{i-\sigma} d_i)(1 - e^{\dagger}_i e_i - p^{\dagger}_{i-\sigma} p_{i-\sigma})^{-1/2}. \tag{7.8}$$

To capture the essence of $CeRhIn_5$ as noted later, we consider the commensurate AF order on a bipartite lattice and consider $\lambda_{ia\sigma}$ and $\lambda_{ib\sigma}$ corresponding to the two sublattice (ia and ib sites) instead of $\lambda_{i\sigma}$, which are expressed as

$$\lambda_{ia\uparrow} = \lambda_{ib\downarrow} \equiv \lambda_i + \delta\lambda_i \tag{7.9}$$

$$\lambda_{ia\downarrow} = \lambda_{ib\uparrow} \equiv \lambda_i - \delta\lambda_i \tag{7.10}$$

respectively. For $H_{U_{\rm fc}}$ in Eq. (7.1), we employ the mean-field decoupling as $n_i^{\rm f} n_i^{\rm c} \simeq \bar{n}_{\rm f} n_i^{\rm c} + \bar{n}_{\rm c} n_i^{\rm f} - \bar{n}_{\rm f}\bar{n}_{\rm c}$ with $\bar{n}_{\rm f} \equiv \sum_i \langle n_i^{\rm f}\rangle/N$ and $\bar{n}_{\rm c} \equiv \sum_i \langle n_i^{\rm c}\rangle/N$. By approximating mean fields and Lagrange multipliers as uniform ones, i.e., $e = \langle e_i \rangle$, $p_\sigma = \langle p_{i\sigma}\rangle$, and $d = \langle d_i \rangle$, $\lambda' = \lambda'_i$, $\delta\lambda = \delta\lambda_i$, and $\lambda = \lambda_i$, respectively, the following set of mean-field equations is obtained under optimal conditions: $\partial\langle H\rangle/\partial x = 0$ for $x = \lambda', \delta\lambda, \lambda, e, d, p_\uparrow$, and $p_\downarrow$:

$$e^2 + p_\uparrow^2 + p_\downarrow^2 + d^2 - 1 = 0, \tag{7.11}$$

$$p_\uparrow^2 - p_\downarrow^2 = \frac{2}{N}\sum_{k}{}' \left[\langle f_{k\uparrow}^\dagger f_{k+{\rm Q}\uparrow}\rangle - \langle f_{k\downarrow}^\dagger f_{k+{\rm Q}\downarrow}\rangle\right], \tag{7.12}$$

$$p_\uparrow^2 + p_\downarrow^2 + 2d^2 = \frac{1}{N}\sum_{k\sigma}{}' \left[\langle f_{k\sigma}^\dagger f_{k\sigma}\rangle + \langle f_{k+Q\sigma}^\dagger f_{k+Q\sigma}\rangle\right], \tag{7.13}$$

$$\frac{V}{N}\sum_{{\rm k}\sigma}{}' \left(\frac{\partial Z_\sigma}{\partial e}\right)\left[\langle f_{k\sigma}^\dagger c_{k\sigma}\rangle + \langle f_{k+Q\sigma}^\dagger c_{k+Q\sigma}\rangle\right] + \lambda' e = 0, \tag{7.14}$$

$$\frac{V}{N}\sum_{k\sigma}{}' \left(\frac{\partial Z_\sigma}{\partial d}\right)\left[\langle f_{k\sigma}^\dagger c_{k\sigma}\rangle + \langle f_{k+Q\sigma}^\dagger c_{k+Q\sigma}\rangle\right] + (U + \lambda' - 2\lambda)d = 0, \tag{7.15}$$

$$\frac{V}{N}\sum_{k\sigma}{}' \left(\frac{\partial Z_\sigma}{\partial p_\uparrow}\right)\left[\langle f_{k\sigma}^\dagger c_{k\sigma}\rangle + \langle f_{k+Q\sigma}^\dagger c_{k+Q\sigma}\rangle\right] + (\lambda' - \lambda - \delta\lambda)p_\uparrow = 0, \tag{7.16}$$

$$\frac{V}{N}\sum_{k\sigma}{}' \left(\frac{\partial Z_\sigma}{\partial p_\downarrow}\right)\left[\langle f_{k\sigma}^\dagger c_{k\sigma}\rangle + \langle f_{k+Q\sigma}^\dagger c_{k+Q\sigma}\rangle\right] + (\lambda' - \lambda + \delta\lambda)p_\downarrow = 0, \tag{7.17}$$

where $\boldsymbol{Q}$ is the antiferromagnetically ordered vector. Here, $\sum_k'$ denotes the summation over the 1st Brillouin zone. The chemical potential μ is determined so as to give the total filling $n = (\bar{n}_{\rm f} + \bar{n}_{\rm c})/2$ with

$$\bar{n}_{\rm f} + \bar{n}_{\rm c} = \frac{1}{N}\sum_{k\sigma}{}' [\langle f_{k\sigma}^\dagger f_{k\sigma}\rangle + \langle f_{k+Q\sigma}^\dagger f_{k+Q\sigma}\rangle + \langle c_{k\sigma}^\dagger c_{k\sigma}\rangle + \langle c_{k+Q\sigma}^\dagger c_{k+Q\sigma}\rangle]. \tag{7.18}$$

These equations are solved self-consistently.

To clarify the mechanism of drastic change of the Fermi surface of CeRhIn$_5$, in which the most typical one is the two-dimensional-like Fermi surface observed as the β_2 branch, we consider $\varepsilon_k = -2t(\cos k_x + \cos k_y)$ on the square lattice. Hereafter, the energy unit is taken as $t = 1$. To simulate the AF order in the heavy-electron state realized in CeRhIn$_5$, we consider the small hybridization $V = 0.2$ case near the half filling $n = 0.9$ and the AF order with $\boldsymbol{Q} = (\pi, \pi)$. Although we here calculated several U cases, we show the result for $U = \infty$, since the result is essentially unchanged even for finite U, as long as U is larger than the bandwidth.

7.1.2.2 Ground-State Phase Diagram

First, we show the ground-state phase diagram determined under the assumption of the paramagnetic states with $p_{\uparrow} = p_{\downarrow}$ and $\delta\lambda = 0$ in Fig. 7.3a. The first-order valence-transition line (solid line with filled triangles) terminates at the QCP (filled circle) at $(\varepsilon_{\rm f}^{\rm QCP}, U_{\rm fc}^{\rm QCP}) = (-0.3623, 1.1778)$. For $U_{\rm fc} > U_{\rm fc}^{\rm QCP}$, $\bar{n}_{\rm f}$ shows a jump as a function of $\varepsilon_{\rm f}$, indicating the first-order transition between the paramagnetic metals with $\bar{n}_{\rm f}$ close to 1 and $\bar{n}_{\rm f} < 1$ in deep and shallow $\varepsilon_{\rm f}$ regions, respectively, since a large $U_{\rm fc}$ forces electrons to pour into either the f level or the conduction band [17–20]. At the QCP, valence fluctuations diverge $\chi_{\rm v} = -\partial\bar{n}_{\rm f}/\partial\varepsilon_{\rm f} = \infty$, and for $U_{\rm fc} < U_{\rm fc}^{\rm QCP}$, $\chi_{\rm v}$ has a conspicuous peak at $\varepsilon_{\rm f}$ represented by the dashed line with open circles in Fig. 7.3a, indicating strong valence fluctuations, as shown in Fig. 7.3b. At the QCP, the characteristic energy scale of the system, the so-called Kondo temperature, is given by $T_{\rm K} \equiv \bar{\varepsilon}_{\rm f} - \mu = 2.93 \times 10^{-3}$ with $\bar{\varepsilon}_{\rm f} = \varepsilon_{\rm f} + \lambda + U_{\rm fc}\bar{n}_{\rm c}$.

When the AF states are taken into account, the AF order parameter defined as $m_{\rm s} \equiv p_{\uparrow}^2 - p_{\downarrow}^2$ decreases as $\varepsilon_{\rm f}$ increases, as shown in Fig. 7.3c for $U_{\rm fc} = 0.5$. When the ground-state energies of this AF state and the paramagnetic state are compared, level crossing occurs at "critical" value of $\varepsilon_{\rm f} = \varepsilon_{\rm f}^{\rm c}$. Then, this AF and paramagnetic phase transition is identified to be of the first order. The phase boundary determined in this manner is shown by the solid line with filled squares in Fig. 7.3a. We find that the AF order terminates in the vicinity of the first-order valence-transition line and the valence-crossover line. These results imply that the suppression of the AF order occurs at the points with strong valence fluctuations.

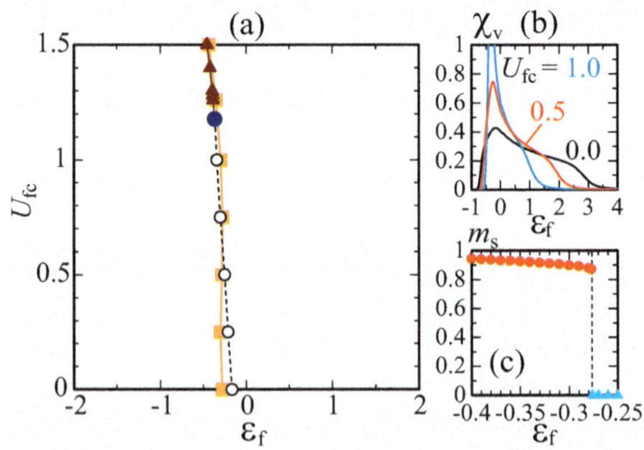

Fig. 7.3 **a** Ground-state phase diagram in the plane of $U_{\rm fc}$ and $\varepsilon_{\rm f}$ for paramagnetic and AF states (see text). The first-order valence-transition line (solid line with triangles) terminates at the quantum critical point (filled circle). Valence crossover occurs at the dashed line with open circles, at which $\chi_{\rm v}$ has a maximum, as shown in **b**. The solid line with filled squares represents the boundary between the AF state and the paramagnetic state (see text). **c** AF order parameter $m_{\rm s}$ versus $\varepsilon_{\rm f}$ for $U_{\rm fc} = 0.5$. All results in **a**–**c** are calculated for $t = 1$, $V = 0.2$, and $U = \infty$ at $n = 0.9$ [15]. Reprinted from Ref. [15]. Copyright 2010 The Physical Society of Japan

These results are favorably compared with those observed in $CeRhIn_5$. Applying pressure to Ce systems corresponds to increasing ε_f, since negative ions approach the tail of the 4f wavefunction at the Ce site. Experimental facts of the sudden disappearance of the AF order at $P = P_c$ and the simultaneous emergence of the residual resistivity ρ_0 peak as well as $\rho \propto T$ near $P = P_c$ seem to be well described by the results shown in Fig. 7.3. The vicinity of ε_f^c for a moderate $U_{fc} (< U_{fc}^{QCP})$ with a well-developed χ_v seems to correspond to the vicinity of $P = P_c$ in $CeRhIn_5$.

7.1.2.3 Fermi-Surface Change

To analyze the Fermi-surface change shown by dHvA measurement in $CeRhIn_5$, we apply the magnetic field to Eq. (7.1) as

$$-h \sum_i (S_i^{fz} + S_i^{cz}). \tag{7.19}$$

The dHvA effect [6] and A coefficient [8] were measured at $H = 12 \sim 17$ T and $H = 15$ T, respectively. The magnetic field of $H = 15$ T is estimated to be $h = 0.0046t$ when the half bandwidth of the conduction band $4t$ of Eq. (7.1) is compared with that of $CeRhIn_5$ by band-structure calculation, about 1.5 eV [23]. Then, we show in Fig. 7.4 the contour plot of the energy band of Eq. (7.1) with $\downarrow$ spin located at the Fermi level μ for $U_{fc} = 0.5$ at $h = 0.005$.

In the AF state for the filling $n = 0.9$, the lower hybridized band of Eq. (7.1) is folded, giving rise to the hole region emerging at the magnetic zone boundary connecting $\boldsymbol{k} = (0, \pi)$ and $(\pi, 0)$, as shown in Fig. 7.4a. Here, in order to make a comparison with the "small" Fermi surface, which consists of only conduction electrons, we plot the Fermi surface of the conduction band $\varepsilon_{\boldsymbol{k}}$ at the filling $\bar{n}_c = 0.8$ in Fig. 7.4a using a dashed line. This Fermi surface corresponds to that obtained when the hybridization between f and conduction electrons is switched off, i.e., $V = 0$ in Eq. (7.1); Since f electrons for $\bar{n}_f = 1$ are located at the localized f level for $V = 0$, extra electrons in $n = 0.9$, i.e., $\bar{n}_c = 0.8$ are in the conduction band. We see that the Fermi surface of the AF state for $V = 0.2$ is nearly the same as the "small" Fermi surface represented by the dashed line. This corresponds to the experimental fact that the dHvA signals of $CeRhIn_5$ are very similar to those of $LaRhIn_5$ in the AF-ordered phase for $P \leq P_c$ [6, 10].

The shape of the Fermi surface close to the "small" Fermi surface remains until ε_f reaches the AF-paramagnetic boundary $\varepsilon_f^c = -0.283$, as shown in Fig. 7.4b, which corresponds to $CeRhIn_5$ at $P \lesssim P_c$. When ε_f exceeds ε_f^c, the Fermi surface drastically changes, as shown in Fig. 7.4c for $\varepsilon_f = -0.280$. The folding of the lower hybridized band disappears and the "large" Fermi surface recovers, which is clearly different from the "small" Fermi surface shown in Fig. 7.4a and b. This "large" Fermi surface remains in the paramagnetic phase, as shown in Fig. 7.4d for $\varepsilon_f = -0.250$.

These results are naturally understood if we draw the energy bands in the paramagnetic phase and the AF phase as in Fig. 7.4e and f, respectively. In Fig. 7.4e, the lower

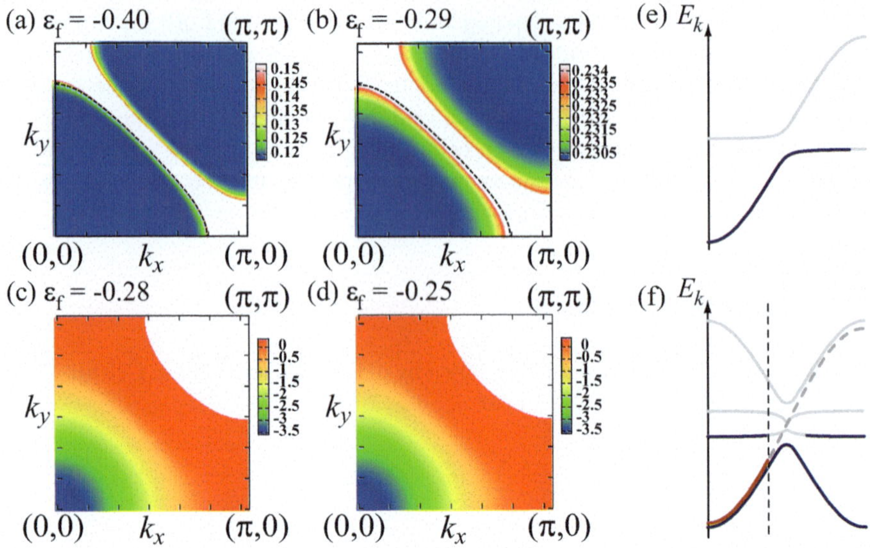

Fig. 7.4 The contour plot of the energy band with ↓ spin located at the Fermi level μ for $t = 1$, $V = 0.2$, $U = \infty$, $U_{\rm fc} = 0.5$, and $n = 0.9$ at $h = 0.005$: **a** $\varepsilon_{\rm f} = -0.40$, **b** $\varepsilon_{\rm f} = -0.29$, **c** $\varepsilon_{\rm f} = -0.28$, and **d** $\varepsilon_{\rm f} = -0.25$. The $E_{\boldsymbol{k}\downarrow} > \mu$ parts are represented by white regions. In **a** and **b**, the dashed line indicates the Fermi surface of the conduction band, $\varepsilon_{\boldsymbol{k}}$ for $\bar{n}_{\rm c} = 0.8$ [15]. Occupied (solid lines) and empty (gray lines) bands in the extended Anderson lattice model at $n = (\bar{n}_{\rm f} + \bar{n}_{\rm c})/2 = 0.9$ are shown in paramagnetic phase **e** and in AF-ordered phase **f**. In **f**, gray dashed line indicates the energy band of the conduction band, $\varepsilon_{\mathbf{k}}$ at $\bar{n}_{\rm c} = 0.8$. Solid dashed line is a guide for the eyes indicating that the Fermi surface in AF-ordered phase with finite **c**–**f** hybridization coincides with the small Fermi surface where f electrons are completely localized [21]. Reprinted with permission from Ref. [21]. Copyright 2011 The Physical Society of Japan

hybridized band of the model (7.1) is occupied as illustrated by the solid line forming the "large" Fermi surface whose volume contains f electrons, i.e., $\bar{n}_{\rm f} + \bar{n}_{\rm c}$. On the other hand, the energy band of conduction electrons is illustrated in Fig. 7.4f by the gray dashed line whose occupied part for $\bar{n}_{\rm c}$ is represented by the solid line, clearly indicating the "small" Fermi surface, in sharp contrast to Fig. 7.4e. Once the AF order occurs, the lower and upper hybridized bands are folded as shown in Fig. 7.4f. Since the lower hybridized band in the magnetic Brillouin zone is fully occupied, the Fermi surface emerging at the folded lower hybridized band shows the area occupied for the filling 0.8 among $n = (1 + 0.8)/2$, giving rise to the same Fermi surface of the conduction band at $\bar{n}_{\rm c} = 0.8$. Namely, the "small" Fermi surface appears in the AF phase even though the c–f hybridization remains finite $\langle f^{\dagger}_{\boldsymbol{k}\sigma} c_{\boldsymbol{k}\sigma}\rangle \neq 0$. Thus, the mechanism of the coincidence of the Fermi surface in the AF phase and the dashed line shown in Fig. 7.4b is naturally understood.

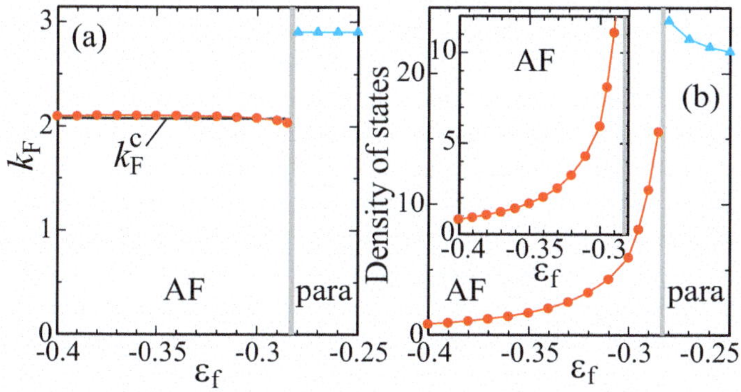

Fig. 7.5 **a** Fermi wave number k_F versus ε_f in the AF state (circles) and the paramagnetic state (triangles) for $t = 1$, $V = 0.2$, $U = \infty$, $U_{fc} = 0.5$, and $h = 0.005$ at $n = 0.9$. The solid line represents k_F for the conduction band ε_k at $\bar{n}_c = 0.8$. **b** $D(\mu)$ versus ε_f for the same parameters as **a**. The inset is enlargement of the AF phase [15]. Reprinted from Ref. [15]. Copyright 2010 The Physical Society of Japan

7.1.2.4 Comparison with the dHvA Measurement

To facilitate the comparison with the dHvA result, we plot the Fermi wave number k_F, defined by the distance between $\boldsymbol{k} = (0, 0)$ and the intersection point of the Fermi surface and the line connecting $\boldsymbol{k} = (0, 0)$ and (π, π), in Fig. 7.5a. Here, for the AF-ordered state $\varepsilon_f < \varepsilon_f^c$, k_F in the 1st Brillouin zone is plotted. We see that k_F is almost unchanged as a function of ε_f, which is nearly the same as the k_F of the conduction band at $\bar{n}_c = 0.8$, k_F^c represented by the solid line. Then, k_F shows an abrupt jump at ε_f^c. This is consistent with the dHvA measurement indicating that the dHvA frequencies, including the β_2 branch, remain almost constant for $0 \leq P < P_c$, and they suddenly jump at $P = P_c$ and remain constant for $P > P_c$ (see Fig. 7.1).

We find that the measured enhancement of the effective mass of electrons from $m^* \sim 6m_0$ at $P = 0$ to $m^* \sim 60m_0$ at $P \to P_{c-}$ [6] is also reproduced. The ε_f dependence of the density of states (DOS) at the chemical potential μ, $D(\mu)$, is shown in Fig. 7.5b, where the DOS is defined by

$$D(\omega) \equiv \frac{1}{2N} \sum_{k\sigma} \delta(\omega - E_{k\sigma}) \tag{7.20}$$

with $E_{\mathbf{k}\sigma}$ being the energy band of Eq. (7.1). The DOS is enhanced about 10 times when ε_f approaches ε_f^c, as shown in the inset in Fig. 7.5b. However, the inverse of the renormalization factor $1/\sqrt{Z_\sigma}$, due to the many-body effect, does not show divergent growth even as ε_f approaches ε_f^c. This implies that the divergent growth of the DOS is mainly due to the band effect. Thus, $D(\mu)$ is proportional to m^* and $\sqrt{A}$, explaining the measured $\sqrt{A}/m^* =$ const. scaling [8, 22]. We note that $D(\mu)$ at $\varepsilon_f = -0.4$ is about 10 times larger than the DOS of conduction electrons at $\bar{n}_c = 0.8$,

$D_c(\mu) = 0.092$, which is also consistent with the enhanced γ of $CeRhIn_5$ [2] from that of $LaRhIn_5$ [10, 11] at $P = 0$. When P increases, i.e., ε_f increases, m_s decreases, as shown in Fig. 7.3c. Since the increase in ε_f tends to increase the renormalization factor $\sqrt{Z_\sigma}$, the energy gap between the original lower hybridized band and the folded band in the AF phase increases. This effect pushes up the folded band in the 1st Brillouin zone, making the f-electron-dominant flat part of the band whose bottom is located at $\boldsymbol{k} = (0, 0)$ start to emerge at the Fermi level (see Fig. 7.4b).

In the paramagnetic phase for $\varepsilon_f > \varepsilon_f^c$, $D(\mu)$'s are larger than those in the AF phase, as shown in Fig. 7.5b. The increase in the DOS toward ε_f^c in the paramagnetic phase is naturally understood since as ε_f decreases, $\bar{n}_f$ increases to approach 1, i.e., the Kondo state, giving rise to the reduction of T_K, i.e., the increase in $D(\mu)$. In $CeRhIn_5$, the dHvA signal of the β_2 branch has not been detected for $P > P_c$, probably because its effective mass is too large, close to $100m_0$ [6]. This is also consistent with our result. For $\varepsilon_f > \varepsilon_f^c$, $D(\mu)$ decreases as h increases. This is also consistent with the field dependence of m^* of the β_2 branch in $CeCoIn_5$ [24], which is expected to correspond to the paramagnetic state of $CeRhIn_5$ for $P > P_c$ [25, 26].

The increase in $D(\mu)$ toward ε_f^c shown in Fig. 7.5b suggests that the total DOS at the 3D Fermi surface of $CeRhIn_5$ gives rise to the measured peak structure of the A coefficient at $P = P_c$ [8]. Our results clearly show that the "small" Fermi surface can be observed by dHvA measurement even without switching off the c–f hybridization, i.e., $\sum_{k\sigma}\langle f_{k\sigma}^\dagger c_{k\sigma}\rangle/N$ is always finite, which reminds us of the elucidation of metamagnetism in $CeRu_2Si_2$ [27–29].

7.1.2.5 Signatures of the Valence Crossover in Ce115 Systems

In $CeIr(In_{0.925}Cd_{0.075})_5$, the NQR measurement was performed under pressure [30]. The ^{115}In NQR frequency exhibits the remarkable change around $P = 2$ GPa, which indicates that the Ce-valence crossover occurs. Interestingly, around $P = 2$ GPa, the AF transition temperature starts to decrease and the superconductivity starts to appear. This suggests that enhanced valence fluctuations suppress the AF order and the valence-fluctuation-mediated superconductivity appears.

It is also noted here that in $CeIrIn_5$ the ^{115}In NQR frequency also exhibits the similar change around $P = 2$ GPa where the superconducting-transition temperature takes a maximum [31]. These experimental observations indicate that the effect of critical valence fluctuations arising from the critical point of the valence transition, which is located at the $T < 0$ side close to $T = 0$ as shown in Fig. 1.4c, play an important role as an underlying mechanism in the Ce115 systems.

7.2 Comprehensive Understanding of the Phase Diagram of the Heavy Fermion Systems

In this section, it is discussed that the phase diagram of the heavy fermion systems can be understood comprehensively on the basis of the extended Anderson lattice model. The key control parameter is the c–f hybridization strength. In Fig. 7.6, the schematic temperature T-pressure P phase diagram with the magnetic transition temperature $T_{\rm mag}$ and the valence-crossover temperature $T_{\rm v}^*$ is shown for $U_{\rm fd}$ smaller than the value of the valence QCP $U_{\rm fd}^{\rm QCP}$. The pink shaded dome represents the superconducting phase caused by the critical valence fluctuations and the gray shaded dome represents the superconducting phase caused by the magnetic fluctuations. In the $U_{\rm fd} > U_{\rm fd}^{\rm QCP}$ case, the valence-crossover temperature $T_{\rm v}^*$ is replaced by the first-order valence-transition temperature $T_{\rm v}$ at least in Fig. 7.6b–e. In that case, the valence-fluctuation mediated superconducting transition temperature is expected to be reduced even if it exists because the critical valence fluctuation is diminished as distanced from the valence QCP.

For large c–f hybridization in the extended Anderson lattice model, the magnetic-transition temperature $T_{\rm mag}$ and the valence-crossover temperature $T_{\rm v}^*$ are sep-

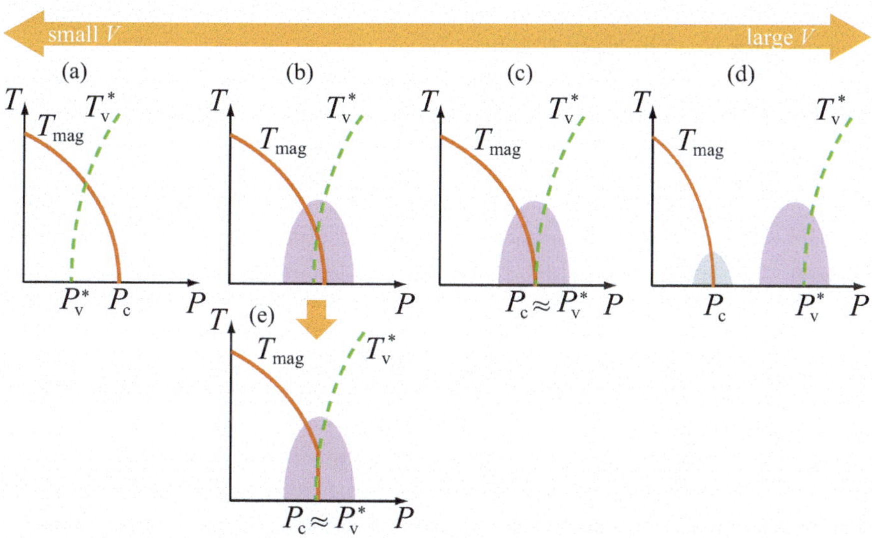

Fig. 7.6 Schematic T-P phase diagram of the extended Anderson lattice model where the **c**–**f** hybridization changes from **a** the small value and increases as **b**, **c**, and **d**. The dashed line represents the valence-crossover temperature $T_{\rm v}^*$ arising from the valence QCP (see Fig. 1.4**c**). The solid line represents the magnetic-transition temperature $T_{\rm mag}$. The pink shaded dome denotes the superconducting phase caused by critical valence fluctuations and the gray shaded dome denotes the superconducting phase caused by magnetic fluctuations. Enhanced valence fluctuation at $P_{\rm v}^*$ in **b** suppress the magnetic order giving rise to the coincidence $P_{\rm c} \approx P_{\rm v}^*$ in **e**

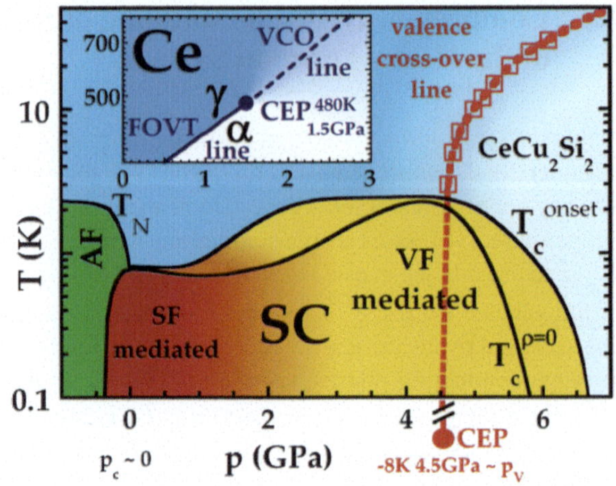

Fig. 7.7 The temperature-pressure phase diagram of $CeCu_2Si_2$. The critical point (red circle) from which the valence-crossover line extends is deduced from the resistivity measurement. SC denotes the superconducting phase and AF denotes the antiferromagnetically ordered phase [33]. Reprinted from Ref. [33]. Copyright 2012 American Physical Society

arated as shown in Fig. 7.6d. This corresponds to the T-P phase diagram of $CeCu_2(Si_{1-x}Ge_x)_2$ [32]. As for $CeCu_2Si_2$, detailed analysis of the temperature and pressure dependence of the electric resistivity, the critical point of the valence transition is identified as $P_v^* = 4.5 \pm 0.2$ GPa, $T_{cp} = -8$ K, as shown in Fig. 7.7 [33]. As the c-f hybridization decreases, the magnetic order expands to the larger (smaller) P direction in the Ce (Yb)-based systems.

At the critical value of the c-f hybridization, the magnetic-transition point P_c and the valence-crossover point P_v^* coincide, as shown in Fig. 7.6c. As the c-f hybridization further decreases, the magnetic-transition temperature T_{mag} and the valence-crossover temperature T_v^* tend to cross as shown in Fig. 7.6b. In case that the magnetic-transition temperature at the crossing point with T_v^* is not so high, the enhanced valence fluctuations suppress the magnetic order, which make coincidence of P_c and P_v^* as shown in Fig. 7.6e. Hence, for a certain range of realistic parameters of the c-f hybridization, coincidence of magnetic-transition temperature and valence-crossover temperature occurs. An important point is that the coincidence $P_c \approx P_v^*$ occurs not only at the critical c–f hybridization but also in rather wide range around the critical value of the c–f hybridization. This is the case in $CeRhIn_5$ discussed in Sect. 7.1 as shown in Fig. 7.3.

In the T-P phase diagram of the 1/1 approximant crystal $Yb_{14}Al_{35}Au_{51}$, the valence QCP is identified at $P = 1.96$ GPa (see Fig. 6.10b) as noted in Sect. 6.6 [35]. For $P > 2$ GPa, the magnetic order appears [35]. Hence, this is the case shown in Fig. 7.6c or Fig. 7.6e.

The phase diagrams shown in Fig. 7.6a–e also hold when the horizontal axis is taken as the magnetic field H or chemical composition x instead of the pressure. In the former case, the T-H phase diagram of $YbAuCu_4$ shown in Fig. 4.23 where the sharp valence-crossover temperature T_v^* appears for $H > H_v^* = 1.3$ T in the temperature-magnetic field phase diagram as noted in Sect. 4.6.3.2 corresponds to Fig. 7.6e. For

$0 \leq H \leq H_v^*$, the antiferromagnetic order is realized. Hence, the coincidence of the magnetic transition and the valence-crossover $H_c \approx H_v^*$ occurs.

As noted in Sect. 6.5.2, in the T-x phase diagram of α-$YbAl_{1-x}Fe_xB_4$, the sharp valence-crossover temperature T_v^* appears at $x = 0.014$ and the magnetic order appears for $x > 0.014$ as shown in Fig. 6.9, which corresponds to Fig. 7.6c or Fig. 7.6e with the pressure-axis direction being inverted.

For $U_{fd} > U_{fd}^{QCP}$, the phase diagrams shown in Fig. 7.6b–e are realized where T_v^* is replaced with the first-order valence-transition temperature T_v. In that case, the T-P phase diagram of $YbInCu_4$ shown in Fig. 1.2, where the first-order valence-transition P_v and the ferromagnetic-transition pressure P_c coincide, corresponds to Fig. 7.6e with the pressure-axis direction being inverted.

In the extended Anderson lattice model, as the c–f hybridization decreases from Fig. 7.6b, P_v^* can be located inside the magnetically ordered phase, i.e., $P_v^* < P_c$ holds. In the T-P phase diagram of $CeRh_{0.5}Ir_{0.5}In_5$, the In-NQR frequency change was detected, which suggests that the Ce-valence crossover occurs [30]. The obtained

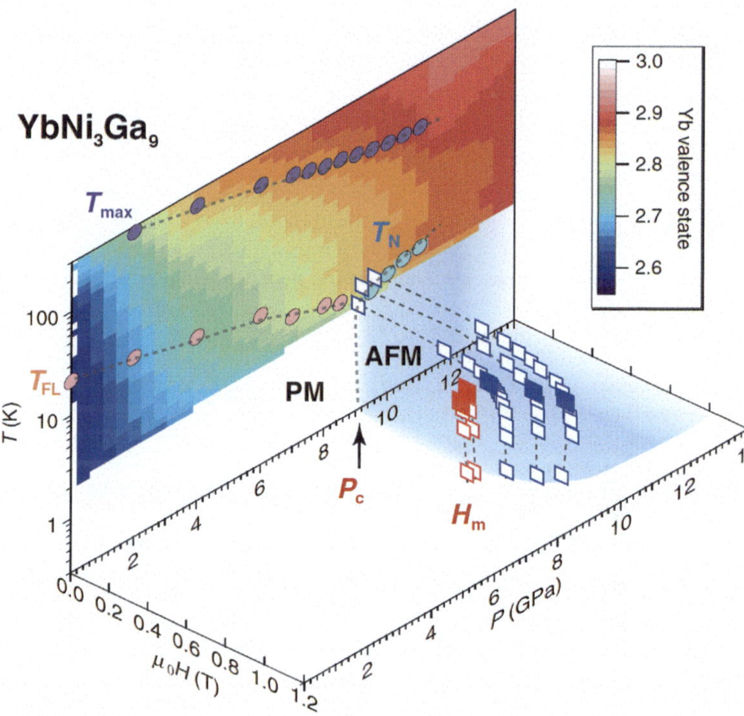

Fig. 7.8 Temperature-pressure-magnetic field phase diagram of $YbNi_3Ga_9$. The magnetically ordered phase appears for $P \geq P_c \approx 9$ GPa and $T \leq T_N$. The first-order transition (red open squares), which exhibits metamagnetic transition, terminates at the critical point (red filled square) at $H \approx 0.7$ T [34]. Reprinted with permission from Ref. [34]. Copyright 2015 American Physical Society

$T_v^*(P)$ temperature crosses the antiferromagnetic transition temperature $T_{mag}(P)$ around $P \approx 1.2$ GPa, which corresponds to the case shown in Fig. 7.6a.

The interplay of the magnetic and valence transition was also observed in $YbNi_3Ga_9$ by the multiple probes of the X-ray absorption spectroscopy and the bulk measurements such as the magnetic susceptibility and resistivity under the extreme conditions of low temperature and high pressure under the magnetic field [34]. At ambient pressure, the paramagnetic metal phase appears in $YbNi_3Ga_9$, which exhibits heavy-fermion behavior at low temperatures. By applying pressure, the magnetic order appears above $P_c = 9$ GPa. Interestingly by applying magnetic field, the metamagnetic behavior was observed along the line denoted by open red squares at $H = 0.7$ T in Fig. 7.8. Moreover, as approaching the point denoted by red filled square, which is located at the upper end of the line denoted by open red squares, from the high-temperature side, the enhancement of the magnetic susceptibility was observed [34]. These measurements are consistent with the diverging uniform magnetic fluctuation at the critical point of the first-order valence transition, as discussed in Sect. 4.6.2. These results indicate that the field-induced valence transition is indeed detected in $YbNi_3Ga_9$. As shown in Fig. 7.8, the critical point of the first-order valence transition is closely located near the magnetic phase boundary, which corresponds to the case of Fig. 7.6e for $U_{fd} > U_{fd}^{QCP}$ with T_v^* being replaced by T_v with the pressure-axis direction being inverted.

References

1. W. Bao, P.G. Pagliuso, J.L. Sarrao, J.D. Thompson, Z. Fisk, J.W. Lynn, R.W. Erwin, Phys. Rev. B **67**, 099903 (2003)
2. H. Hegger, C. Petrovic, E.G. Moshopoulou, M.F. Hundley, J.L. Sarrao, Z. Fisk, J.D. Thompson, Phys. Rev. Lett. **8**, 4986 (2000)
3. T. Mito, S. Kawasaki, G.-Q. Zheng, Y. Kawasaki, K. Ishida, Y. Kitaoka, D. Aoki, Y. Haga, Y. Ōnuki, Phys. Rev. B **63**, 220507 (2001)
4. T. Muramatsu, N. Tateiwa, T.C. Kobayashi, K. Shimizu, K. Amaya, D. Aoki, H. Shishido, Y. Haga, Y. Ōnuki, J. Phys. Soc. Jpn. **70**, 3362 (2001)
5. G. Knebel, M.-A. Méasson, B. Salce, D. Aoki, D. Braithwaite, J.P. Brison, J. Flouquet, J. Phys.: Condens. Matter **16**, 8905 (2004)
6. H. Shishido, R. Settai, H. Harima, Y. Ōnuki, J. Phys. Soc. Jpn. **74**, 1103 (2005)
7. T. Park, F. Ronning, H.Q. Yuan, M.B. Salamon, R. Movshovich, J.L. Sarrao, J.D. Thompson, Nature **440**, 65 (2006)
8. G. Knebel, D. Aoki, J.P. Brison, J. Flouquet, J. Phys. Soc. Jpn. **77**, 114704 (2008)
9. T. Park, V.A. Sidorov, F. Ronning, J.-X. Zhu, Y. Tokiwa, H. Lee, E.D. Bauer, R. Movshovich, J.L. Sarrao, J.D. Thompson, Nature **456**, 366 (2008)
10. H. Shishido, R. Settai, D. Aoki, S. Ikeda, H. Nakawaki, N. Nakamura, T. Iizuka, Y. Inada, K. Sugiyama, T. Takeuchi, K. Kindo, T.C. Kobayashi, Y. Haga, H. Harima, Y. Aoki, T. Namiki, H. Sato, Y. Ōnuki, J. Phys. Soc. Jpn. **71**, 162 (2002)
11. N.F. Phillips, R.A. Fisher, F. Bouquet, M.F. Hundley, P.G. Pagliuso, J.L. Sarrao, Z. Fisk, J.D. Thompson, J. Phys.: Condens. Matter **15**, S2095 (2003)
12. T. Moriya, *Spin Fluctuations in Itinerant Electron Magnetism* (Springer, Berlin, 1985)
13. J.A. Hertz, Phys. Rev. B **14**, 1165 (1976)
14. A.J. Millis, Phys. Rev. B **48**, 7183 (1993)

15. S. Watanabe, K. Miyake, J. Phys. Soc. Jpn. **79**, 033707 (2010)
16. G. Kotliar, A.E. Ruckenstein, Phys. Rev. Lett. **57**, 1362 (1996)
17. S. Watanabe, M. Imada, K. Miyake, J. Phys. Soc. Jpn. **75**, 043710 (2006)
18. S. Watanabe, A. Tsuruta, K. Miyake, J. Flouquet, J. Phys. Soc. Jpn. **78**, 104706 (2009)
19. Y. Onishi, K. Miyake, J. Phys. Soc. Jpn. **69**, 3955 (2000)
20. S. Watanabe, A. Tsuruta, K. Miyake, J. Flouquet, Phys. Rev. Lett. **100**, 236401 (2008)
21. S. Watanabe, K. Miyake, J. Phys.: Conf. Ser. **273**, 012063 (2011)
22. K. Miyake, T. Matsuura, C.M. Varma, Solid State Commun. **71**, 1149 (1989)
23. D. Hall, E.C. Palm, T.P. Murphy, S.W. Tozer, C. Petrovic, E.M. Ricci, L. Peabody, C.Q.H. Li, U. Alver, R.G. Goodrich, J.L. Sarrao, P.G. Pagliuso, J.M. Wills, Z. Fisk, Phys. Rev. B **64**, 064506 (2001)
24. R. Settai, H. Shishido, S. Ikeda, Y. Murakawa, M. Nakashima, D. Aoki, Y. Haga, H. Harima, Y. Ōnuki, J. Phys.: Condens. Matter **13**, L627 (2001)
25. K. Miyake, J. Phys.: Condens. Matter **19**, 125201 (2007)
26. L.D. Pham, T. Park, S. Maguion, J.D. Thompson, Z. Fisk, Phys. Rev. Lett. **97**, 056404 (2006)
27. S. Watanabe, J. Phys. Soc. Jpn. **69**, 2947 (2000)
28. R. Daou, C. Bergemann, S.R. Julian, Phys. Rev. Lett. **96**, 026401 (2006)
29. K. Miyake, H. Ikeda, J. Phys. Soc. Jpn. **75**, 033704 (2006)
30. S. Kawasaki, T. Oka, A. Sorime, Y. Kogame, K. Uemoto, K. Matano, J. Guo, S. Cai, L. Sun, J.L. Sarrao, J.D. Thompson, G.-Q. Zheng, Commun. Phys. **3**, 148 (2020)
31. M. Yashima, N. Tagami, S. Taniguchi, T. Unemori, K. Uematsu, H. Mukuda, Y. Kitaoka, Y. Ōta, F. Honda, R. Settai, Y. Ōnuki, Phys. Rev. Lett. **109**, 117001 (2012)
32. H.Q. Yuan, F.M. Grosche, M. Deppe, C. Geibel, G. Sparn, F. Steglich, Science **302**, 2104 (2003)
33. G. Seyfarth, A.-S. Rüetschi, K. Sengupta, A. Georges, D. Jaccard, S. Watanabe, K. Miyake, Phys. Rev. B **85**, 205105 (2012)
34. K. Matsubayashi, T. Hirayama, T. Yamashita, S. Ohara, N. Kawamura, M. Mizumaki, N. Ishimatsu, S. Watanabe, K. Kitagawa, Y. Uwatoko, Phys. Rev. Lett. **114**, 086401 (2015)
35. S. Matsukawa, K. Deguchi, K. Imura, T. Ishimasa, N. K. Sato, J. Phys. Soc. Jpn. **85**, 063706 (2016)

Chapter 8
Instead of Epilogue—Ubiquity of Critical Valence Fluctuations

In this chapter, the ubiquity of critical valence fluctuations is briefly discussed by introducing the cases already discussed in the previous chapters and new cases where the relevance to those phenomena is realized recently. In particular, sharp and pronounced enhancement of the residual resistivity is emphasized as the characteristic of critical valence fluctuations.

8.1 Candidate Materials in Which Critical Valence Transition or Sharp Valence Crossover Seems to Exhibit

Close relation of anomalous critical phenomena and valence fluctuations has been also indicated in $Ce_{0.9-x}Th_{0.1}La_x$ [1] and in $YbAuCu_4$ [2]. Both are related to typical valence-transition materials: One is Ce metal, well known as γ-α transition [3], and the other is $YbInCu_4$ [4], both showing a discontinuous valence jump of a Ce and Yb ion, respectively, when T and P are varied. Since the first-order valence transition is basically an isostructural transition, the critical point exists in the T-P (and chemical composition) phase diagram, as in the liquid-gas transition.

At $x \approx 0.1$ in $Ce_{0.9-x}Th_{0.1}La_x$, at which the critical point is almost suppressed and is close to $T = 0$ K [1], critical phenomena arising from the quantum critical point were revealed: T-linear resistivity emerges prominently and uniform magnetic susceptibility is enhanced at low T giving rise to a large Wilson ratio, $R_W \sim 3$.

In $YbAuCu_4$, the uniform magnetic susceptibility is enhanced as $\chi(T) \sim T^{-0.6}$ as T decreases in spite of the fact that the AF transition takes place at $T_N = 0.8$ K [5], similarly to $YbRh_2Si_2$. Furthermore, the sharp Yb-valence crossover temperature $T_v^*(H)$ is induced by applying a magnetic field [2], suggesting that $YbAuCu_4$ is located in the vicinity of the quantum critical point of the valence transition (see Fig. 4.23 in Sect. 4.6.3.2). It should also be noted that the T-H phase diagram of

S. Watanabe and K. Miyake, *Quantum Critical Phenomena of Valence Transition*, Springer Tracts in Modern Physics 289,
https://doi.org/10.1007/978-981-99-3518-5_8

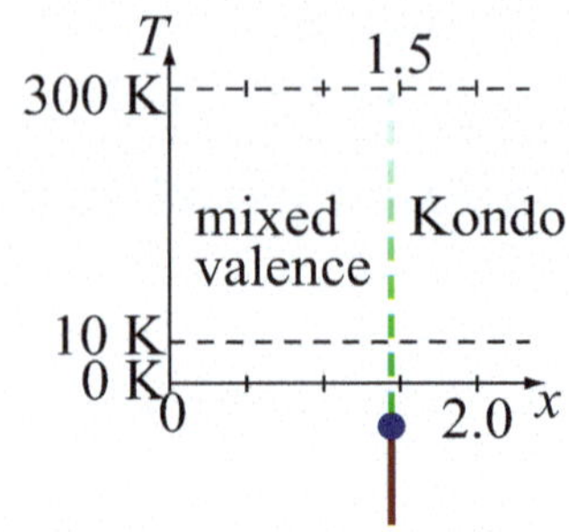

Fig. 8.1 Schematic phase diagram of $YbCu_{5-x}Al_x$ [7–9]. Near $x = 1.5$ sharp Yb-valence crossover occurs between the Kondo regime and mixed valence regime at low temperatures $T < 10$ K [10]. Reprinted from Ref. [10]. Copyright 2012 IOP Publishing

$YbAuCu_4$ [2] closely resembles that of $YbRh_2Si_2$ [6]. The H dependence of the crossover temperature $T^*(H)$, emerging in several physical quantities [5], whose origin is unclear in $YbRh_2Si_2$ is quite similar to the $T_v^*(H)$ in $YbAuCu_4$.

8.1.1 $YbCu_{5-x}Al_x$

In the heavy fermion metal $YbCu_{5-x}Al_x$, at $T = 300$ K, a gradual decrease in the Yb valence occurs as x decreases from 2.0 to 0.0. However, at $T = 10$ K, a sharp decrease in the Yb valence occurs near $x = 1.5$ [7], which has been also supported by recent detailed measurements of the Yb valence [8, 9]. These results indicate that a sharp Yb-valence crossover occurs near $x = 1.5$ with strong Yb-valence fluctuations. The schematic T-x phase diagram of $YbCu_{5-x}Al_x$ is shown in Fig. 8.1. At $x = 1.5$, as T decreases, logarithmic divergence in the specific-heat coefficient $C/T \sim -\log T$ and the power-law divergence in the uniform magnetic susceptibility $\chi \sim T^{-2/3}$ were observed in 1997 by Bauer et al. [7], which well agree with the quantum valence criticality (see Table 6.1).

The shape of $T_v^*(g)$, i.e., the control parameter dependence of the valence crossover temperature with g being the control parameter such as a magnetic field, pressure, and chemical substitution, depends on where the material is located in the phase diagram in Fig. 8.1. Detailed comparison of $T_v^*(g)$ between the theory and experiments will be interesting future studies.

8.1.2 Direct Observation of 4f-5d Coulomb Repulsion in Rare-Earth Atom

As discussed in Chaps. 4 and 6, the microscopic origin for causing the valence transition is the inter orbital Coulomb repulsion between 4f and 5d electrons U_{fd} at the rare-earth atom in the heavy fermion systems. Hence, it is important to develop the method to evaluate the value of U_{fd} experimentally. Such an attempt has been performed by using the X-ray absorption spectroscopy (XAS) combined with the theoretical analysis [11].

So far, the evaluation of the interaction between electrons by the XAS measurement has been performed mostly on the onsite Coulomb repulsion for f electrons U in the heavy fermion systems on the basis of the single-impurity Anderson model [12]. The evaluation has usually been performed in a phenomenological way, which carries two problems. One is the difficulty in determining the parameters uniquely because of many adjustable parameters such as the transfers of conduction electrons and hybridizations between the 4f and conduction electrons. The other problem is the difficulty in considering the realistic valence band structure since, in most analyses of the single-impurity Anderson model, the simplified conduction band such as a constant density of states is employed. Recently, Uozumi and Mizumaki and coworkers have proposed a new method for evaluating not only U but also $U_{\rm fd}$ which overcomes these problems [11].

The partial fluorescence yield (PFY), which corresponds to a high-resolution version of the XAS, is experimentally observed at the Ce L_3 edge in CeO_2 [11]. The fine spectral features observed experimentally are well reproduced by the XAS spectrum calculated by the single-impurity Anderson model considering the realistic band structure. A key point of this approach is that by performing the local density approximation (LDA) calculation taking into account the effect of the onsite Coulomb repulsion for f electrons U, i.e., the LDA+U calculation, the values of the Slater-Koster parameters are estimated. For the estimation of the 4f-5d Coulomb repulsion $U_{\rm fd}$, the number of adjustable parameters is drastically reduced by using the Slater-Koster parameters. Owing to the characteristic and sensitive dependence of the XAS spectrum on $U_{\rm fd}$, the value of $U_{\rm fd}$ is evaluated to be $U_{\rm fd} = 3.0$ eV in CeO_2.

This approach has broad applicability to the heavy fermion systems. In particular, it is interesting to perform the direct evaluation of $U_{\rm fd}$ in the materials which exhibit the quantum valence criticality (see Table 6.1) and also in the candidate materials noted in Sects. 7 and 8.1 in order to confirm the microscopic origin quantitatively.

8.1.3 $CeCu_6$

$CeCu_6$ is a typical heavy fermion metal as $CeCu_2Si_2$ and its effective mass *monotonically* decreases under pressure without the magnetic field in contrast to the case of $CeCu_2Si_2$. However, it exhibits the critical aspect characteristic of the critical valence fluctuations under the magnetic field. Namely, as discussed in Fig. 4.17 in Sect. 4.6, the magnetic field is a good tuning parameter for searching the QCP of valence transition. In particular, the QCP can be precisely hit by simultaneously changing the magnetic field h and the pressure P if the system is located near the QCP in the $\varepsilon_{\rm f}$-$U_{\rm fc}$ plane at the ambient state, as shown in Fig. 8.2a. Indeed, since both $U_{\rm fc}$ and $\varepsilon_{\rm f}$ increase as P increases in the Ce-based compounds, the point indicating the location of the system shifts toward right and upper direction as shown in Fig. 8.2a. Therefore, the locus of the QCP due to applying h intersects with that due to applying P.

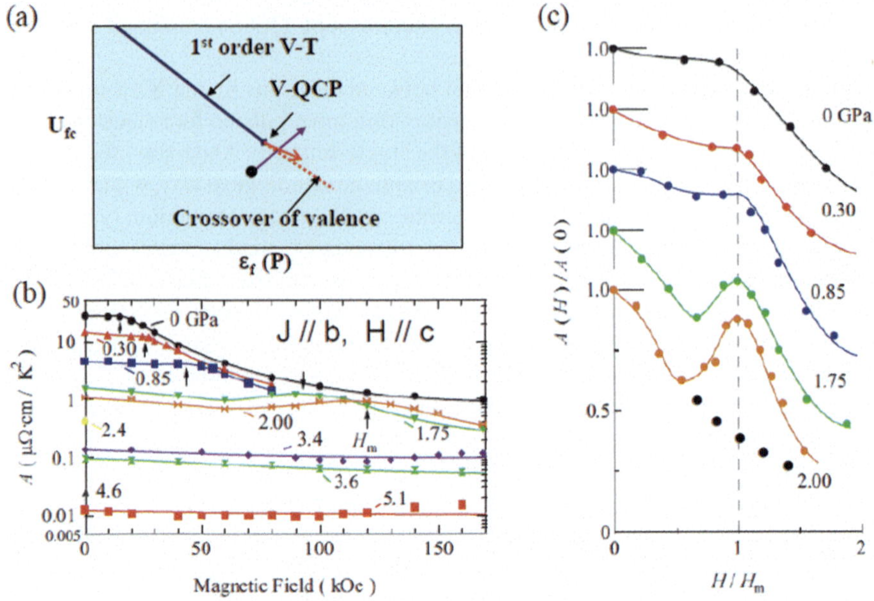

Fig. 8.2 **a** Schematic view how the critical end point of $CeCu_6$ is tuned by pressure P and magnetic field H. Large and small closed circles represent the position of the system and the QCP, respectively. A line with arrow from the system point indicates the locus of the system point as P increases, and a *red* line with arrow from the QCP indicates the locus of QCP as h increases. **b** The magnetic field (H) dependence of the A coefficient of the T^2 term in the resistivity $\rho(T)$ in the logarithmic scale under a series of pressures Ps [13]. Vertical arrows indicate the metamagnetic field H_m for each pressure. **c** Scaled relation between the A coefficient and the magnetic field H for a series of pressures Ps [13]. A series of closed circles at $P = 2.00$ GPa is a guide to the eye showing a possible background variation of A not related to the valence crossover. Reprinted with permission from Ref. [13]. Copyright 2012 The Physical Society of Japan

A symptom of this phenomenon was observed in $CeCu_6$ [13]. Figure 8.2b shows the magnetic field dependence of the A coefficient of the T^2 term in the resistivity of $CeCu_6$ under a series of pressures Ps. At ambient pressure, A is an almost monotonically decreasing function of H. By increasing pressure, it begins to show a clear maximum at H_m where the metamagnetic sharp increase in the magnetization is observed. Note that A is shown in a logarithmic scale. Variations of the A coefficient are rather prominent if it is plotted in a linear scale, as shown in Fig. 8.2c where A is scaled by $A(0)$ at ambient pressure and H is scaled by the metamagnetic field H_m. The peak structure becomes sharper and sharper as P increases, which strongly suggests that A diverges at the QCP because the T dependence of the resistivity should exhibit the T-linear dependence there. Experiments at higher pressure and magnetic field are highly desired.

Indeed, it was reported that $CeCu_6$ is located near the QCP of valence transition and a sharp valence crossover occurs at $P \simeq 5$ GPa, although the decrement in the A coefficient against the pressure is slightly moderate compared to that in $CeCu_2Si_2$ and

$CeCu_2Ge_2$ [16, 27]. A characteristic temperature T_F^* (or the effective Fermi energy ϵ_F^*) of $CeCu_6$ is very low of the order of 10 K reflecting the heaviness of its effective mass. This suggests that the QCP is recovered by the magnetic field of the order of T_F^*, i.e., $H \gtrsim 10$ T and under a certain pressure $P_c > 2$ GPa, which is inconsistent with the experiment reported in Ref. [13]. This physical picture is also consistent with the disappearance of magnetic correlations in the region $H \gtrsim 2.5$ T, which was observed by the inelastic neutron scattering experiment [21].

A direct evidence of the sharp crossover in the Ce valence can be seen in the pressure dependence of the coefficient A, the residual resistivity ρ_0, and T_{max} where the resistivity $\rho(T)$ takes a maximum, was reported by Raymond and Jaccard [16]. In Fig. 8.3, the relation between A and T_1^{max} (measure of the pressure P) of $CeCu_6$ for a series of pressures is displayed, together with those of $CeCu_2Si_2$ and $CeCu_2Ge_2$ [22]. The dashed curve is for $CeCu_6$. We can see that the valence changes rather sharply at around $T^{max} = 100$ K which corresponds to $P = 5$ GPa [16]. This clearly shows that $CeCu_6$ is a system near the QCP of valence transition although its position is a bit distant from its QCP compared with the case of $CeCu_2Si_2$ and $CeCu_2Ge_2$. It is noted that AF-QCP of $CeCu_6$ is induced by replacing Cu by Au [18], and seems to be located at $P = P_c \simeq -0.3$ GPa [19]. Then, the pressure for QCP of valence transition is again about 5 GPa higher than that for AF-QCP as in $CeCu_2(Si,Ge)_2$. The characteristic temperature T_F^* (or the effective Fermi energy ϵ_F^*) of $CeCu_6$ is very low of the order of 10 K reflecting the heaviness of its effective mass. Therefore, it is expected that the QCP is recovered by the magnetic field of the order of T_F^*, i.e., $H \sim 10$ T.

Related to this case, it is to be noted that the heavy fermion compound $CeRu_2Si_2$, exhibiting the metamagnetic behavior, is also expected to exhibit a valence transition

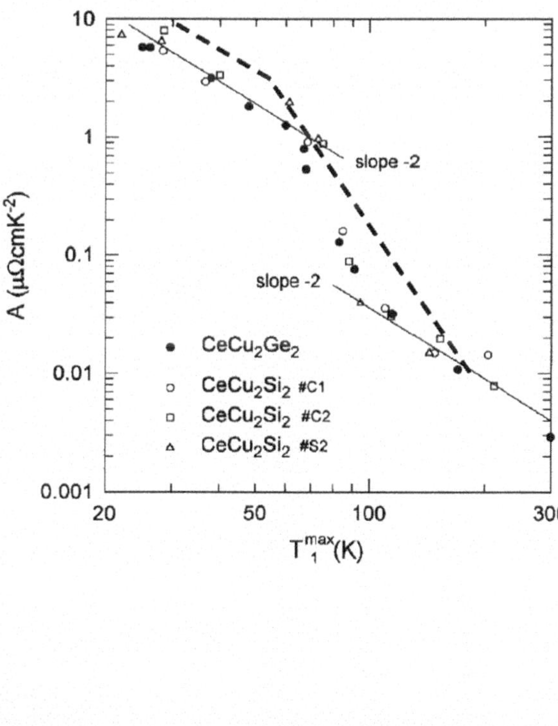

Fig. 8.3 A versus T_1^{max} of $CeCu_6$ ($CeCu_2Si_2$ and $CeCu_2Ge_2$) [27]. The dashed lines are for $CeCu_6$ [16], filled circles are for $CeCu_2Ge_2$, and open symbols are for $CeCu_2Si_2$ [17]. Reprinted from Ref. [27]. Copyright 2007 IOP Publishing

at $P \sim 4$ GPa, offering us another possible candidate for investigating the effect of magnetic field on valence transition [20].

8.1.4 Ce(Ir,Rh)Si_3

In this subsection, signature of possible critical valence transition in $CeIrSi_3$ and $CeRhSi_3$ is discussed. Figure 8.4a shows the $P - T$ phase diagram of $CeIrSi_3$ [22], in which the pressure $\bar{P}_c$, where the smooth extrapolation of the Néel temperature T_N in the SC phase vanishes, and the pressure P_c^*, where the SC transition temperature T_{sc} takes a broad maximum, are markedly different. Furthermore, the uppercritical field H_{c2} exhibits a sharp and huge enhancement around $P = P_c^*$, as shown in Fig. 8.4b. This implies that the pairing interaction is increased sharply as the magnetic field H is increased, suggesting that the QCP of valence transition is induced by the magnetic field at around $P = P_c^*$, as in the case of $CeRhIn_5$ at $P = P_c$, as shown in Fig. 8.4, although the definition of $P = P_c$ and $P = P_c^*$ are different in these two

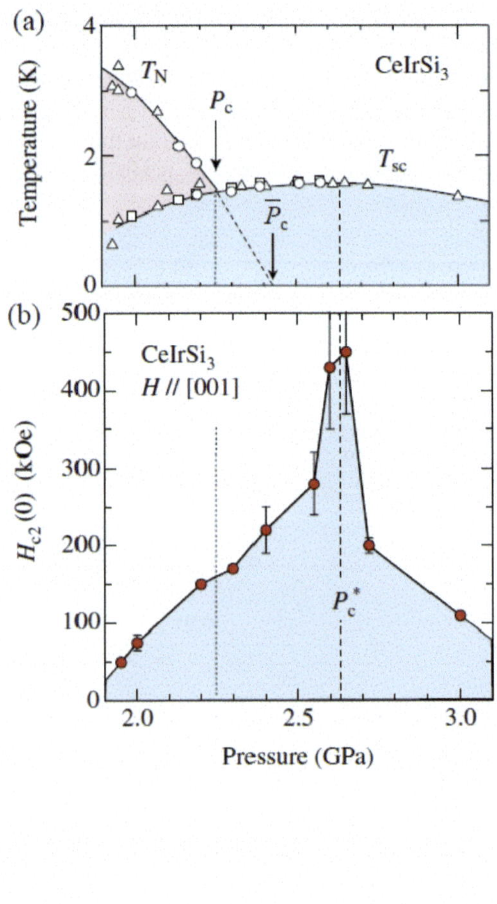

Fig. 8.4 **a** $P - T$ phase diagram of $CeIrSi_3$ without magnetic field, i.e., $H = 0$. P_c is defined as the pressure where the Néel temperature T_N coincides with the superconducting transition temperature T_{sc} [22], and $\bar{P}_c$ is a hypothetical critical pressure where the T_N extrapolated into the SC phase vanishes. **b** Upper-critical field $H_{c2}(T \to 0)$ as a function of P. $H_{c2}(T \to 0)$ takes sharply enhanced maximum at P_c^*, where the T_{sc} takes a broad maximum at $H = 0$ [22]. Reprinted with permission from Ref. [22]. Copyright 2008 The Physical Society of Japan

[H]

references [22, 29]. We also note that a similar enhancement of H_{c2} was observed in UGe_2 at $P \gtrsim P_x$ [23], which was explained nicely by the effect that the pairing interaction is enhanced by the effect of field-induced magnetic transition between two types of ferromagnetic states in UGe_2 [24].

8.1.5 Ce(Co,Rh,Ir)In_5

In previous sections, it has been shown that the critical valence fluctuations of Ce ion are the single origin for a variety of anomalous properties of $CeCu_2(Si,Ge)_2$ under pressure. The superconductivity realized at $P \sim P_v$ is in a *high-T_c state* in the sense that T_c is of $\mathcal{O}(10^{-1})$ of the renormalized Fermi energy. In this sense, the critical-valence-fluctuation mechanism may offer one of the routes to the *room temperature superconductivity*. A natural question is whether there are other Ce-based systems in which the critical valence fluctuations play an important role. In this section, some candidates are briefly discussed.

The properties of $CeTIn_5$ (T=Co, Rh, Ir) have the aspect similar to $CeCu_2(Si,Ge)_2$. Indeed, the P-T phase diagram of $CeRhIn_5$ is quite similar to that shown in Fig. 4.5 (Sect. 4.3) although the peak structure of T_c is less prominent. The $\rho(T = T_c)$ exhibits sharp peak structure at around $P = P_{max}$ where T_c takes the maximum, and the temperature dependence of $\rho(T)$ follows an approximate T-linear behavior [25]. This similarity can be seen more clearly in Fig. 8.5 in which T_N and T_c are shown in P-T-x space, x being the concentration of $CeCo_{1-x}Rh_xIn_5$ $(0 \le x \le 1)$, or $CeRh_{2-x}Ir_{x-1}In_5$ $(1 \le x \le 2)$, or $CeIr_{3-x}Co_{x-2}In_5$ $(2 \le x \le 3)$ [26]. It is noted that the line of mag-

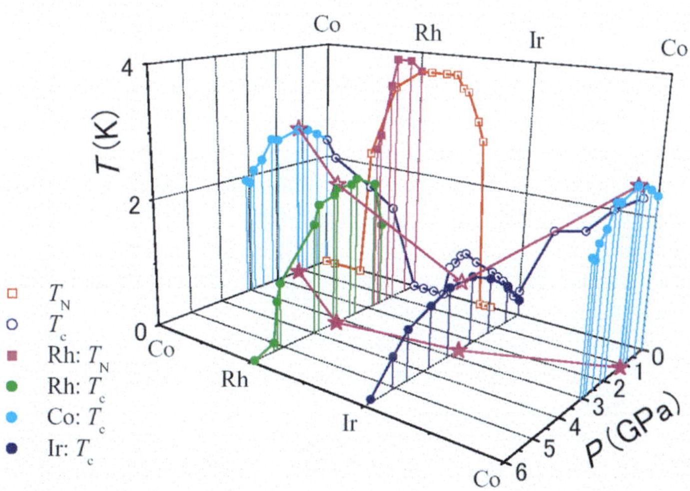

Fig. 8.5 Phase diagram of Ce(Co,Rh,Ir)In_5 in P-T-x space. (Fig. 15 in [27]). Reprinted from Ref. [27]. Copyright 2007 IOP Publishing

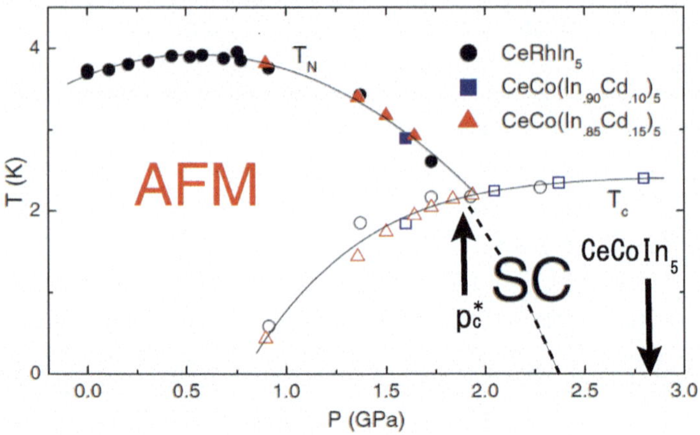

Fig. 8.6 Phase diagram of Ce(Co,Rh)(In,Cd)$_5$ under pressure [27, 28]. The dashed line represents a hypothetical Néel temperature realized if the superconducting state were absent [29, 30]. Consulting the phase diagram of CeCo(In,Cd)$_5$ in Ref. [28], CeCoIn$_5$ at ambient pressure is estimated to locate at $P = 2.8$ GPa in the CeRhIn$_5$ phase diagram as indicated by the down arrow. $p_c^* \simeq 1.9$ GPa denotes the pressure where T_N and T_c coincides [29]. Reprinted from Ref. [27]. Copyright 2007 IOP Publishing

netic QCP (P_c) around CeRhIn$_5$ is surrounded by the line of T_c maximum around which $\rho \propto T$ behavior is observed. This suggests that the T_c maximum is accompanied by the drastic valence change of Ce ion.

CeCoIn$_5$ is located off the magnetic critical point which corresponds to CeCo$_{0.4}$Rh$_{0.6}$In$_5$ [26]. Results of solid solution CeCo(In,Cd)$_5$ under pressure also suggest that the AF-QCP for CeCoIn$_5$ is located at negative pressure side around $P = -0.6$ GPa [28]. Indeed, it can be easily understood if we realize the fact that 6.5 pressure by 0.9 GPa, and that AF-QCP is located at a pressure higher by 0.43 GPa than that where T_N and T_c conicide [28]. This situation can be seen in Fig. 8.6.

Importance of the critical valence fluctuations can be seen more clearly in the case of CeIrIn$_5$ in which the maximum of T_c is attained in the pressure region where the AF fluctuations are suppressed as has been revealed by In^{115}-NQR measurement [31]. Two separated T_c domes have been observed under pressures in CeIr$_{3-x}$Co$_{x-2}$In$_5$ ($x = 0.8$) [32] as in CeCu$_2$(Si$_{0.9}$Ge$_{0.1}$)$_2$ [33]. These facts strongly suggest that the critical valence fluctuations are developed in Ir-based compounds under pressure and the origin of their *unconventional* behaviors.

It is also of interest to note that Pu-based superconductors PuCoGa$_5$ [34] and PuRhGa$_5$ [37] exhibit a trace of enhanced valence fluctuations. From a general point of view, 5f electrons in Pu element itself are located near the boundary between the localized state with nearly an integral valence and the itinerant state with a fractional valence. Indeed, the systematic variation of the Wigner-Seitz radius for a series of actinide elemental metals clearly shows that the ionic radius changes drastically between Pu and Am metals, suggesting the existence of the developed valence fluctuations in metallic compounds including such elements associated with the "valence transformation" [36].

A circumstantial evidence supporting this point of view is that the ratio A/γ^2, A being a coefficient of T^2-term in the resistivity and γ the Sommerfeld coefficient, of those Pu115 compounds $\{(A/\gamma^2) \simeq 3.4 \times 10^{-6}[\mu\Omega\text{cm}(\text{K}^2\text{mole/mJ})^2]$ for $PuCoGa_5$ and $(A/\gamma^2) \simeq 2 \times 10^{-6}[\mu\Omega\text{cm}(\text{K}^2\text{mole/mJ})^2]$ for $PuCoGa_5$ [37]$\}$ are between that for strongly correlated metals (the Kadowaki-Woods ratio, $(A/\gamma^2)_{\text{KW}} \simeq 1.0 \times 10^{-5}[\mu\Omega\text{cm}(\text{K}^2\text{mole/mJ})^2]$) and that for weakly correlated metals ($(A/\gamma^2) \simeq 0.4 \times 10^{-6}$ $[\mu\Omega\text{cm}(\text{K}^2\text{mole/mJ})^2]$ [38]). This also suggests that Pu115 compounds are not strongly correlated metals near magnetic instability but are subject to the valence fluctuations of Pu ion, although the Sommerfeld coefficient ($\gamma \sim 77$ [mJ/K^{2}mole] for $PuCoGa_5$ [34] and $\gamma \sim 95$ [mJ/K^{2}mole] for $PuRhGa_5$ [37]) is moderately enhanced. The value of A-coefficient is that of zero temperature limit obtained where the superconductivity is destroyed by the magnetic field. Another circumstantial evidence is that the T-dependence of the resistivity ρ is nearly linear as far as one can see in Fig. 3 of Ref. [34] although it is claimed in that paper that ρ increases approximately as $T^{1.35}$ at $T_c < T < 50$ K.

These anomalous aspects observed in Pu115 superconductors suggest a possibility that the high transition temperature ($T_c = 18.5$ K for $PuCoGa_5$ and $T_c = 9$ K for $PuRhGa_5$) is promoted at least in part by the enhanced valence fluctuations of Pu ion. It should be also noted that the AF fluctuations do not develop at all in these compounds, judging from the fact that NMR & NQR relaxation rates $1/T_1$ follow the Korringa law in rather wide temperature range [39, 40].

8.1.6 Compounds with Sharp Enhancement of Residual Resistivity ρ_0

As discussed physically and intuitively in Sect. 2.1.3, the residual resistivity ρ_0 exhibits sharp and pronounced peak around the critical valence transition or sharp valence crossover on the variation of external parameters such as the pressure or the magnetic field. In this sense, the observation of the sharp peak of ρ_0 is one of good signatures of sharp valence change or the critical valence transition with enhanced valence fluctuations. In this subsection, such candidate materials are introduced.

Here, as a typical case exhibiting such an anomaly, the case of $EuCu_2Ge_2$ is discussed below. First, the sharp peak of ρ_0 around $P = P_c$ is really observed as shown in Fig. 8.7. This P_c corresponds to the pressure where the antiferromagnetic order disappears discontinuously (i.e., as the first order transition), as shown in Fig. 2 in Ref. [41]. It is crucial here that P_c coincides with P_v^* where the sharp crossover in the valence occurs.

Figure 8.8 exhibits the exponent n of the T dependence of the resistivity $\rho(T)$ $(= \rho_0 + BT^n + \cdots)$, implying that $\rho(T) \propto T$ at $P = P_c$, which is one of the signals for the critical valence fluctuations.

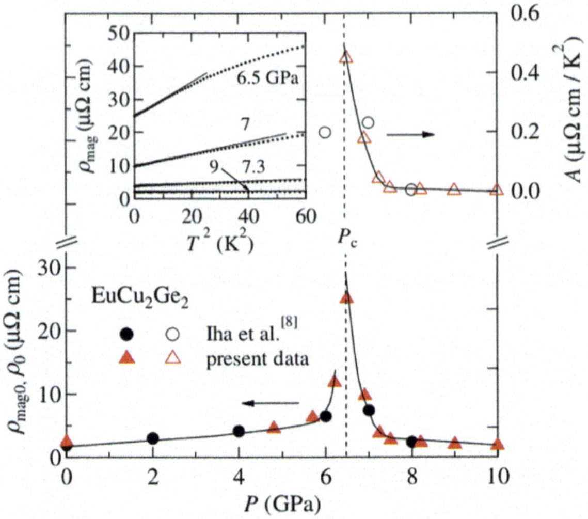

Fig. 8.7 Pressure dependence of the coefficient A of the T^2 dependence of the electrical resistivity $\rho(T)$, the residual resistivity ρ_0, and the magnetic residual resistivity ρ_{mag0} [41]. The solid lines serve as a visual guide. The circular and triangular symbols indicate our previous and present data, respectively. The open and closed symbols indicate the coefficient A and ρ_0 or ρ_{mag0}, respectively. Note that the closed circles and triangles indicate residual resistivity ρ_0 and ρ_{mag0}, respectively. The inset shows the T^2 dependence of ρ_{mag} at low temperatures. Reprinted with permission from Ref. [41]. Copyright 2020 The Physical Society of Japan

Fig. 8.8 Pressure dependence of exponent n of the power law $\rho_{\mathrm{mag}} = \rho_{\mathrm{mag0}} + BT^n$ of the magnetic resistivity at $T = 0.05$ K for $EuCu_2Ge_2$ [41]. The solid curves are guides to the eye. Reprinted with permission from Ref. [41]. Copyright 2020 The Physical Society of Japan

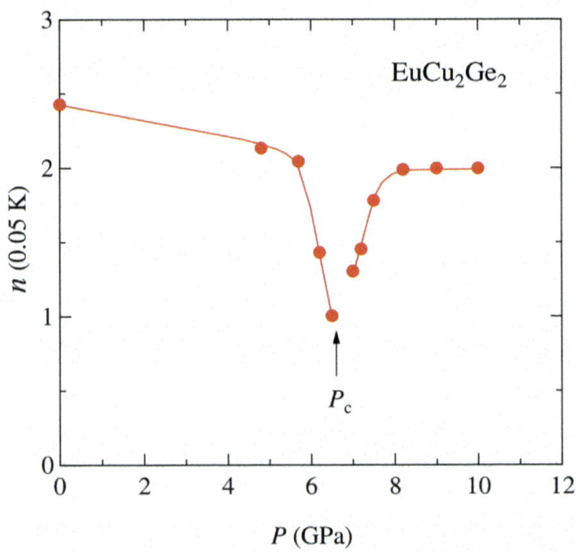

Exactly the same behaviors were observed not only in a sister compound $EuPt_3Al_{15}$ [42] but also in $CeAl_2$ [43, 44] and $CeRhIn_5$ [14]. These four examples reinforce the fact that the sharp peak of the residual resistivity ρ_0 is the smoking gun for the critical valence transition with enhanced valence fluctuations. In other words, these phenomena are rather ubiquitous ones which are expected to be observed in a rather wide class of materials near and long-term future.

References

1. J.C. Lashley, A.C. Lawson, J.C. Cooley, B. Mihaila, C.P. Opeil, L. Pham, W.L. Hults, J.L. Smith, G.M. Schmiedeshoff, F.R. Drymiotis, G. Chapline, S. Basu, P.S. Riseborough, Phys. Rev. Lett. **97**, 235701 (2006)
2. S. Wada, A. Yamamoto, K. Ishida, J.L. Sarrao, J. Phys.: Condens. Matter **20**, 175201 (2008)
3. K.A. Gschneidner, L. Eyring, *Handbook on the Physics and Chemistry of Rare Earths* (North-Holland, Amsterdam, 1978)
4. I. Felner, I. Nowik, Phys. Rev. B **33**, 617 (1986)
5. J.L. Sarrao, C.D. Immer, Z. Fisk, C.H. Booth, E. Figueroa, J.M. Lawrence, R. Modler, A.L. Cornelius, M.F. Hundley, G.H. Kwei, J.D. Thompson, F. Bridges, Phys. Rev. B **59**, 6855 (1999)
6. P. Gegenwart, T. Westerkamp, C. Krellner, Y. Tokiwa, S. Paschen, C. Geibel, F. Steglich, E. Abrahams, Q. Si, Science **315**, 969 (2007)
7. E. Bauer, R. Hauser, L. Keller, P. Fischer, O. Trovarelli, J.G. Sereni, J.J. Rieger, G.R. Stewart, Phys. Rev. B **56**, 711 (1997)
8. K. Yamamoto, H. Yamaoka, N. Tsujii, A.M. Vlaicu, H. Oohashi, S. Sakakura, T. Tochio, Y. Ito, A. Chainani, S. Shin, J. Phys. Soc. Jpn. **76**, 124705 (2007)
9. H. Yamaoka, I. Jarrige, N. Tsujii, N. Hiraoka, H. Ishii, K.-D. Tsuei, Phys. Rev. B **80**, 035120 (2009)
10. S. Watanabe, K. Miyake, J. Phys.: Condens. Matter **24**, 294208 (2012)
11. H. Tonai, N. Sasabe, T. Uozumi, N. Kawamura, M. Mizumaki, J. Phys. Soc. Jpn. **86**, 093704 (2017)
12. F. de Groot, A. Kotani, *Core Level Spectroscopy of Solids* (CRC Press, Boca Raton, 2008)
13. Y. Hirose, J. Sakaguchi, M. Ohya, M. Matsushita, F. Honda, R. Settai, Y. Ōnuki, J. Phys. Soc. Jpn. **81**, SB009 (2012)
14. G. Knebel, D. Aoki, J.-P. Brison, J. Flouquet, J. Phys. Soc. Jpn. **77**, 114704 (2008)
15. T. Park, V.A. Sidorov, F. Ronning, J.-X. Zhu, Y. Tokiwa, H. Lee, E.D. Bauer, R. Movshovich, J.L. Sarrao, J.D. Thompson. Nature **456**, 366 (2008)
16. S. Raymond, D. Jaccard, J. Low Temp. Phys. **120**, 107 (2000)
17. D. Jaccard, H. Wilhelm, K. Alami-Yadri, E. Vargoz, Phys. B **259–261**, 1 (1999)
18. H. von Löhneysen, J. Phys. C **8**, 9689 (1996)
19. J. Flouquet, Private communication
20. J. Flouquet, Y. Haga P. Haen, D. Braithwaite, G. Knebel, S. Raymond, S. Kambe, J. Magn, Magn. Mater **272–276**, 27 (2004)
21. J. Rossat-Mignod, L.P. Renault, J.L. Jacoud, C. Vettier, P. Lejay, J. Flouquet, E. Walker, D. Jaccard, A. Amato, J. Magn. Magn. Mater. **76–77**, 376 (1988)
22. R. Settai, Y. Miyauchi, T. Takeuchi, F. Lévy, I. Sheikin, Y. Ōnuki, J. Phys. Soc. Jpn. **77**, 073705 (2008)
23. I. Sheikin, A. Huxley, D. Braithwaite, J.-P. Brison, S. Watanabe, K. Miyake, J. Flouquet, Phys. Rev. B **64**, 220503(R) (2001)
24. S. Watanabe, K. Miyake, J. Phys. Soc. Jpn. **71**, 2489 (2002)
25. T. Muramatsu, N. Tateiwa, T.C. Kobayashi K. Shimizu, K. Amaya, D. Aoki, H. Shishido, Y. Haga, Y. Ōnuki, J. Phys. Soc. Jpn. **70**, 3362 (2001)

26. P.G. Pagliuso, Private communication
27. K. Miyake, J. Phys.: Condens. Matter **19**, 125201 (2007)
28. L.D. Pham, T. Park, S. Maquilon, J.D. Thompson, Z. Fisk, Phys. Rev. Lett. **97**, 056404 (2006)
29. G. Knebel, D. Aoki, D. Braithwaite, S. Salce, J. Flouquet, Phys. Rev. B **74**, 020501 (2006)
30. T. Park, F. Ronning, H.Q. Yuan, M.B. Salamon, R. Movshovich, J.L. Sarrao, J.D. Thompson, Nature **440**, 65 (2006)
31. S. Kawasaki, G.-Q. Zheng, H. Kan, Y. Kitaoka, H. Shishido, Y. Ōnuki, Phys. Rev. Lett. **94**, 037007 (2005)
32. M. Nicklas, V.A. Sidorov, H.A. Borges, P.G. Pagliuso, J.L. Sarrao, J.D. Thompson, Phys. Rev. B **70**, 020505 (2004)
33. H.Q. Yuan, F.M. Grosche, M. Deppe, C. Geibel, G. Sparn, F. Steglich, Science **302**, 2104 (2003)
34. J.L. Sarrao, L.A. Morales, J.D. Thompson, B.L. Scott, G.R. Stewart, F. Wastin, J. Rebizant, P. Boulet, E. Colineau, G.H. Lander, Nature **420**, 297 (2002)
35. F. Wastin, P. Boulet, J. Rebizant, E. Colineau, G.H. Lander, J. Phys.: Condens. Matter **15**, S2279 (2003)
36. N.G. Cooper (ed.), *Challenges in Plutonium Science*. Los Alamos Science No. 26 (2000)
37. F. Wastin, Private communication (2002)
38. K. Miyake, T. Matsuura, C.M. Varma, Solid State Commun. **71**, 1149 (1989)
39. N.J. Curro, T. Caldwell, E.D. Bauer, L.A. Morales, M.J. Graf, Y. Bang, A.V. Balatsky, J.D. Thompson, J.L. Sarrao, Nature **434**, 622 (2005)
40. H. Sakai, Y. Tokunaga, T. Fujimoto, S. Kambe, R.E. Walstedt, H. Yasuoka, D. Aoki, Y. Homma, E. Yamamoto, A. Nakamura, Y. Shiokawa, K. Nakajima, Y. Arai, T.D. Matsuda, Y. Haga, Y. Ōnuki, J. Phys. Soc. Jpn. **74**, 1710 (2005)
41. J. Gouchi, K. Miyake, W. Iha, M. Hedo, T. Nakama, Y. Ōnuki, Y. Uwatoko, J. Phys. Soc. Jpn. **89**, 053703 (2020)
42. T. Koizumi, F. Honda, Y.J. Sato, D. Li, D. Aoki, Y. Haga, J. Gouchi, S. Nagasaki, Y. Uwatoko, Y. Kaneko, Y. Ōnuki, J. Phys. Soc. Jpn. **91**, 043704 (2022)
43. H. Miyagawa, G. Oomi, M. Ohashi, I. Satoh, T. Komatsubara, M. Hedo, Y. Uwatoko, Phys. Rev. B **78**, 064403 (2008)
44. E. Pugh, Private communication

Appendix A
Distribution Function of Fermions

Grassmann number η_i satisfies anticommutation relation$(\eta_i\eta_j = -\eta_j\eta_i)$. In general, the distribution function of the fermion system described by the Hamiltonian $H(c^\dagger, c)$ is expressed as

$$Z = \mathrm{Tr}e^{-\beta H} = \int \prod_{i=1}^{M} d\eta_i^* d\eta_i \exp\left[-\sum_{j=1}^{M}\left\{\eta_j^*(\eta_j - \eta_{j+1}) + \frac{\beta}{M}H(\eta_j^*, \eta_{j+1})\right\}\right], \quad (A.1)$$

where the integral is performed under the condition $\eta_{M+1} = -\eta_1$. By introducing $\Delta\tau = \beta/M$ and taking $M \to \infty$, we obtain

$$Z = \lim_{M\to\infty}\int \prod_{i=1}^{M} d\eta_i^* d\eta_i \exp\left[-\int_0^\beta d\tau \left\{\eta^*(\tau)\frac{\partial\eta(\tau)}{\partial\tau} + H(\eta^*(\tau), \eta(\tau))\right\}\right]. \quad (A.2)$$

The multiple integral $\lim_{M\to\infty}\int \prod_{i=1}^{M} d\eta_i^* d\eta_i = \int \mathcal{D}\eta^*\mathcal{D}\eta$ is referred to as functional integral. The $\{\cdots\}$ part is the Lagrangean L. The Eq. (5.3) in the main text is the result of this formulation applied to the Hubbard model (5.1).

S. Watanabe and K. Miyake, *Quantum Critical Phenomena of Valence Transition*, Springer Tracts in Modern Physics 289,
https://doi.org/10.1007/978-981-99-3518-5

Appendix B
Coefficient of the Gaussian Term—Derivation of C_q in Eq. (5.17) in Chap. 5

In this appendix, the coefficient C_q in Eq. (5.17) is derived. Here, C_q is given by $C_q = C/q$ for the ferromagnetic case and $C_q = C$ for the antiferromagnetic case.

For $|\omega_l| \le v_F q$ with the Fermi velocity being v_F, we have

$$\chi_0(\boldsymbol{q}, \mathrm{i}\omega_l) - \chi_0(\boldsymbol{q}, 0) \approx -b\chi_0(\boldsymbol{q}, 0)\frac{|\omega_l|}{v_F q} \tag{B.1}$$

where b is given by $b = \pi/2$ in three spatial dimenstion and $b = 1$ in two spatial dimension. For the ferromagnetic case, the expansion of $\chi_0(\boldsymbol{q}, \mathrm{i}\omega_l)$ around $\boldsymbol{q} = \boldsymbol{0}$ gives

$$C = \frac{bU^2 N(\varepsilon_F)}{2v_F}. \tag{B.2}$$

For the antiferromagnetic case, the expansion of $\chi_0(\boldsymbol{q}, \mathrm{i}\omega_l)$ around $\boldsymbol{q} = \boldsymbol{Q}$ gives

$$C = \frac{bU^2 \chi_0(\boldsymbol{Q}, 0)}{2v_F Q}. \tag{B.3}$$

Here, $N(\varepsilon_F)$ is the density of states at the Fermi level ε_F per spin

$$N(\varepsilon_F) = \lim_{\boldsymbol{q}\to\boldsymbol{0}} \chi_0(\boldsymbol{q}, 0). \tag{B.4}$$

S. Watanabe and K. Miyake, *Quantum Critical Phenomena of Valence Transition*, Springer Tracts in Modern Physics 289,
https://doi.org/10.1007/978-981-99-3518-5

Appendix C
Quantum Criticality in the SCR Theory

In this appendix, the quantum criticality in the SCR theory is derived for each class in the spatial dimensions of $d = 3$ and 2. In Sect. C.1, the criticality in $d = 3$ is derived from the SCR equation. In Sect. C.2, the solution of the SCR equation is derived for the dynamical exponent $z = 3$ in $d = 2$. In Sect. C.3, the criticality is analyzed for $z = 2$ in $d = 2$. The result is shown to be equivalent to the result of the renormalization group theory.

C.1 Quantum Criticality in $d = 3$

In this appendix, it is explained that the quantum criticality at the QCP in $d = 3$ is given by Eq. (5.56) by analyzing the solution of the SCR equation [Eq. (5.47)] at low temperatures. The x integral in the r.h.s. of Eq. (5.47) is defined by Eq. (5.90) as

$$L \equiv \int_0^{x_c} dx x^{d+z-3} \left\{ \ln u - \frac{1}{2u} - \psi(u) \right\}, \tag{C.1}$$

where d is the spatial dimension and z is the dynamical exponent. By changing the integral variable as $x' = x/t^{\frac{1}{z}}$, Eq. (C.1) is expressed as

$$L = t^{\frac{d+z-2}{z}} \int_0^{\frac{x_c}{t^{\frac{1}{z}}}} dx'(x')^{d+z-3} \left\{ \ln u - \frac{1}{2u} - \psi(u) \right\}, \tag{C.2}$$

where u is given by

$$u = (x')^{z-2} \left\{ \frac{y}{t^{\frac{2}{z}}} + (x')^2 \right\}. \tag{C.3}$$

S. Watanabe and K. Miyake, *Quantum Critical Phenomena of Valence Transition*, Springer Tracts in Modern Physics 289,
https://doi.org/10.1007/978-981-99-3518-5

We see that at low temperatures $t \ll 1$ for

$$\frac{y}{t^{\frac{2}{z}}} \to 0, \tag{C.4}$$

the x' integral in $d = 3$ converges in Eq. (C.2) where the upper bound of the integral is set to be ∞. Hence, the t dependence of L is evaluated as

$$L \propto t^{\frac{z+1}{z}}. \tag{C.5}$$

Then, from the SCR equation [Eq. (5.47)], the following solution:

$$y \propto t^{\frac{z+1}{z}} \tag{C.6}$$

is immediately obtained at the QCP where $y_0 = 0.0$ is set in Eq. (5.47). It is confirmed that this result satisfies the condition of Eq. (C.4), i.e., $y/t^{\frac{2}{z}} \ll 1$, for $t \ll 1$.

C.2 Solution of SCR Equation for $z = 3$ in $d = 2$

The derivation of the solution of the SCR equation [Eq. (5.47)] for $z = 3$ in $d = 2$ is shown in this appendix [1].

By using the approximation formula

$$\ln u - \frac{1}{2u} - \psi(u) \approx \frac{1}{2u(1+6u)}, \tag{C.7}$$

in Eq. (C.1), the x integration can be performed and the leading terms are evaluated as

$$L \approx \frac{t}{4} \ln \left[\frac{1}{y} \left(\frac{t}{6} \right)^{\frac{2}{3}} \right] \tag{C.8}$$

for $y \ll t^{\frac{2}{3}} \ll 1$. Then, the solution of the SCR equation [Eq. (5.47)] $y = y_0 + y_1 L$ at the QCP with $y_0 = 0.0$ is obtained as follows.

$$y = -\frac{y_1}{12} t \ln t. \tag{C.9}$$

Since the correlation length ξ is given by $y \propto \xi^{-2}$, this coincides with the result of the renormalization group theory for $z = 3$ in $d = 2$ in Ref. [2].

C.3 Equivalence of SCR Solution and Renormalization Group for $z = 2$ in $d = 2$

In this appendix, it is shown that the solution of the SCR equation [Eq. (5.48)] for $z = 2$ in $d = 2$ coincides with the result of the renormalization group theory by Millis [2].

By using the approximation formula Eq. (C.7) in the SCR equation [Eq. (5.48)], the x integration can be performed as

$$y = y_0 + \frac{y_1}{2}\left(y\ln y + \frac{t}{2}\left\{\ln\frac{x_c^2 + y}{y} - \ln\frac{x_c^2 + y + \frac{t}{6}}{y + \frac{t}{6}}\right\}\right). \tag{C.10}$$

At the QCP with $y_0 = 0.0$, the leading terms are evaluated as

$$-\ln 2y \approx \frac{t}{2y}\ln\frac{t}{2y}, \tag{C.11}$$

for $y \ll t \ll 1$. Then, one finds that

$$y = -t\frac{\ln(-\ln t)}{2\ln t} \tag{C.12}$$

is the solution of Eq. (C.11) up to the order of $\ln t$, which coincides with the result of the renormalization group theory for $z = 2$ in $d = 2$ in Ref. [2].

Appendix D
Thermal-Expansion Coefficient and Grüneisen Parameter

The magnetic volume effect is discussed on the basis of the SCR theory for spin fluctuations in Sects. 5.3.5 and 5.3.6. The valence volume effect is discussed on the basis of the SCR theory for valence fluctuations in Sects. 6.4.4 and 6.4.5. In both cases, the specific heat, the thermal-expansion coefficient, and the Grüneisen parameter are derived from the SCR free energy, which gives the thermodynamically consistent expressions. In this appendix, the definitions of the thermal-expansion coefficient α and and the Grüneisen parameter Γ are summarized. The equivalent expressions of α and Γ are also derived by using the thermodynamic relations.

D.1 Thermal-Expansion Coefficient

The thermal-expansion coefficient is defined as

$$\alpha = \frac{1}{V}\left(\frac{\partial V}{\partial T}\right)_P, \tag{D.1}$$

where P is the pressure. By using the relation

$$\left(\frac{\partial V}{\partial T}\right)_P = -\frac{\left(\frac{\partial P}{\partial T}\right)_V}{\left(\frac{\partial P}{\partial V}\right)_T}, \tag{D.2}$$

Equation (D.1) is expressed as

$$\alpha = \kappa_T \left(\frac{\partial P}{\partial T}\right)_V, \tag{D.3}$$

S. Watanabe and K. Miyake, *Quantum Critical Phenomena of Valence Transition*, Springer Tracts in Modern Physics 289,
https://doi.org/10.1007/978-981-99-3518-5

where the isothermal compressibility κ_T is defined as

$$\kappa_T = -\frac{1}{V}\left(\frac{\partial V}{\partial P}\right)_T . \tag{D.4}$$

On the other hand, by using the Maxwell relation $(\partial V/\partial T)_P = -(\partial S/\partial P)_T$ in Eq. (D.1), α can be expressed as

$$\alpha = -\frac{1}{V}\left(\frac{\partial S}{\partial P}\right)_T . \tag{D.5}$$

D.2 Grüneisen Parameter

The Grüneisen parameter Γ is defined by [3]

$$\Gamma = \frac{\alpha V}{C_V \kappa_T}, \tag{D.6}$$

where C_V is the specific heat at a constant volume

$$C_V = T\left(\frac{\partial S}{\partial T}\right)_V . \tag{D.7}$$

With the use of Eq. (5.83) for α in Eq. (D.6), the Grüneisen parameter Γ is expressed as follows:

$$\begin{aligned}
\Gamma &= -\frac{\partial(S,T)}{\partial(P,T)} \frac{1}{T\frac{\partial(S,V)}{\partial(T,V)}} \frac{(-V)}{\frac{\partial(V,T)}{\partial(P,T)}} \\
&= -\frac{V}{T}\frac{\partial(S,T)}{\partial(S,V)} \\
&= -\frac{V}{T}\left(\frac{\partial T}{\partial V}\right)_S .
\end{aligned} \tag{D.8}$$

If the entropy is expressed as $S = k_B S(T/T^*)$ with a single characteristic temperature scale T^* where k_B is the Boltzmann constant as in the Fermi-liquid region of metals, $(\partial T/\partial V)_S$ is given in a form as

$$\left(\frac{\partial T}{\partial V}\right)_S = \frac{T}{T^*}\left(\frac{\partial T^*}{\partial V}\right)_S . \tag{D.9}$$

Then, the Grüneisen parameter Γ is expressed as a conventional form as [4, 5]

$$\Gamma = -\left(\frac{\partial \ln T^*}{\partial \ln V}\right)_S. \tag{D.10}$$

Appendix E
Electric Quadrupole Susceptibility Near the Valence QCP

In Sect. 6.6.4, the temperature dependence of the elastic constant in the quantum critical Yb-Al-Au approximant crystal and the quasicrystal is discussed. The elastic constant reflects the electric quadrupole fluctuations at low temperatures. In this Appendix, the mechanism of the temperature dependences of electric quadrupole susceptibility χ_Γ for $\Gamma = x^2 - y^2, 2z^2 - x^2 - y^2, xy, x^2 + y^2 + z^2$, and x^2 shown in Fig. 6.21b is explained [6].

The explicit form of $\langle i\ell \pm | \hat{O}_{x^2} | i\ell\pm \rangle$ in Eq. (6.127) is given by

$$\langle i\ell \pm | \hat{O}_{x^2} | i\ell\pm \rangle = \frac{a_\ell}{2}(\cos^2\alpha_i \cos^2\beta_i + \sin^2\alpha_i) + a_\ell^2 \cos^2\alpha_i \sin^2\beta_i, \quad \text{(E.1)}$$

where i denotes the Yb site on the icosahedron shown in Fig. 6.19c and ℓ indicates the 4f state for $\ell = 1$ and the 5d state for $\ell = 2$. The $\pm$ in the left hand side denotes either of the Kramers states.

In Table E.1, the expectation values of $\langle i1 \pm | \hat{O}_\Gamma | i1\pm \rangle$ on the 4f state in Eq. (6.127) for $\Gamma = x^2 - y^2, 2z^2 - x^2 - y^2, xy, x^2 + y^2 + z^2$, and x^2 at the ith Yb site are listed.

In Table E.2, the values of the RPA susceptibility of the 4f orbital with the Kramers state $\eta = +$, i.e. $\lim_{\boldsymbol{q}\to 0} \chi^{ii'}_{1+1+1+1+}(\boldsymbol{q}, \omega = 0)$ [see Eq. (6.121)], are listed for the $i = 1$ and $i' = 1$–12 at $T = 1.0 \times 10^{-5}$, 3.0×10^{-5}, and 2.0×10^{-4} as a representative case. The diagonal component $(i, i') = (1, 1)$ has a positive value, which increases sharply on cooling below $T = 3.0 \times 10^{-5}$. On the other hand, in the off-diagonal component for $i \neq i'$, negative values appear as seen in Table E.2 at $T = 2.0 \times 10^{-4}$. As T decreases, these negative values change to positive values finally as seen at $T = 1.0 \times 10^{-5}$.

The electric quadrupole susceptibility is calculated in Eq. (6.128), which is reflected in the elastic constants as discussed in the main text. Since $\langle i1 \pm | \hat{O}_{x^2+y^2+z^2} | i1\pm \rangle$ are all positive at the ith Yb site, the above negative-to-positive change in the off-diagonal components all contribute to increase in $\chi_{x^2+y^2+z^2}$ on cooling from $T = 2.0 \times 10^{-4}$ to $T = 1.0 \times 10^{-5}$. This results in the convex curve

S. Watanabe and K. Miyake, *Quantum Critical Phenomena of Valence Transition*, Springer Tracts in Modern Physics 289,
https://doi.org/10.1007/978-981-99-3518-5

Table E.1 $\langle i1 \pm | \hat{O}_\Gamma | i1\pm \rangle$ for $\Gamma = x^2 - y^2$, $2z^2 - x^2 - y^2$, xy, $x^2 + y^2 + z^2$, and x^2 at the ith Yb site (see Fig. 6.19c)

i	$x^2 - y^2$	$2z^2 - x^2 - y^2$	xy	$x^2 + y^2 + z^2$	x^2
1	−2.00	−21/8	8.17	63/4	4.69
2	−1.27	3.04	0	63/4	4.69
3	−2.00	−21/8	−8.17	63/4	4.69
4	−1.27	3.04	0	63/4	4.69
5	3.27	−0.42	0	63/4	1.75
6	3.27	−0.42	0	63/4	1.75
7	−1.27	3.04	0	63/4	4.69
8	3.27	−0.42	0	63/4	1.75
9	3.27	−0.42	0	63/4	1.75
10	−1.27	3.04	0	63/4	4.69
11	−2.00	−21/8	−8.17	63/4	4.69
12	−2.00	−21/8	8.17	63/4	4.69

Table E.2 RPA susceptibility $\lim_{\boldsymbol{q}\to\boldsymbol{0}} \chi^{ii'}_{1+1+1+1+}(\boldsymbol{q}, \omega = 0)$ at $T = 1.0 \times 10^{-5}$, 3.0×10^{-5}, and 2.0×10^{-4}. The i and i' denote the Yb site on the icosahedron shown in Fig. 6.19c

i	i'	$T = 1.0 \times 10^{-5}$	3.0×10^{-5}	2.0×10^{-4}
1	1	27.10	21.74	21.08
1	2	3.41	−0.70	−0.99
1	3	7.08	3.07	2.84
1	4	3.41	−0.70	−0.99
1	5	3.41	−0.70	−0.99
1	6	3.41	−0.70	−0.99
1	7	2.78	−0.99	−1.17
1	8	2.78	−0.99	−1.17
1	9	2.78	−0.99	−1.17
1	10	2.78	−0.99	−1.17
1	11	5.22	0.37	−0.15
1	12	3.94	0.31	0.19

of the $a_{\mathrm{B}}^4 e^2/\chi_{x^2+y^2+z^2}$ vs. T plot in Fig. 6.21b. The same applies to χ_{x^2} since $\langle i1 \pm | \hat{O}_{x^2} | i1\pm \rangle$ are all positive, as shown in Table E.1.

For $\Gamma = xy$, the form factor is zero at the $i = 2, 4, 5, 6, 7, 8, 9, 10$th Yb site (see Fig. 6.19c), as shown in Table E.1. This implies that the above negative-to-positive change in the off-diagonal component in the RPA susceptibility (see Table E.2) does not contribute to increase in χ_{xy}, which is in sharp contrast to $\chi_{x^2+y^2+z^2}$ and χ_{x^2}. Hence, the main contribution comes from the diagonal component for $i = i'$ in Table E.2, which results in almost flat-T dependence of χ_{xy} in Fig. 6.21b.

For $\Gamma = x^2 - y^2$ and $2z^2 - x^2 - y^2$, the form factors are not zero but have positive and negative values depending on the ith Yb site (see Fig. 6.19c), as shown in Table E.1. Hence, the off-diagonal components in the RPA susceptibility are canceled partly as T decreases, which results in the gradual increase in χ_Γ around the lowest temperatures in Fig. 6.21b.

References

1. M. Hatatani, T. Moriya, J. Phys. Soc. Jpn. **64**, 3434 (1995)
2. A. J. Millis, Phys. Rev. B **48**, 7183 (1993)
3. E. Grüneisen, Ann. Phys. **39**, 257 (1912)
4. R. Takke, M. Niksch, W. Assmus, B. Lüthi, R. Pott, R. Schefzyk, D.K. Wöhlleben, Z. Phys. B **44**, 33 (1981)
5. J. Flouquet, Prog. Low. Temp. Phys. **15**, 149 (2005)
6. S. Watanabe, Solid State Commun. **336**, 114425 (2021)

Index

S. Watanabe and K. Miyake, *Quantum Critical Phenomena of Valence Transition*, Springer Tracts in Modern Physics 289,
https://doi.org/10.1007/978-981-99-3518-5

The manufacturer's authorised representative in the EU is Springer Nature Customer Service Centre GmbH, Europaplatz 3, 69115 Heidelberg, Germany. If you have any concerns regarding our products, please contact ProductSafety@springernature.com

Printed and bound by CPI Group (UK) Ltd, Croydon, CR0 4YY

30/09/2026

02225317-0002